RIGHTING AMERICA
AT THE
CREATION MUSEUM

MEDICINE, SCIENCE, AND RELIGION
IN HISTORICAL CONTEXT
Ronald L. Numbers, *Consulting Editor*

RIGHTING AMERICA

AT THE

CREATION MUSEUM

Susan L. Trollinger
and William Vance Trollinger, Jr.

JOHNS HOPKINS UNIVERSITY PRESS   BALTIMORE

Johns Hopkins University Press
2715 North Charles Street
Baltimore, Maryland 21218-4363
www.press.jhu.edu

Library of Congress Cataloging-in-Publication Data

Names: Trollinger, Susan L., 1964–
Title: Righting America at the Creation Museum / Susan L. Trollinger and
    William Vance Trollinger, Jr.
Description: Baltimore : Johns Hopkins University Press, 2016. |
    Series: Medicine, science, and religion in historical context | Includes
    bibliographical references and index.
Identifiers: LCCN 2015030255| ISBN 9781421419510 (hardcover : alk. paper)
    | ISBN 9781421419534 (electronic) | ISBN 1421419513 (hardcover : alk.
    paper) | ISBN 142141953X (electronic)
Subjects: LCSH: United States—Church history—21st century. |
    Creationism—United States. | Evangelicalism—United States. |
    Fundamentalism—United States. | Creation Museum (Petersburg, Ky.)
Classification: LCC BR526 .T76 2016 | DDC 277.3/083—dc23 LC record
    available at http://lccn.loc.gov/2015030255

A catalog record for this book is available from the British Library.

*Special discounts are available for bulk purchases of this book. For
more information, please contact Special Sales at 410-516-6936 or
specialsales@press.jhu.edu.*

Johns Hopkins University Press uses environmentally friendly book materials,
including recycled text paper that is composed of at least 30 percent post-
consumer waste, whenever possible.

For our children

CONTENTS

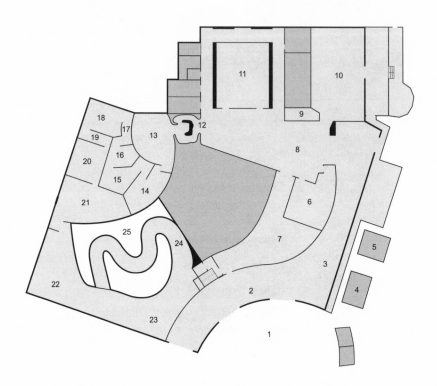

UPPER LEVEL:

1. Grand Plaza
2. Main Entrance and Tickets
3. Portico
4. Zip Line Tower
5. Ark Encounter Exhibit
6. Stargazer's Planetarium
7. Dragon Hall Bookstore
8. Main Hall
9. Coffee and Ice Cream Shop
10. Noah's Café
11. Special Effects Theater
12. Entrance to Bible Walkthrough Experience
13. Dig Site
14. Starting Points
15. Biblical Authority
16. Biblical Relevance
17. Graffiti Alley
18. Culture in Crisis
19. Time Tunnel
20. Six Days Theater
21. Wonders of Creation
22. Garden of Eden
23. Cave of Sorrows
24. Ramp to Lower Level
25. Ark Construction

RIGHTING AMERICA
AT THE
CREATION MUSEUM

INTRODUCTION

On May 28, 2007, the Creation Museum in Petersburg, Kentucky, opened its doors, its purpose to "point today's culture back to the authority of Scripture and proclaim the gospel message." It does this by demonstrating that "the account of origins presented in Genesis is a simple but factual presentation of actual events." Thus, throughout the seventy-five-thousand-square-foot museum, visitors encounter exhibits that assert claims such as these: the God of the Bible created the universe in six consecutive twenty-four-hour days less than ten thousand years ago; "The various original life forms (kinds) . . . were made by direct creative acts of God"; these acts of creation included the "special creation of Adam . . . and Eve," whose "subsequent fall into sin" resulted in "death (both physical and spiritual) and bloodshed enter[ing] this world"; and the global flood was an actual historic event that accounts for geological strata and the fossil record.[1] The museum explains these claims to visitors by way of more than one hundred and fifty exhibits featuring animatronic human figures and dinosaurs (sometimes appearing in the same display), numerous explanatory placards and diagrams, and several miniature dioramas depicting a global flood, as well as a re-creation of a portion of the Garden of Eden that includes many life-size animal figures placed among artificial plant life, a waterfall with pool below, a life-size reproduction of the Tree of Life, multiple scenes with Adam and Eve, and the serpent.

It is tempting to dismiss the Creation Museum as a surreal oddity, an inexplicable and bizarre cultural site. But to imagine that the museum is a wacky but essentially irrelevant outpost on the far outskirts of American life is a huge mistake. As peculiar as it may seem, the Creation Museum lies squarely within the right side of the American cultural, political, and religious mainstream. That is to say, the museum exists and thrives not because it is so preposterous—although some people are surely drawn to it precisely for this reason—but because it represents and speaks to the religious and political commitments of a large swath of the American population. More than this, and more important, the Creation Museum seeks to shape, prepare, and arm millions of American Christians as uncompromising and fearless warriors for what it understands to be the ongoing culture war in America.

In short, the Creation Museum matters, and all Americans ought to understand what is going on there. Hence this book. And to truly understand the museum, the only place to start is with fundamentalism, that remarkable movement that shows no sign of disappearing from the American landscape. As a quintessential fundamentalist institution, the museum shares and promotes the movement's core commitments: biblical inerrancy, premillennialism, patriarchy, political conservatism, and (of course) creationism.[2] Thus follows a brief history of fundamentalism and how the Creation Museum fits within this story.[3]

Fundamentalism, Creationism, and Answers in Genesis

Fundamentalism finds its origins in the mid- and late nineteenth century, when Darwinism (*On the Origin of Species* appeared in 1859) and historicism (or "higher criticism") challenged traditional understandings of the Bible. The former raised questions about the Genesis story of creation (six days? all those separate creations?), not to mention larger theological questions about God's role in creation and the nature of human beings. The latter, because of its recognition that time and place shape texts and because of its determination to evaluate the Bible as one of these historical texts, raised serious questions about the supernatural character and literal authenticity of the biblical record. Who really were the authors of the sixty-six books of the Bible? How does one reconcile the inconsistencies and errors in the texts? What about the borrowings from the stories of other cultures?

Many American Protestants responded to these intellectual challenges by accommodating Darwinism and by coming to understand the Bible as an errant document that human beings, living in the stream of time, wrote. But other Protestants responded quite negatively to the threats posed by Darwinism and historicism. The most significant theological response was the doctrine of biblical inerrancy. First formulated by Princeton theologians in the late nineteenth century, inerrancy emphasizes that the original "autographs" are the infallible product of the Holy Spirit's guidance. As such, they contain no errors of any sort; they are factually accurate in all that they have to say, including when they speak on matters of history, science, and the like. While the texts and their translations that have come down to us have a few errors, since only the original autographs are truly inerrant, the mistakes are understood to be so few and so minor that we can trust the Bible in our hands as the Word of God.

Not only is the Bible errorless, but it also foretells the future. A second set of ideas intimately connected to inerrancy and developed in the nineteenth century made even clearer the Bible's supernatural character: dispensational premillennialism. According to this eschatological system, a literal reading of the Bible (particularly the books of Daniel and Revelation) provides a sure guide to the past, present, and future of human history. Dispensational premillennialism divides history into (generally seven) segments, or dispensations; in each dispensation God tests humans, they fail, and God imposes a divine judgment (e.g., the Genesis flood). The current dispensation, the "church age," displays the increasing apostasy of the church and the increasing decadence of civilization. But at the end of the church age, which would be preceded by the return of the Jews to Palestine, Christ will return in the air (the "rapture") to retrieve the faithful. This will be followed by a time of "tribulation" that will include the reign of the antichrist, followed by the return of Christ and the saints, who will annihilate the enemy and establish the millennial kingdom of God.

Thanks in good part to a series of Bible and prophecy conferences, by the turn of the century many American evangelicals were strongly committed to biblical inerrancy and dispensational premillennialism. Then in 1909 Cyrus Scofield published his *Reference Bible* (a second edition appeared in 1917), which became the Bible of choice for conservative Protestants in the United States and which (with its heavy-handed dispensationalist gloss of the biblical text) cemented inerrancy and premillennialism in the evangelical consciousness.

These premillennialist evangelicals were alarmed but not surprised, given the dire "end times" predictions, by the spread of theological liberalism in Protestantism. In response, Lyman and Milton Stewart, who were wealthy evangelical oilmen, funded the publication and distribution (three million copies mailed to Protestant ministers, editors, seminary students, and others) of *The Fundamentals*, a twelve-volume series on the "fundamentals of the faith" that appeared between 1910 and 1915. These volumes articulated a conservative theology—biblical inerrancy was at the center—that was designed to serve as the doctrinal rock upon which "orthodox" Christians would do battle with the liberal enemy. But while these volumes were suffused in a culture war binary, and while they provided the name of the crusading movement to come, the approach was not consistently militant. For example, while the Stewarts and the editors embraced dispensational premillennialism and took it to be a given, the volumes failed to proclaim it to be a "fundamental" of the faith. More striking, *The Fundamentals* did not treat Darwinism as anathema, with one essay even suggesting the possibility of theistic evolution.

This moderate (especially in retrospect) tone would be swept away—never to return—in the wake of World War I, when many Americans became convinced that the war against the barbarous "Huns" threatened Christian morality and Western civilization. Conservative evangelicals explained Germany's devolution into amoral savagery in terms of the nation's widespread acceptance of Darwinian evolution and biblical higher criticism (which they took to be a German invention). More than this, dispensationalists saw the British capture of Jerusalem in 1918 as thrilling evidence of the rapidly approaching end of history. In this atmosphere, further charged by the Red Scare, evangelicals gathered in Philadelphia in May of 1919 to create the World's Christian Fundamentals Association (WCFA). Presided over by the ardent Baptist premillennialist William Bell Riley, this interdenominational organization set forth two primary goals: to promote the "fundamentals of the faith" (including biblical inerrancy and dispensational premillennialism) among American Protestants and to purge the major Protestant denominations of liberals and modernists.

While a fundamentalist understanding of the Bible continued to spread rapidly among American evangelicals in the 1920s, and while many Protestant denominations experienced a fundamentalist "controversy," the fundamentalist movement failed miserably in its efforts to capture control of the major Protestant denominations. Aggressive fundamentalist campaigns among the Northern Baptists and the Northern Presbyterians did not succeed in imposing fundamentalist creedal statements on the denominations, nor did they succeed in removing theological liberals from seminaries and mission fields.

By 1922, the fundamentalist movement had turned much of its attention to ridding public schools of Darwinian evolutionism. After all, evolution rejected the Genesis creation account, emphasized natural processes, and seemingly regarded human beings as nothing more than highly developed animals. The movement found the moral results of the latter notion in World War I and the dastardly aggression of a Germany fully committed to a Darwinian "survival of the fittest." In response to this deadly threat, the WCFA and other fundamentalists embarked on a campaign designed to pressure state governments to pass legislation outlawing the teaching of human evolution in the public schools. Tennessee passed such legislation in 1925, making it illegal "to teach any theory that denies the Story of Divine Creation of man as taught in the Bible, and to teach instead that man has descended from a lower order of animal." When Dayton science teacher John Thomas Scopes and the American Civil Liberties Union challenged the law, the result was the media circus known as the Scopes trial. Although Scopes was convicted, a prominent segment of the national media held the fundamentalist movement up to great ridicule, and that ridicule fueled the notion among academics and journalists that the fundamentalist movement not only lost the trial, but its death was a foregone conclusion.

Yet states, particularly in the South, continued to introduce anti-evolution laws, and three states maintained such laws into the 1960s. More important, the fundamentalist movement continued to advance at the grassroots level, with a network of local churches (independent, affiliated with a fundamentalist denomination, or nominally mainline) across the nation that flourished thanks to a rapidly expanding web of nondenominational publishing houses, mission agencies, radio stations, and Bible institutes. Less than two decades after the Scopes trial, the movement reappeared on the national stage. In recognition of the damage done to the word "fundamentalist" in the 1920s, many of these fundamentalists drew upon their nineteenth-century heritage and renamed themselves "neo-evangelical," or, eventually, just "evangelical."

Some conservative Protestants rejected the name change, defiantly holding on to the word "fundamentalist." But in good part this was a squabble over labels: notwithstanding the name change, many "evangelicals" continued (and continue) to maintain their commitments to biblical inerrancy, premillennialism, patriarchy, political conservatism, and creationism. Actually, in the 1930s and 1940s political conservatism had become much more pronounced. Strongly committed to unfettered capitalism, fundamentalists were horrified by the New Deal, viewing Franklin D. Roosevelt's activist state as a clear sign that the one world government of the Antichrist was just around the corner. Such apocalyptic concerns intensified with the onset of the Cold War and the threat of atheistic communism at home and abroad, a threat intensified by the very real fear of nuclear warfare. By the 1950s, fundamentalists and evangelicals clearly and loudly occupied the right end of the American political spectrum. Fervently pro-business, militarist, and anticommunist, they passionately opposed both the expansion of the welfare state and all but the mildest threats to white privilege. While the tumultuous 1960s and 1970s saw the emergence of a small but influential evangelical left, the vast majority of evangelicals and fundamentalists remained staunch political conservatives who decried the antiwar protests and the civil rights movement, opposed the Johnson administration's expansion of the New Deal, adamantly condemned the "sexual revolution" and feminism, attacked U.S. Supreme Court decisions prohibiting institutionalized school prayer and legalizing abortion, and blasted the Internal Revenue Service's efforts to remove tax-exempt status from Christian schools that discriminated on the basis of race.[4]

Tapping into this sense of outrage, television evangelists and shrewd Republican Party operatives in the late 1970s combined forces to mobilize these Protestant conservatives (most of whom were already Republican) to "take back" America by electing "pro-family, pro-life, pro-Bible morality, pro-America candidates" to office. Led by Jerry Falwell and his "Moral Majority," the Christian Right made a substantive contribution to the elections of Ronald Reagan in 1980 and 1984. In the post-Reagan years, the Christian Right became a political powerhouse with an intricate web of local evangelical churches and national organizations, including Focus on the Family and Concerned Women of America. Perhaps most significant, with George W. Bush, the Christian Right had one of their own as president of the United States for the first eight years of the twenty-first century.

In short, over the past four decades the Christian Right has become the most reliable and perhaps the most important constituency within the Republican Party. Demonstrating how important it had become to the GOP, as of 2015 not one of the many candidates for the 2016 Republican presidential nomination publicly affirmed that he or she believed in evolution. While some sought to dodge the question, many flatly rejected evolution, with a number of them suggesting that creationism should be taught in public schools.[5]

Interestingly, when people in the early twenty-first century use the word "creationism," they generally do not mean the "creationism" of William Jennings Bryan and other early fundamentalists. That is to say, what passes as "creationism" in much of fundamentalism and evangelicalism has changed. Oddly enough, this change has its origins far outside American evangelicalism and fundamentalism, in Seventh-day Adventism (SDA). In 1864 Ellen C. White, prophet and (along with her husband, James) SDA founder, had a vision in which she witnessed God's creation of the world in six days (God rested on the seventh, an important point for the fledgling organization because of its focus on the importance of the seventh day as the Sabbath). Not only did White confirm that the Earth was approximately six thousand years old, but she declared that Noah's Flood had reconfigured the Earth's surface and produced the fossil record. No one outside of Adventism seems to have attended to White's proclamations regarding the creation of the Earth until the early twentieth century, when SDA convert George McCready Price embarked on a writing career devoted to explaining and publicizing White's pronouncements. In books such as *Outlines of Modern Christianity and Modern Science* (1902), *The Fundamentals of Geology* (1916), and (most important) *The New Geology* (1923), Price attacked evolution while providing the "scientific" evidence for an understanding of the Earth's past that confirmed Ellen White's vision of a catastrophic global flood. As Price saw it, his "flood geology" not only explained the fossil record but also resolved all questions raised by modern science about the Genesis account of creation.[6]

At the Scopes trial, William Jennings Bryan referred to Price as one of two scientists he respected when it came to the history of the Earth. But Bryan and almost all early fundamentalists were old Earth creationists who had made their peace with mainstream geology. They either interpreted the days in Genesis 1 as allowing for a gap of time between the creative act of Genesis 1:1 and the remainder of the creation process, or they understood the word "day" as not a day of twenty-four hours, but as an "age," that is, a large but unspecified amount of time. Bryan held to the latter "day-age" understanding of Genesis, a point he made clear at the trial under Clarence Darrow's interrogation.[7]

Bryan's betrayal (which is how Price understood it) notwithstand-

ing, Price's flood geology made inroads among American fundamentalists in the first few decades after the Scopes trial. Then, in 1961, John
C. Whitcomb Jr., a theologian and professor of Old Testament at Grace
Seminary in Indiana, joined forces with Henry M. Morris, a PhD in
hydraulic engineering and chair of the civil engineering department
at Virginia Tech, to write *The Genesis Flood: The Biblical Record and
Its Scientific Implications.* Borrowing heavily from Price (while signif-
icantly downplaying their indebtedness to this Seventh-day Adventist,
in order not to alienate their fundamentalist and evangelical audience),
Morris and Whitcomb claimed—as indicated in the book's title—a
"twofold purpose" for *The Genesis Flood.* First, convinced as they were
of the "complete divine inspiration," "verbal inerrancy," and "perspicuity of Scripture," they sought "to ascertain exactly what the Scriptures
say concerning the Flood and related topics." Second, they sought to
delineate the "scientific implications of the Biblical record of the Flood,
seeking if possible to orient the data of these sciences within this Biblical framework." In 489 pages they made their case: the Bible asserts
that Noah's Flood, a global event, lasted one year; science confirms that
this global flood produced the geological strata that can be seen today;
ergo, Morris and Whitcomb demolished the case for evolution and an
old Earth. While all of this did little more than reiterate Price's flood geology (albeit reworked for an evangelical and fundamentalist audience),
Whitcomb and Morris did go beyond the Adventists in one important
detail: they claimed that God created not simply the Earth in six twenty-
four hour days, but, instead, the entire universe, which "must have had
an 'appearance of age' at the moment of creation."[8]

Morris and Whitcomb produced one of the most important books
in twentieth-century American religious history. Like the *Scofield Reference Bible* before it, *The Genesis Flood* and the ideas it promoted
swept through conservative Protestantism with extraordinary speed.
Vast numbers of American evangelicals and fundamentalists enthusiastically accepted the notion that a commitment to reading the Bible
"literally" necessarily required a commitment to a six twenty-four-hour-
day creation; they were reinforced in their commitment by the apparent
scientific apparatus of *The Genesis Flood* (which was replete with footnotes, photographs, and even the occasional mathematical equation). A
host of organizations popped up to spread the young Earth creationist
word throughout the United States and beyond. Among the most important were two organizations with which Morris had direct ties: the

Creation Research Society (CRS), established in 1963, and the Institute for Creation Research (ICR), founded in 1972. While these organizations conducted very little in the way of scientific "research," they argued that "creation science," a legitimate endeavor, deserved equal status with evolutionary science.[9]

ICR's likely greatest long-term contribution to the creationist cause can be found in the fact that it provided the auspices under which Ken Ham made his American debut. Born in 1951 in Cairns, Australia, Ham's father (Mervyn) was a school principal who served at various institutions throughout Queensland, and who inculcated Ham and his siblings in the conviction that the Bible (including the book of Genesis) had to be read literally. Armed with this knowledge, Ham secured a bachelor's degree in applied science from the Queensland Institute of Technology and a diploma in education from the University of Queensland. In 1975 he began work as a science teacher in the town of Dalby, where he later reported to have been appalled by the fact that some of his students assumed their textbooks that taught evolutionary science successfully proved the Bible to be untrue. According to Ham, this experience "put a 'fire in my bones' to do something about the influence that evolutionary thinking was having on students and the public as a whole." Having just read *The Genesis Flood*, which thrilled him, Ham began delivering well-received talks to local churches in behalf of young Earth creationism.[10]

In 1977 Ham moved to a school in Brisbane, where he continued his presentations on young Earth creationism. Soon he joined with another teacher who shared his young Earth creationist views, John Mackay, to begin selling creation science materials to Queensland public schools, which by law taught both evolution and creationism. In 1979, Ham left his job to found with Mackay what eventually became known (after merging in 1980 with the Creation Science Association, a similar group from South Australia headed by Carl Wieland) as the Creation Science Foundation (CSF). The CSF ministry of spreading the young Earth creationist gospel expanded rapidly in Australia; it even ventured into the United States, in the form of speaking tours. In January 1987, Ham moved to the United States to work with Henry Morris and ICR as a traveling creation science evangelist. The next month an extraordinarily strange conflict erupted between Ham (who was still a co-director of CSF) and Mackay. The latter accused Ham's personal secretary, Margaret Buchanan, of being a "broomstick-riding, cauldron-stirring witch," a "frequent attender of seances and satanic orgies" who engaged in "necrophilia." In response to a request for evidence, Mackay claimed that he had received divinely inspired "spiritual discernment." CSF eventually pushed Mackay out of the organization, and (after a few months with CSF scientist Andrew Snelling as temporary manager) Wieland replaced Mackay as the organization's co-director (and later married Buchanan). Despite all of this, Mackay eventually resumed work with Ken Ham.[11]

Ham remained in America, working in behalf of ICR as an evangelist for young Earth creationism, touring the nation and delivering his popular "Back to Genesis" seminars. Unlike ICR, which sought to develop and publicize a "creation science," Ham bypassed research and instead concentrated on reaching Christian laypersons with a simple, three-pronged message. He argued that evolutionary teaching was evil and had produced almost unspeakable cultural decadence; the first eleven chapters of Genesis, read literally, revealed both the truth of the origins of the universe and a guidebook for the proper organization of society; and, finally, true Christians should join the culture war against the forces of atheistic humanism. This message proved to be wildly popular with evangelicals and fundamentalists. In contrast with the generally paltry crowds that attended ICR presentations, people flocked to hear the charismatic Australian creationist. In the wake of his remarkable success, and with Morris's blessing, Ham and a few colleagues left ICR in 1994 to establish Answers in Genesis (AiG) as an outreach of CSF. In 1997 CSF itself became Answers in Genesis, reflecting both the success of the American organization and a commitment to emphasizing biblical creationism. In 2005, Ham and Wieland not-so-amicably parted ways over, to quote AiG's official history, "organizational and philosophical differences" (and not over "doctrinal issues"). Ham retained control of AiG activities in the United States and United Kingdom, while Wieland remained in charge of what was now called Creation Ministries International in Australia, with connections to ministries elsewhere in the English-speaking world.[12]

According to AiG, its purpose is "to provide seminars, lectures, debates, books, along with other forms of media, museums, facilities, and exhibitions that uphold the authority and inerrancy of the Bible as it relates to origins and history." Its website (www.answersingenesis.org), which it launched in 1995, forms the center of all this activity; the popularity of the website (in 2014 it received 14.4 million visits and 43.9

million "pageviews") gives credence to the organization's claim to be "the world's largest apologetics organization."[13]

The website[14] has links to a number of online AiG magazines, including *Answers* (a quarterly magazine started in 2006 for which there is also a print version), which seeks "to illustrate the importance of Genesis in building a creation-based worldview, and to equip readers with practical answers so they can confidently communicate the gospel and biblical authority with accuracy and graciousness," and *Answers in Depth,* also started in 2006, which "provides Christians with powerful apologetic answers, careful critiques, and close examinations of the world around them." One also finds on the AiG website links to various AiG blogs, with Ken Ham's most prominent. Under "Media" visitors can find a link to *Answers with Ken,* the daily sixty-second radio program that Ham started in 1994, which, according to AiG, "is now heard on more than 700 stations." Also under "Media" are: *Answers Conversation,* weekly fifteen-minute podcasts that discuss "the objective propositional truth revealed to us by God through . . . His infallible, inerrant, and inspired Word"; *Answers Mini-Dramas,* sixty-second radio plays on topics such as "Aliens and the Bible," "Dad: Spiritual Leader," and "Halloween Evangelism"; video clips of various lengths on topics such as "Age of the Earth," "Evolution," and "Worldview"; and, a plethora of creationist cartoons attacking (to mention just a few targets) evolution and its social effects, the idea of global warming, and the myth of liberal tolerance. Under "Outreach," one finds a list of conferences and activities (including "Embrace: Answers for Women 2015," "Answers Mega Conference," "Dealing with Compromise: Answers for Pastors," "Children's Ministry Conference," and "Grand Canyon Raft Trips") plus a calendar of large and small conferences and a roster of more than thirty speakers (including some in the United Kingdom) who are available for those seeking to organize an Answers in Genesis event. Finally, a link to the AiG "Store," offers an abundant supply of creationist apparel, books, curricular material, digital downloads, DVDs, and more.[15]

In short, Answers in Genesis is a creationist juggernaut. Strikingly, a relatively small group of people (the same names repeatedly appear) produces a mind-boggling flood of print, media, and social media material. Such production testifies to the missionary zeal of this cadre of young Earth creationists and to the fact that this cadre is relentlessly "on message," presenting the same set of propositions again and again. This is even true for the online *Answers Research Journal (ARJ),* which from its inception in 2008 has been edited by AiG's director of research (and young Earth geologist), Andrew Snelling. *ARJ* advertises itself as "a professional, peer-reviewed technical journal for the publication of interdisciplinary scientific and other relevant research" that produces "cutting-edge creation research." Certainly the titles of many of the articles suggest that *ARJ* is not a typical research publication, for example, "Fungi from the Biblical Perspective," "Where in the World Is the Tower of Babel?," "An Initial Estimate toward Identifying and Numbering the Ark Turtle and Crocodile Kinds," and "A Proposed Bible-Science Perspective on Global Warming," which claims that "there is no reason either biblically or scientifically to fear the exaggerated and misguided claims of catastrophe as a result of increasing levels of man-made carbon dioxide."[16] Moreover, when one looks closely at *ARJ*, one notices that authors are often not identified in the table of contents, and just a few individuals contribute a large percentage of the articles. For example, in 2012, Callie Joubert, whose credentials are not provided, contributed almost 50 percent of the articles published that year, including one in which he uses philosopher of science Michael Ruse to make the point that "a fear of God and the afterlife play a major role in shaping the thinking and behavior of the so-called atheist." In 2013, Joubert only contributed one article; however Simon Turpin, identified on the AiG website as having a "BA degree in theology and intercultural studies," and Danny Faulkner, AiG's resident young universe astronomer, combined to contribute eleven of the thirty articles published that year. The next year *ARJ* editor Snelling (five articles), Joubert (three articles), and Faulkner (six articles, including "Interpreting Craters in Terms of the Day Four Cratering Hypothesis") produced 45 percent of the 2014 volume.[17]

One explanation for the small number of contributors to the *Answers Research Journal* could be that, for the past half-century, "creation science" has produced meager results. But from the beginning Ken Ham and AiG focused not on scientific research but on making the case for biblical creationism. And this meant building a museum. According to Ham, this dream went back to his days in Australia: "standing near an 'ape-man' exhibit" in a "secular . . . museum," he "overheard a father telling his young son, 'This was your ancestor' . . . My heart ached [and] my cry to the Lord was: 'Why can't we have a creation museum that teaches the truth?'" When Ham and his colleagues founded AiG in 1994, they set up shop just south of Cincinnati, a location "chosen because almost 2/3 of America's population lives within 650 miles," and thus perfect for the future Creation Museum. Despite local and national opposition, AiG succeeded in 1999 in getting a forty-seven-acre plot just west of the Cincinnati Airport rezoned for a museum, and then secured the final purchase of the land in May 2000. Just seven years later, the $27 million museum was finished, funded by donations and AiG funds, and without need for a mortgage.[18]

Within a year of its opening, 404,000 visitors had toured the museum.[19] On April 26, 2010, less than three years after opening, the Creation Museum welcomed its one-millionth guest; by the summer of 2015 2.4 million people had visited the museum. The average visitor had a "college or advanced degree," a "household income of $67, 500," and had traveled over 250 miles to get to the museum. With these admission numbers and the museum's ticket prices—as of 2015 adults paid $29.95 for a one-day admission ticket, with an additional $7.95 for a ticket to the Stargazer's Planetarium—the Creation Museum has generated significant tax-free revenues (tax free because AiG is a religious nonprofit and tax-exempt organization). As indicated by AiG's 2013 tax return, the Creation Museum generated nearly $4.8 million in total revenue during the fiscal year that ended June 30, 2013.[20] Visitors keep arriving at the Creation Museum, and AiG expects their numbers to increase by 50 percent annually once the organization's latest project, the construction of a life-size replica of Noah's Ark (the Ark Encounter project), comes to completion at a location just forty-five minutes away from the Creation Museum.[21]

Questions and Method

The Creation Museum is the crown jewel of the AiG apologetics enterprise, an impressive and sophisticated visual argument on behalf of young Earth creationism and a highly politicized fundamentalism. Given its ongoing popularity, given that the kinds of claims it makes increasingly appear in American culture and politics, the Creation Museum demands close examination. This book—the first full-scale scholarly treatment of the museum—offers precisely such an examination. Three central questions animate this examination: What is the message of the Creation Museum? How does the museum convey this message to its visitors? and How (in conveying this message) does the museum constitute its visitors as Christians and as Americans? Just under the surface throughout the book is a fourth question: What does all this mean for American religion and politics?

To answer these questions, the authors engaged in participant observation. Accordingly, we each visited the museum at least seven times since 2007, taking field notes and photographs. In addition, we purchased and watched the videos on display in the museum and read materials by AiG representatives on the museum and its construction. Beyond attending to the museum itself, we read hundreds of books, online articles, and blog entries written by Ken Ham and his AiG colleagues. We also attended an AiG event at a church in eastern Ohio, which provided a window into AiG's outreach programs.

The Creation Museum makes a sophisticated and complex argument by way of state-of-the-art communication technology. AiG further advances these and other arguments on its website, through its books and DVDs, and by way of off-site events. Because a visit to the museum or some other AiG location—real or virtual—can be an overwhelming experience, one of the chief aims of this book is, quite simply, to slow it all down. That is to say, this book contains a close content analysis of the exhibits, placards, dioramas, videos, and images on display in the museum and of the flood of words produced by the small band of AiG young Earth creationists. Only by slowing it down, by taking the time to analyze precisely what is being conveyed, can anyone hope to understand what the Creation Museum is saying and doing and why it matters.

To capture and analyze all that the museum conveys requires an interdisciplinary approach. So this book draws upon several scholarly fields, including religious and political history, museum studies, visual rhetoric, argumentation, biblical studies, and history of science. That said, the focus of the book remains the same throughout: to see as clearly as possible what the Creation Museum displays and says, and to see what it does with what it displays and says.

In this regard this book takes very seriously what the museum says about itself and what AiG spokespersons—especially Ken Ham—say about what the museum is doing. This is particularly true in the first three chapters, where we examine the museum as a museum, the museum's treatment of science, and the museum's treatment of the Bible. These chapters start with what the museum/AiG assert—about the museum as a state-of-the-art museum, about the museum displaying lots of "real science," and about the museum as committed to upholding biblical authority—and then compare these claims with what is going on at the museum.

Suffice to say here that there is a significant gap between what the museum and AiG promise and what the museum actually delivers. The final two chapters—on politics and on judgment—provide a more fully developed explanation of what is happening at the museum. To do this requires more time outside of the confines of the museum and in the larger world of AiG. While both the museum and AiG are public— anyone can visit the museum, visit the AiG website, and/or attend AiG events—much of the messaging in the world of AiG is directed toward the true young Earth creationist believers. Attending carefully to what is being said here, to insiders as it were, reveals much about what is actually happening at the museum.

The Creation Museum seeks to shape Christianity and Christians in powerful ways that will have a lasting impact on American life. All of us have a stake in understanding what is happening at the museum and its role in preparing and arming crusaders for the ongoing culture war that polarizes and poisons U.S. religion and politics. Put simply, as bizarre as the museum may seem to many Americans, what happens inside its doors matters to all of us.

Let us enter.

1

MUSEUM

I n its very name the Creation Museum makes an important claim—
namely, that it is a museum. That claim matters because museums
enjoy a special status "as one of the most important sources for educating our children and as one of the most trustworthy sources of objective
information." People hold museums in high regard for good reason. For
more than two hundred years, museum directors, curators, designers,
trustees, and the like have endeavored to develop the museum as an
institution that not only collects, safeguards, studies, presents, and explains objects of interest but also seeks to improve the people it serves
by, among other things, providing them the most up-to-date and authoritative knowledge about those objects.[1] In its name, then, the Creation Museum claims to be one of those special institutions worthy of the
public's trust and high regard. In this way, the Creation Museum invites
visitors to receive all of the substantive material they encounter as current, reliable, and a resource for self-improvement.

Of course, the Creation Museum goes well beyond merely claiming
that it is a museum. Locked glass cases display what appear to be rare
objects such as fossils, mineral samples, and bones. Inside those cases,
labels identify skulls and other bones by providing both the common
and scientific names of the creatures to which the items once belonged.
Large placards placed throughout the museum feature photographs of
natural phenomena. Other signage displays diagrams of natural and
historical processes including the separation of continents, the transformation of animal species over time, and the movement of peoples
about the Earth. An authoritative-sounding male voice heard by way of
speakers placed throughout the museum provides information for visitors about the contents of displays. Life-size models and skeletal recon-

structions of dinosaurs stand ready for visitors' inspection. Room-size dioramas offer visitors the opportunity to walk through scenes depicting ancient people and animals in what appear to be their natural settings. In short, the Creation Museum does not just say it is a museum. It also looks like one.[2]

That said, not everyone is convinced. Indeed, a number of commentators, both popular and scholarly, have taken issue with this claim. A columnist argues that "the Creation Museum isn't really a museum at all," and a journalist claims that "it all resembles just another Disney-style magic kingdom." A physicist calls it a "museum of misinformation," a geologist calls it an "anti-museum," a philosophy professor says it is a "rebuttal museum," and a biologist characterizes it as a "haunted house."[3] Other journalists who do not explicitly contest the claim that the Creation Museum is a museum nevertheless comment that it may be "one of the weirdest museums in the world," that it offers a "disorienting mix of faith and reason," and that it "induces an eerie vertigo." One journalist recounts that upon entering the Creation Museum he "felt like [he] had entered a world that had gone mad" but then wondered at the close of his tour whether he "was the crazy one after all."[4]

Why does the Creation Museum prompt questions about whether it really is a museum? The quick answer is that although the Creation Museum looks like a museum, it puts forth an argument at odds with the museums it resembles—namely, that the Earth, along with the rest of the universe, is only about six thousand years old. Moreover, in support of that claim, it mobilizes a literal interpretation of the Bible along with scientific evidence. In short, the Creation Museum, as one commentator put it, "stands the natural history museum on its head."[5]

It makes sense that AiG wants to claim that the Creation Museum is a museum and that certain scientists and others energetically contest that claim.[6] However that question gets resolved, the Creation Museum certainly functions as a museum for many, if not most, of its visitors. That being the case, the question that drives this book is not about its proper name but about what it does as it claims to be a museum.

Of course, all museums make arguments throughout their exhibits. Further, whatever its reputation and however well researched its information, no museum provides an objective perspective on its contents. Any claim that a museum makes is open to contestation. In this sense all museums, just like the Creation Museum, are rhetorical—that is, their purposes are persuasive, and their claims are never neutral.[7]

Moreover, museums are rhetorical in more ways than one might expect. For instance, museums make claims about the objects they display. Museums name them, define them, and put them into categories.[8] Often museums place them within historical narratives telling visitors about the causes that produced them, which, for example, might include a brilliant artist, a physical process in nature, or a culture's development over time. A narrative might also include the effects that something has or has had on other things, people, or history. In all sorts of ways, museums tell visitors not only how to think about items on display but even how to look at them. Thus, they provide ways to make meaning of objects whether they are familiar or strange. Indeed, just by displaying an object museums give it special status.[9]

By making meaning of the items displayed, museums also shape the visitors who encounter them. Not only do museums shape the perspective that visitors have or develop of the things in view, but they also shape who the visitor is in relation to them. The museum might position the visitor as the subject of knowledge of items on display—that is, as one who knows what something is, what its behavior is likely to be, what its significance is, and how it ranks in a hierarchy of value. Or a museum might convey to the visitor that he or she is unable to know the object because only individuals with specialized knowledge can really understand it.[10] Museums often situate items on display within big discourses such as the history of art, evolution, the history of human civilizations, scientific knowledge, and so forth. When museums exhibit their objects within these expansive narratives, they provide powerful ways of making sense not just of the objects on display but of any apparently similar object beyond their displays. Thus, museums offer big stories that visitors may adopt to make sense of both the content of museums and their world.

When the Creation Museum is taken seriously on its own terms— that is, as a museum that functions like other museums—what can be seen is that the entire museum is rhetorical. Like any museum, the Creation Museum produces its rhetorical effects with words and objects on display. But that is not all. It also uses glass cases, dioramas, diagrams, drawings, photographs, videos, walking paths, lighting, colors, and so forth to rhetorical effect.[11] Tony Bennett, renowned historian of museums and especially natural history museums, calls these strategies "the how of a museum" and argues that studying the "how" of a museum is at least as important as paying attention to the "what of a museum" or its contents. Only when attention is given to how a museum presents its contents, he argues, can the impact that any museum has on its visitors be discerned.[12]

In this chapter, which focuses on the Creation Museum as a museum, we attend to the "how" of the museum or, to use Bruce Ferguson's term, its material speech. Specifically, we study the material means by which the museum aims to alter or refine how visitors see objects and their relationship to them, how the museum encourages visitors to think about the place of those objects within a bigger story, how the museum invites visitors to think about their own place in that story, and how all that works to shape visitors' convictions about themselves and how that story should unfold. Put simply, the material speech of the Creation Museum is, as it is in all other museums, political.[13] Indeed, it seeks to shape not just how visitors think about young Earth creationism but how they understand themselves as subjects in the world and, in particular, as Christian agents in a world into which they should intervene.

The material speech of the Creation Museum, like any museum, has not arisen in a vacuum. Rather, it has a history, and it is its historical nature—the fact that it borrows the forms of its material speech from museums that have come before it—that makes it intelligible to visitors. Take just one display strategy—the glass case. When visitors enter a museum and see items displayed in a glass case, they know that those items matter. They are important, valuable, worthy of consideration. Why? Because for centuries museums have put important, valuable, and worthy objects in glass cases. But why did museums do that? Even more importantly, what kinds of effects do display cases typically have on visitors? How do they position visitors in relationship to the things they hold?

What follows is a brief history of museums that pays particular attention to the display strategies that museum designers created over time, the purposes for which they were developed, and the effects they tended to have. The focus is on natural history museums (with some attention to science museums), given that the Creation Museum borrows its overall design and display techniques from natural history museums, given that the Creation Museum engages and relies in part on the discourse of science to make its arguments, and given that the ostensive subject matter of the Creation Museum—the creation of the universe and all that is in it—is the kind of subject matter typically encountered in a natural history museum.

A Brief History of Natural History Museums

Before there were museums there were curiosity cabinets of the sixteenth and seventeenth centuries that held rare or exotic objects expected to inspire awe and wonder because of some exceptional quality they possessed. Typically the property of princes and seen by invitation only, these cabinets contained objects that, taken together, represented miniature versions of the world and, if interpreted properly, could reveal the hidden hierarchies put in place by God at the creation. Beyond gaining access to such hidden truths, additional pleasures attendant to owning a curiosity cabinet included displaying one's power to own such extraordinary objects. The increase in travel, exchange, and communication that came with the establishment of colonial empires brought forth an astonishing explosion in the West of objects and artifacts that made this power possible.[14]

The modern or traditional natural history museum emerged in the late eighteenth and early nineteenth centuries[15] and, like the curiosity cabinet, served as a repository for valued objects and artifacts. But whereas the curiosity cabinet displayed items deemed exceptional, the modern museum presented things that were considered typical or common. Moreover, the modern museum displayed a collection much more extensive than that of a curiosity cabinet. Indeed, the modern museum aimed to gather into one place and for all time as exhaustive a collection of objects and artifacts as possible. Another distinctive characteristic of the modern museum was that it was public. Often created by an action of the state, the modern museum was designed to make what had once been held in private collections available for all to see. Given the state's new role in improving its citizens, it is not surprising that the purpose of creating a public space within which important objects and artifacts would be seen went well beyond simply making collections public. The modern natural history museum sought to remake for the better the masses of people who passed through it by displaying its contents for visual inspection, telling a particular kind of history, and controlling how visitors moved through its interiors.[16]

Displaying Objects

Modern natural history museums depended on the glass case and the diorama to display their objects. Of course, the glass case closely resembled the curiosity cabinet in that it too enabled items to be purposefully arranged for view. But while the curiosity cabinet displayed extraordinary things that inspired reverence and awe, the glass case of the modern natural history museum encouraged a scientific gaze upon common or typical items for purposes of close observation. Provided that perspective by the museum, museumgoers could experience a sense of mastery over the objects arrayed before them.[17]

The diorama represented a more significant departure from the curiosity cabinet than did the glass case because it displayed objects and artifacts "in situ." Whereas the curiosity cabinet decontextualized its contents, the diorama seemed to display its contents as if they remained in their natural habitat. Moreover, while the items in the curiosity cabinet were to be revered for their extraordinary qualities, the specimens in a diorama stood in for something much larger such as a whole species or a culture.[18]

To mount animals for a diorama meant capturing, killing, and skinning them in the wild. Then taxidermy, which demanded extensive knowledge of anatomy and tremendous skill in sculpture, was used to produce figures that appeared lifelike. But when museumgoers looked upon a completed diorama, they did not see all that effort. Instead, they saw beautiful animals that looked as though they had been plucked from the wild and deposited in an authentic re-creation of their natural habitat. With all its seemingly effortless realism, the diorama served as a "peephole"—a direct line of sight—into the wild or into some distant culture. Not surprisingly, given the power of vision that dioramas afforded visitors, dioramas became a centerpiece of natural history museums in the nineteenth century.[19]

Of course, the diorama did not provide an objective view of its contents. Rather, it consisted of a painstakingly and thoughtfully constructed scene that always told a story with a certain perspective or agenda driving it. In the case of Carl Akeley's famous mammalian dioramas, which almost always featured a dominant male specimen with smaller female specimens nearby, the story told was that the patriarchal family, with its corresponding sexual division of labor, was *the* natural, necessary, and proper organizing unit for all social creatures.[20] Thus, much more than a peephole, the diorama constructed a story about certain social relations that naturalized them.

The diorama also served as a "time machine" that transported museumgoers with every visit to, as Donna Haraway puts it in her insightful analysis of Akeley's dioramas, a veritable "Garden of Eden" that

ordered everything as it should be. At a time when Americans understood themselves to be facing troubling forms of social and cultural
decay at the hands of monopoly capitalism and technological change,
Haraway argues, these modern dioramas provided encouragement especially for men about the dominant role that they needed to fulfill so
that things might be right once again in America.[21]

Both the glass case and the diorama privileged sight over the other
senses. First and foremost, visitors were to observe the collection of
objects and artifacts in the museum. That the modern natural history
museum put a premium on sight is not at all surprising given the centrality of sight for the production of knowledge in the nineteenth century in the West. Whether through the camera lens, the telescope, the
microscope, or controlled observation, sight revealed the truth of all
things. Undergirding the idea that sight revealed truth was, of course,
empiricism—the notion that human beings gain knowledge through experience via the senses and especially sight.[22]

Objects and artifacts were presented in the glass case and diorama
complete with labels to provide museumgoers with information that
would render those objects and artifacts (to borrow a term from Bennett) immediately "legible." In this way, modern natural history museums enabled visitors to see and therefore to know something important
about those objects and artifacts. Moreover, the realism employed in the
construction of the dioramas discouraged museumgoers from seeing
the diorama as constructing meaning and encouraged them, instead, to
see the diorama as providing a direct and reliable experience of its contents. Given that the objects and artifacts presented were often drawn
from places and cultures considered strange by Western museumgoers,
it is fair to say that the modern natural history museum afforded its
visitors a "specular dominance," or profound sense of power through
vision, over a world that otherwise was likely to seem confusing and
bewildering.[23]

Telling History

With the general acceptance of Darwin's thesis regarding evolution
among the various modern disciplines such as history, geology, biology,
and anthropology came the development of a particular grand narrative of natural history. According to that grand narrative, all life shared
a common origin and developed progressively through time by way of
evolutionary processes. Situated at the outermost point of that narra-

tive and, thus, at the apex of all life, history, and civilization was modern man. Gendered as he was, modern man represented not only the
best that had ever been but also the promise of further progress into the
future.[24]

Modern museums assisted in the construction and dissemination
of this grand narrative by fundamentally altering the logic that had
governed the display of objects. Whereas in the eighteenth century, objects were typically displayed on tables within categories constructed
from the similarities and differences observed among the objects, in
the nineteenth century they were arranged according to their place
within an ordered development. Thus, the modern museum served as
a "backteller" or intellectual force that saw the past in the objects it col-
lected, discerned the universal laws that ordered them, and then transformed that knowledge into a series of exhibits that made that history
and the logic of progress that drove it unmistakable and memorable for
visitors.[25]

Controlling Movement

With the birth of the modern museum and other public spaces such
as railway stations, department stores, and public exhibitions, concern
emerged about the conduct of crowds. How might a large gathering of
people from a wide range of social and economic classes be kept from
devolving into chaos? Moreover, was there a way to control such crowds
that would be "both unobtrusive and self-perpetuating?" A solution
emerged with the crucial insight that at the heart of the visitors' experience in museums (and other public spaces) was their movement. Once
visitors were seen as "minds on legs," museums focused on constraining
their movement not only for purposes of crowd control but also so that
visitors could not help but internalize the narratives and, thus, the messages of museums.[26]

Instead of consisting of one or more large open spaces within which
visitors were free to move about (or not) at will, the modern natural
history museum was structured by a series of rooms that constituted
a unidirectional path. Within the series of rooms, objects and artifacts
were sequenced according to that linear grand narrative of developmental progress. As those "minds on legs" moved from one room to the
next, they seemed to witness evolutionary processes as those processes
unfolded. Structured in this way, museums disincentivized any movement that transgressed the prescribed path. Outside that preset order

and its corresponding narrative, the contents of the museum had little meaning.[27]

As museumgoers moved along the path set before them, they not only learned that evolutionary processes governed all life. They also adopted the scientific perspective of the time such that they saw themselves as also subject to those laws. Thus they came to observe their own behavior as an object of knowledge and came to assess it according to its compliance with the progressive laws of nature. If they found defects in their behavior, they could rectify them.[28]

The modern natural history museum provided opportunities for such self-improvement. Through its architectural features, like the promenade that was borrowed from the arcade, the modern natural history museum made it easy for visitors to observe one another. Thus, lower class visitors could watch the behaviors of "their betters" and imitate their "more highly developed" mannerisms. In addition, as museumgoers observed others on the promenade they also learned that they too were being watched. With that awareness, they became intensely self-monitoring and self-regulating.[29]

In all of these ways, the modern natural history museum taught its visitors that they were the subjects of a powerful knowledge that could reveal the hidden truth of all things and also that they were the objects of that same knowledge. In the end, the modern natural history museum educated its Western visitors in the incontrovertible march of progress that lay beneath all visible evidence from nature. Moreover, it invited them to adopt a social position within that grand narrative or script from which they would recognize their elevated location within the hierarchical order of nature and better embody the truth of their rank through self-monitoring and emulation of others.[30]

A famous exhibit that employed strategies typical of the modern natural history museum and that illustrates an additional effect beyond those already mentioned was the Races of Mankind exhibit that was on display at the Field Museum in Chicago from 1936 to 1966.[31] Commissioned in 1929 by Stanley Field, then president of the Field Museum, the exhibit consisted of 104 busts, heads, and life-size bronze figures intended to represent an array of human races from around the world. Designed and cast by artist Malvina Hoffman, the figures were realistic looking—like animal figures in a diorama—as if they had just been plucked out of their natural surroundings. In addition, the sculptures were arranged in a developmental sequence from "least" to "most" ad-

vanced racial type. The figure that appeared at the end of the developmental series and, thus, as the culmination of evolutionary development
was "American/Nordic Man." And, whereas the other life-size figures in
the exhibit were depicted in a costume of their culture, and while each
also carried an everyday implement, American/Nordic Man was naked
and carried nothing. Thus, he appeared to transcend racial typography.
As Tracy Lang Teslow puts it, "He recalls no moment of life, but rather
classical statuary. . . . 'American' or 'Nordic' man represents a classless,
raceless human pinnacle."[32]

Teslow argues that the exhibit was popular because the figures
looked so realistic, because (unlike real human beings) they did not resist the close scrutiny of museumgoers, and because their arrangement
in a developmental series seemed to prove that the white European (and
especially male) visitor represented the apex of all history and development. Thus, the Field Museum and so many other natural history
museums with similar exhibits taught their visitors that the nation that
was home to the museum represented the apex of all human and cultural development.[33]

The Contemporary Natural History Museum

In September 1969, a new kind of museum opened in the once-empty
Palace of Fine Arts in San Francisco's Marina District. The brainchild of
Frank Oppenheimer, member of the Manhattan Project and brother to
its director (J. Robert Oppenheimer), the Exploratorium took a different approach than had the modern museum to presenting science to its
visitors. Rather than send visitors down predetermined paths designed
to reveal the hidden laws of the universe, Oppenheimer's museum was
constructed as an open space, absent any borders or boundaries between
exhibits, within which visitors could wander as they pleased. And rather
than display objects in glass cases and dioramas for museumgoers to
gaze upon, Oppenheimer designed (with the help of many others) a
museum that put science and technology directly into the hands of the
visitor by offering a collection of interactive exhibits that encouraged
visitors to employ all of their senses toward an exciting experience in
which they made their own scientific and technological discoveries.[34]

Oppenheimer's new museum responded to two problems from
which science and technology suffered in the mid to late twentieth century. The first was that in the twentieth century, science had become a

discourse most people could not understand. Its focus was too specialized, its vocabulary too esoteric, its objects of study either too big or too
small to comprehend. The second problem was that, with the profound
changes in industrial warfare witnessed during World War II, Americans had come to fear some forms of technology and their implications
for life on Earth. As an innovative and dedicated science educator, Oppenheimer responded to these challenges by creating a museum organized around themes of human perception that enabled laypeople to have
a direct and comprehensible experience of science and technology as
positive forces in their lives.[35]

Oppenheimer's creative redesign of the modern museum quickly
spread to museums across the country.[36] As the demands of a consumer
economy became more pressing, natural history museums (along with
all other kinds of museums) were seen less as educational institutions
serving a national public and more as independent interests that endeavored to offer an entertaining product (the exhibit) to a selective
consumer (the museumgoer) in a competitive market of leisure in-
dustries. Moreover, museums were responding to an important critique of the modern museum—namely, that it had taken a paternalis-
tic and therefore antidemocratic approach to the museum visitor.[37] By
determining the path of the visitor, by putting objects out of reach, by
staging dioramas to instruct visitors in the proper way to do family,
by telling grand stories about the ineluctable laws of the universe, and
so forth, so the argument went, modern museums treated visitors as
though they were merely passive recipients of the wisdom of invisible
experts.

In addition to adopting Oppenheimer's borderless, interactive, and
multisensory museum design, contemporary museums also changed the
way they represented knowledge. Rather than construct exhibits that
put forth the supposedly one-and-only definitive word on any matter,
contemporary museums often stage exhibits that raise questions about
the authority of scientific knowledge. Rather than tell the visitor what to
think, such exhibits more likely pose challenging questions about what
scientists think they know on a matter. Moreover, instead of providing
a single and presumably correct answer to some question, exhibits tend
to present multiple views. In these ways, contemporary exhibits have
become self-reflexive both about the production of knowledge and its
authority as well as about the power that museums exercise in presenting that knowledge. The hope in making these changes is that museum-

Stone and metal gate located at the entrance of the Creation Museum that signals the importance of dinosaurs at the museum, both as source of entertainment and as evidence of a young Earth. Photograph taken by Susan Trollinger. For more images of and information about the museum, go to creationmuseum.org.

goers will become active participants in the production and discussion of knowledge.[38]

A Tour of the Creation Museum

In the course of their history, natural history museums have developed a rich array of forms and strategies that new museums may draw upon for their own designs. Not surprisingly, the Creation Museum has drawn on those forms and strategies, but which of the many choices available does it employ? Do the strategies and forms used at the Creation Museum tend to derive more from the modern museum or the contemporary museum, and more importantly, what is their likely impact? Answers to these questions begin with a brief tour of the Creation Museum.[39]

As indicated in the introduction, the Creation Museum is located about twenty miles west of Cincinnati in a rural area just off Interstate

Outdoor dining area overlooking a pond and extensive landscaped grounds that surround the Creation Museum and that underscore the claim made within the museum that beauty such as this must have been created by the God of the Bible. Photograph taken by Susan Trollinger. For more images of and information about the museum, go to creationmuseum.org.

275. It is housed in a $27 million facility that sits on forty-nine acres of land. Much of the land immediately surrounding the museum has been transformed into meticulously landscaped areas that include a manmade lake, a pond, three themed gardens, and a petting zoo. Large parking lots provide ample room for cars, RVs, and school buses. The building itself is 75,000 square feet and takes the shape of an irregular polygon that roughly resembles a pentagon. The two front sides of the museum consist of floor-to-ceiling tinted windows that are about two stories high and that are interrupted by large beige cement columns. Other exterior walls are covered in what looks like locally quarried stone.[40] A life-size model of a Sauropod often stands near the main entrance.

Just inside the main entrance, visitors find themselves in the Portico with its forty-foot ceiling and tinted windows in the exterior wall. Interior walls appear to be made of the same stone-like material as the

A child plays near two small *Tyrannosaurus rex* dinosaurs in this life-size diorama in the Main Hall of the Creation Museum providing a provocative visual representation of the argument that human beings cohabitated the Earth with dinosaurs. Photograph taken by Susan Trollinger. For more images of and information about the museum, go to creationmuseum.org.

exterior walls. The floor consists of tumbled pavers. Inside the main entrance visitors can peruse placards explaining how the dragons that appeared in legend and literature represent a shared cultural memory of the time when humans and dinosaurs walked the Earth together. Just past the Dragon Legends exhibit visitors can purchase tickets to gain access to the museum's exhibits ($29.95 for adults) and to the Stargazer's Planetarium ($7.95 for adults). Having purchased tickets, visitors are encouraged to have their picture taken in front of a green screen that yields a souvenir photograph that makes it appear as though they were standing near Noah's Ark, or in the Garden of Eden, or—most dramatically—next to what looks like a hungry dinosaur.

Visitors then enter the Main Hall, a large open space. The high ceilings continue, and the walls and floor are done in what appears to be stone tile. Immediately to the right of the Main Hall is Noah's Café, an Ark-themed restaurant that serves American fare. Beyond the café is a series of hallways that are lined with glass cases that display skeletal objects, fossils, and mineral deposits. At the end of those hallways, visitors enter a multimedia theater featuring the film *Men in White*.[41] Across the Main Hall from Noah's Café is located both the Stargazer's Planetarium and the Dragon Hall Bookstore, where visitors can purchase books, DVDs, and souvenirs.

At the center of the Main Hall is a long hallway, on the right side of which (if one is coming from the Portico) is a series of inset display cases that are home to live birds, frogs, and chameleons. On the left side of the hallway is a large diorama of a rocky setting in which two animatronic boys play near two small animatronic *Tyrannosauras rex* dinosaurs. Above visitors' heads, a forty-foot-long animatronic Sauropod sits atop the wall that houses the glassed-in cases and munches on green leaves.

At the end of the Main Hall sits the entrance to the Bible Walk-through Experience. The walls of the entrance resemble the curved and layered rock of the Grand Canyon, and the floor looks like a well-worn stone path. An authentic-looking "Grand Canyon National Park" sign marks the opening. At the end of the main path (a second path splits off that is designed especially for children), visitors enter a room called the Dig Site. At the center of this room is a life-size diorama that features two figures who sit on sandy ground that seems to be yielding a dinosaur skeleton to their brushes and chisels. To explain the scene, a white man dressed just like one of the figures in the diorama appears on flat-screen monitors to say that one of the men at the dig is an evolutionist and the other is a creationist but that both are scientists.

As visitors enter the next room, called Starting Points, they are confronted by a light gray wall upon which a question has been stenciled in bold white and charcoal gray letters: "Same Facts, but Different Views . . . Why?" When they turn the corner into this modern-looking room with its blond wood floors, gray walls, and black ceiling, the answer is provided on a large placard that opposes "Human Reason" (represented by a stack of books) and "God's Word" (represented by a scroll). Above the placard, words stenciled on the wall tell the visitor that there are "Different Views because of Different Starting Points." A small sign attached to this placard instructs the visitor, "Individuals must choose God's Word as the starting point for all their reasoning, or start with their own arbitrary philosophy as the starting point for evaluating everything around them, including how they view the Bible."[42] This opposition between "Human Reason" and "God's Word" is then reiterated on a series of placards that contrast the different conclusions that evolutionists and creationists draw from the evidence they encounter in the world. Following this series of signs, another presents a timeline from 4004 BC to the present. On that timeline are placed the "Seven Cs" or "God's intervention at key periods of history [that] explains most of the world we see today."[43] At the center of this room stands a large glass case that holds a model of a small ape. All about the case are placards that contest the scientific argument that the skeleton discovered in 1974, which was named "Lucy," provides a crucial link in the evolutionary development of human beings.[44]

From here, visitors move into the Biblical Authority room wherein another sharp contrast is drawn, this time between the unity of God's Word and challenges that human beings have made to it over time. When visitors enter this room, they encounter life-size models of Old Testament figures Isaiah, Moses, and David, as well the New Testament apostle Paul. At the feet of these figures, a long placard displays what appear to be words spoken by them (and other biblical figures) regarding the Creation, the Flood, the Resurrection, and a new creation. On a wall opposite these figures, a large flat screen displays a video of individuals speaking short excerpts from scripture. On the far wall a huge placard reports the many times and ways in which human beings (beginning with Adam and Eve and including ancient philosophers, the medieval church, linguists, geologists, and archaeologists) have contested "God's Word." Visitors follow this sign to the Biblical Relevance room in which, among other things, a life-size figure of Luther appears to be posting his ninety-five theses, a wood printing press produces copies of the Bible, and the Scopes trial unfolds on a video screen. Placed in a prominent spot on the largest wall in the room is a timeline that depicts the fall of the church into disobedience from God's Word at the hands of modern philosophers and scientists along with misguided theologians and evangelists.

Turning the corner from this room into the next, the visitor enters Graffiti Alley, a narrow, dark hallway lined with what appear to be exterior brick walls covered in the sort of graffiti visitors have learned to associate with street gangs. Plastered on those bricks are pages torn from newspapers and news magazines about such topics as stem-cell research, the terrorist attacks of 9/11, euthanasia, abortion, gay teens, Islamic terrorism, evolution, gay marriage, school shootings, and the decline of Christian America. As visitors move through the hall, they hear sounds of the city. Turning the corner of Graffiti Alley, visitors enter another dark "street" depicting a Culture in Crisis. Along the left side of the street is what appears to be the wood-sided exterior of a suburban home. Along that wall, windows protrude from the siding and reveal, by way of video loops, activities going on within the home. Opposite these windows stands the brick exterior of what looks to be a mainline Protestant church. The same individuals who appear in the windows of the house also appear in the windows of the church. Outside, a wrecking ball with "Millions of Years" carved into its side looks like it has just smashed into the foundation of the church.

When visitors leave the Culture in Crisis room they enter the Time Tunnel, a dark hallway with small white lights all around, that delivers them to the Six Days Theater. There, visitors watch a computer-generated animation, projected on a curved wall in front of them, of the story of Creation based in Genesis 1. From here, visitors enter the room devoted to the Wonders of Creation. This is a large, brightly lit space with off-white walls that are interrupted here and there by fat, white pillars. On these walls are hung brilliant color photographs that are about the size of a large poster. These photographs are framed and lit from behind by a bright light so that they seem to glow from within. The photographs depict various creatures and plants, the Earth as seen from space, the solar system, the double helix, and so forth. Above many of these photographs are hung flat screens that play fifteen videos in a continuous loop on topics such as plants, the Sun, the solar system, DNA, and the mechanics of bird flight that aim to prove that "there has to be a powerful Intelligence behind the universe."[45] Benches set parallel to these walls provide an opportunity for visitors to sit as they view these images and videos. Floor-to-ceiling signage at the center of the room provides an outline of the six days of Creation and excerpts from scripture. As visitors exit this large open space, they enter a smaller one, with similar décor, wherein they encounter a computer-generated animation of the creation of Adam.

From here, visitors pass between two more white pillars and enter a completely different environment—a lush, life-size re-creation of a portion of the Garden of Eden complete with trees, huge rock formations, a waterfall and pool, various animal figures (including animatronic dinosaurs), and Adam and Eve. These human figures appear in multiple biblical scenes such as when Adam names the animals, just after Eve

Along the path of the Bible Walkthrough Experience, a child visitor looks on as an animatronic *Tyrannosaurus rex* stands over its bloody kill signaling the entrance of carnivorous diets and violence among animals that came as a result of the Fall. Photograph taken by Susan Trollinger. For more images of and information about the museum, go to creationmuseum.org.

was created from Adam's rib, and when Eve presents the forbidden fruit to Adam.

Visitors then enter the Cave of Sorrows, two rooms that depict scenes of sin that follow from Adam and Eve's transgression. In the first room, large black-and-white images are hung on the walls to the left and projected on concrete gray walls to the right. These images depict a man preparing to inject himself with a hypodermic needle, a mushroom cloud emerging from the explosion of an atomic weapon, and a huge pile of human skulls. In the next room, visitors come upon more life-size re-creations of mostly biblical scenes depicting the effects of Adam and Eve's sin. These include the first animal sacrifice, Adam cultivating a garden for food, Cain's murder of Abel, and a *Tyrannosauras rex* devouring a smaller dinosaur.[46]

After visitors pass through what looks like an ancient tent and hear an animatronic Methuselah talk about Adam's descendants and the coming catastrophe, they enter Ark Construction, which portrays Noah's worksite for the Ark. Much of the room is occupied by an installation of a full-scale wood reproduction of a tiny portion of the Ark that visitors can pass through. Nearby, an animatronic Noah stands talking to his foreman while animatronic craftsmen work on the Ark and craftswomen make baskets. In the adjoining room, Voyage of the Ark, visitors enter a re-creation of the interior of the Ark with its wood floors, wood-paneled walls, beamed ceiling, and warm lighting. Here, visitors can listen to another animatronic Noah answer preprogrammed questions about life on the Ark, study the inner workings of animal waste removal systems on the Ark, and look at miniature dioramas depicting various scenes within and outside the Ark. On a composite set of four large flat screens, visitors watch a computer-generated animation of the Flood as it subsumes the Earth.

The next room, Flood Geology, is modern looking with high black ceilings, dark carpet, and dark walls. Some of the walls are painted to resemble rock formations or the shapes of the continents. Lighting systems illuminate the many exhibits that explain flood geology. A miniature diorama depicting Noah and his family worshipping God after the Flood welcomes the visitor into the area. Placards nearby discuss the dramatic effects of the Mount Saint Helen's eruption, which are said to provide "clues" to the power of the Flood to alter the face of the Earth. The Natural Selection room includes descriptive placards and glass case exhibits that argue that natural selection does not lead to evolution. Added in May 2014, the Facing the Allosaurus room contains a thirty-foot-long and ten-foot-high dinosaur skeleton standing behind a glass partition.[47] Nearby, digital animations describe the skeletal remains that were found and the dig site, as well as placards describing the Allosaurus's physical dimensions along with his demise in the Flood.[48] A final room in the flood geology area features placards explaining how the Flood caused a gigantic landmass to break apart into the continents known today and how the Flood carved canyons out of rock and formed new lakes. They also explain how the cooling of North America following the Flood brought changes through natural selection such that the animals on the Ark produced descendants who became the horses and dogs known today.

From here, visitors enter the last room of the Bible Walkthrough, the Babel room, which looks like a scene from an ancient Babylonian city. The walls are covered in a mélange of light blue, medium blue, and green brick and tile. Decorative gold tile forms a border at the ceiling. Golden ram heads mounted on one of the brick walls deliver water through their mouths into a tile pool below. A lamassu, or winged bull with a human head, stands guard at the center of the room. In this room, visitors learn that after God thwarted the plan of Noah's descendants to build a tower, they spread across the Earth and developed unique physical traits according to the processes of natural selection. Placards explain the emergence of religions that departed from Noah's and the emergence of racism.

As visitors exit this room, they pass by a timeline that begins with the "First Family" and finishes with "Rome" and that recounts biblical examples of God's "Judgment of the Nations" as well as God's "Blessing of Nations." An adjoining room, with walls that look like the stone and brick exterior walls of an ancient street, serves as a waiting area for entrance into the Last Adam Theater. Inside the theater, a twenty-minute video displayed on three screens offers a dramatization of Jesus's death and resurrection as seen through the eyes of his mother, Mary. Across the theater is a room that originally served as a chapel "built to represent a first-century synagogue."[49]

From here visitors exit the Walkthrough and enter Palm Plaza, a large open area "designed with an Egyptian architectural influence."[50] The walls, which look like the exterior walls in an ancient village, appear to be made of large slabs of light-colored stone with matching round pillars that support deep ledges. The floor is finished in a mix of natural-toned tile. A series of fake, lighted palm trees stand in a single row down the center of the room. To the right can be found a traveling exhibit on loan from the Bible Museum, which typically showcases rare books and/or other artifacts. To the left is Dr. Crawley's Insectorium, which was added in May 2013.[51] Therein visitors may look at case upon case of preserved insects, ask an animatronic scientist questions about bugs and their origins, or play a video game featuring a dragonfly. Visitors who take the stairs to an area above the Insectorium enter Buddy Davis's Dino Den where they can examine the skeletal structure of a Triceratops along with sculpted models of various other dinosaurs. Placards near each specimen identify the name of the dinosaur and provide information about the creature such as its height, weight, length, the location where its fossil was found, and what it ate after the Fall.

Next, visitors pass through a hallway, climb a set of stairs, and

enter the Dragon Hall Bookstore from its back entrance, above which
a statue of Beowulf stands guard. According to AiG, the bookstore was
"[t]hemed after a medieval castle's great hall" with its high ceiling,
stone facades, and medieval-styled arches and lighting fixtures.[52] Home
schooling materials, Sunday school resources, DVDs, and books line
the stained wood bookcases that cover most of the walls. Developing the
theme of the bookstore, an engraved faux stone façade tells the story of
Saint George, who, by slaying a beast (which looks a lot like a dinosaur),
saved the king's daughter and converted the townspeople to Christianity. As visitors exit the bookstore and re-enter the Main Hall where their
tour began, they pass beneath an animatronic Pteranodon that resembles the image of a dragon that appears just below it and the name of
the bookstore.

The Creation Museum: Contemporary or Modern?

The Creation Museum incorporates an impressive array of new technologies: a multimedia theater with special effects, a planetarium project-
ing images of the universe on a thirty-foot-diameter dome, numerous
flat-screen monitors, many computer-generated animations and images, sound effects to accompany life-size dioramas, five animatronic
dinosaurs, additional animatronic human figures, and many interactive displays. Just considering its impressive incorporation of advanced
technology, the Creation Museum looks like it belongs among those
museums considered contemporary.

Both popular and scholarly commentators on the Creation Museum
tend to agree. ABC News describes the Creation Museum as a "high-tech
sensory experience with animatronic dinosaurs and a movie theater
with seats that shake." Graphic designer and journalist Jandos Rothstein
notes that "AiG has co-opted the latest trends in information design,
with the same sophisticated graphics, displays, and attitude of any large
science museum." And architect Joseph Clark, writing for an on-line
magazine says that "its high-tech immersive dioramas and media-savvy
exhibits are wholly contemporary." Even Lawrence Krauss, a professor
of physics who vigorously disputes the claim that the Creation Museum
is a museum, grants that the Creation Museum "takes advantage of the
technical wizardry of science to get its point across, with a series of dazzling animatronic displays and explanations."[53]

But its use of advanced technology is not the only similarity between the Creation Museum and a contemporary natural history museum. The Creation Museum presents two perspectives on the origins of Earth: one based in Genesis and the authority of God's Word, and another based in reason and the scientific method. Moreover, the museum makes a point of instructing visitors that the choice between these two perspectives is entirely theirs. Thus, the Creation Museum positions visitors, like good consumers, as subjects who decide between competing perspectives. Finally, even as the Creation Museum mobilizes the discourse of science in support of its arguments for a young Earth, it repeatedly raises questions about the authority of science. Thus, not just its use of advanced technology but also its incorporation of multiple perspectives and its positioning of the visitor as the one who chooses among them suggests that the Creation Museum is best understood as a contemporary natural history museum.

Not surprisingly, AiG encourages visitors to consider the Creation Museum a cutting-edge, state-of-the-art natural history museum. On the home page of the Creation Museum, its description of the museum begins: "The state-of-the-art 75,000 square foot museum brings the pages of the Bible to life." A similar claim also appears in AiG's companion volume, *Journey through the Creation Museum*, which says, "This journey through history is a multimedia adventure—incorporating the latest in cutting-edge technology to communicate effectively on multiple levels, a memorable presentation visitors of any age will enjoy."[54]

Moreover, the fact that the Creation Museum presents not one but two perspectives on the origins of the Earth is something spokespersons for the museum like to bring to people's attention. According to Barbara Bradley Hagerty, Ken Ham had this to say in her interview with him on National Public Radio: "'We actually do give both sides as people walk in,' he says, explaining that a fossil exhibit has 'a creation paleontologist' and 'an evolutionary paleontologist' offering different interpretations of the same fossil." Likewise, on the AiG website and in response to a critic who accused the museum of "promot[ing] isolation in thought," Bodie Hodge, a writer and researcher for AiG, argues that "we present the major evolutionary beliefs inside the museum, and in a fair way—as even many of our detractors have told us." He continues, "The museum has major exhibits that examine natural selection, the proposed cornerstone of Darwinian evolution." Finally, in an interview with Stephen Bates, a journalist for *The Guardian*, Ken Ham underscores the point with a comparison between the Creation Museum and a world-renowned natural history museum: "We give both sides, which is more than the Science Museum in London does."[55]

That a museum putting forth a case for a young Earth based in a literal reading of the Bible would not just mention evolution but actually dedicate a whole diorama to making the point that young Earth creationism is not the only perspective available startles some visitors, including journalist and *Discover* columnist Bruno Maddox:

> The disarming and unexpected thing about the Answers in Genesis folks, however, is this: They don't pretend to be right. Yes, we get to wander through a full-scale section of the Garden of Eden and gawp at the Tree of Knowledge. . . .
>
> But every time I feel my scientific hackles start to rise, I turn a corner and there's a reminder . . . that everything I've seen is merely a theory, a possible scenario, a best guess. The exhibit of which the museum is most proud is a simple, life-size diorama in which two khaki-clad paleontologists are squatting and peering down at the bones of a fossilized dinosaur. Beside one is an open Bible; beside the other is a standard paleontology textbook—the message being . . . that everyone's view of reality is inescapably colored and distorted by that person's "starting assumptions." In other words, truth is an illusion, and no one can ever really know anything. I had heard tell that creationists had a problem with Darwin, but I had no idea they idolized Jacques Derrida.[56]

If it is right that the Creation Museum favors the forms and conventions of the contemporary natural history museum, then it would be reasonable to expect that the Creation Museum would have democratizing effects on its visitors. But is this the case? Does it employ the museum design and display strategies of the contemporary natural history museum such that its visitors are empowered to make their own discoveries, draw their own conclusions, and contest the authoritative voice of the museum—even perhaps to the point of inviting them to consider the possibility that "truth is an illusion"?

Displaying Objects at the Creation Museum

Even as the Creation Museum employs some of the most advanced technologies of contemporary museum display, it also makes ample use of traditional forms. One that is used throughout the museum, but especially in and around the Main Hall, is the glass case. Six inset glass cases appear in the Main Hall and offer visitors an opportunity to inspect finches, frogs, fossils, and a placard that outlines AiG's Seven Cs. Another series of floor-to-ceiling glass cases lines one side of the three hallways through which visitors pass on their way to the Special Effects Theater, in which numerous specimens of fish jaws, animal skulls and teeth, mineral deposits, and fossils are mounted on the back wall of the case with labels identifying their scientific and common names. Other glass cases located in various places throughout the museum display such items as a replica of the Lucy skeleton along with a mannequin depicting what she might have looked like according to AiG's biblical perspective, a facsimile of a Codex or fourth-century Greek Bible, and a copy of Charles Darwin's *Descent of Man*.

Like glass cases in a modern natural history museum, the Creation Museum's glass cases make it possible for the visitor to examine objects, such as fossils, skeletons, and animal teeth, which they may have never before encountered. Labels next to items in the cases alert visitors who are likely to be unfamiliar with this or that fish jaw or mineral deposit as to what exactly they are examining. Placards provide additional information and perspective on the contents of every case. In short, the Creation Museum's glass cases position the visitor as the subject of knowledge of their contents while simultaneously alerting the visitor to the fact that the creators of the Creation Museum, who assembled the items as well as produced all those labels and placards, have considerably more expertise than visitors do about the contents of the cases.

Importantly, however, the locked glass doors also present obstacles to the visitors, limiting how close they can get to the objects and insuring that they cannot use any of their other senses, especially touch, to explore the objects. Of course, by limiting visitors' access to objects in this way, the glass provides security for its contents. Perhaps even more importantly, it conveys a clear message that the objects held within are important, perhaps rare or valuable, and worthy of visitors' consideration. In this way, these glass cases also convey, just as they did in the modern natural history museum, that the Creation Museum is the sort of special institution that has possession of valuable specimens and artifacts.

More prominent than the glass case at the Creation Museum is the diorama. The first one that a visitor is likely to see is in the Main Hall. It is a life-size depiction of two children playing near two dinosaurs. Many other life-size dioramas follow this one that feature scenes such as the two paleontologists uncovering the same dinosaur skeleton, the three

Old Testament figures, the Apostle Paul writing 2 Timothy 3:15–17 in Greek, and a *Tyrannosaurus rex* standing amidst rocks and foliage. Of course, the most impressive use of the diorama at the Creation Museum comes in the form of a series of life-size scenes that flow one into the next and that depicts the Garden of Eden, the Fall, and the construction of the Ark (the latter of which is followed by numerous miniature dioramas that provide visitors both interior and exterior three-dimensional views of Noah's boat).

Although taxidermy had no role in the creation of the many animal and other figures that appear throughout the Creation Museum's dioramas, realism nevertheless is the ruling aesthetic. By way of other visual arts such as sculpture and painting and new technologies such as electronics and robotics, all of the figures on display at the Creation Museum are meant, as AiG puts it in comments regarding the turtles in the Main Hall, to "look so real."[57] Indeed, elaborate detail was used, especially in the dioramas depicting scenes from Genesis. Just one example of the fruit of such effort is the Tree of Life. Rising twenty feet out of the floor of the Garden and with a canopy thirty feet wide at the top, the Tree of Life is, according to AiG, "stunningly detailed, [with] over 31,000 leaves . . . each placed by hand upon the tree during its construction."[58]

Such attention to detail, which was a hallmark of modern museum dioramas, is coupled at the Creation Museum with another signature strategy of the modern museum's diorama—namely, the creation of scenes that look like snapshots of the featured figures in action. Indeed, every biblical scene (and other scenes besides) that appears in the museum is depicted in this way. Adam reaches his hand toward an animal he is naming while other animals look on with apparent interest. Just after Eve has been created, Adam rises from his induced sleep to see her for the first time. Adam runs his fingers through Eve's hair and looks at her intently as the two of them stand in a pool near the base of a waterfall. Adam takes his hoe to the vegetable garden as one child harvests carrots, another child carries a basket of grapes, and a pregnant Eve descends stone stairs nearby. Cain raises his arms and cries out in anguish as his brother lies on the ground, his blood staining the sand.

Miniature dioramas have the same snapshot look as the life-size versions. One offers a cut-away of the interior of the Ark that shows a woman descending a set of stairs into the lower level where two big cats are pawing at a couple of turtles in one bay and a pair of dinosaur-like creatures rest on a bed of straw in another. Another depicts the Ark finally resting on high ground as receding waters cascade down cliffs nearby. Still another captures Noah and his family gathered on a hilltop as they present an animal sacrifice to God. Still another shows the Ark floating safely on the fast-rising waters as numerous animals and humans who have scrambled to the top of nearby mountains confront their impending doom.[59]

Just as the realism of the modern museum's dioramas functioned like a peephole into the life of animals and people who appeared strange to Western visitors, so too the Creation Museum provides its visitors a peephole but this time into stories found in the first eleven chapters of the book of Genesis. Moreover, the Creation Museum's adaptation of the diorama makes it all seem even more real. The fact that the museum's life-size dioramas are not enclosed in glass, as the modern museum's dioramas typically were, affords the visitor the feeling not just of viewing biblical scenes but of walking among them. The absence of the glass, the flow of one scene into another, and the accompanying sound effects all strongly encourage visitors to feel as though they are witnessing these stories as they unfold.

Although the Creation Museum has eliminated the glass for most of its life-size dioramas to powerful effect, it has put in its place other barriers so that its dioramas remain, like their modern counterparts, off limits to visitors. Faux rock formations, railings, fences, ropes, and the like, along with placards that say "Thou Shalt Not Touch! Please," keep visitors at a distance from the three-dimensional scenes.[60] Moreover, glass is used as a barrier for the many miniature dioramas that appear throughout the museum. The point here is that while the Creation Museum has adapted the diorama in order to intensify the effects of its realism, the diorama remains here, as it was in the modern museum, a form that privileges sight over the other senses and keeps the visitor at a distance.

With its realism and emphasis on sight, the Creation Museum offers its visitor the same specular dominance that the modern museum did. But this time, visual command is not over some seemingly strange animal or exotic culture. Instead, visitors hold within their vision important objects like skeletal remains, mineral deposits, and rare fossils that promise to offer evidence of the historical truth of those ancient stories taken off the biblical page and brought to three-dimensional life. Thus, the Creation Museum positions its visitor as the subject of knowledge of those objects and stories.

Telling a Story at the Creation Museum

The Creation Museum does not just tell individual stories that appear in Genesis 1–11. It also tells a much larger story, a story that encompasses all of history from the beginning to the end of time, but a story that is dramatically different from the progressive one characteristic of the modern natural history museum.

The visitor first encounters the Creation Museum's story in the Main Hall, in the first glass case of the series of inset glass cases that displays poison dart frogs, finches, and chameleons. There one finds a placard containing the "Seven Cs in God's Eternal Plan," which represent the "seven pivotal events from the beginning to the end of time."[61] These Seven Cs reappear on a much larger placard in the Starting Points room, which at the top states that "God's Word *is the key* to the past, present, and future" (emphasis in original). Below this statement the Seven Cs are positioned on a timeline that begins at 4004 BC and ends in the present, and are divided into three groups: "God's Perfect Creation," "God's Judgments for the Rejection of God's Word," and "God's Restoration of Creation."

Each of the Seven Cs then appears singly on a large placard, placed at spots among the museum's exhibits. Each placard identifies the C, explains the C, and offers a supporting quote from the Bible. Not surprisingly, the placard for the first C, "Creation," appears at the entrance to the Garden of Eden. The title of this placard is "Creation by God's Word." Under the title, additional text explains, "In the beginning—in six, twenty-four-hour days—God made a perfect creation (~4000 BC)"; the scenes that follow include Adam naming the animals and Adam and Eve just after her creation, and Adam and Eve in the pool at the base of the waterfall. Right after the waterfall is the placard for the second C, "Corruption," which is entitled "Rejection of God's Word Led to Corruption," and which describes the Fall: "The first man, Adam, disobeyed the Creator, bringing death and corruption into the creation. His disobedience explains the catastrophes, disease, suffering, and death in the present world." After this placard comes inset glass cases, the first with a red snake that appears to be coming toward the visitor, and the second with Eve presenting the fruit to Adam.

The third C, "Catastrophe," appears after visitors leave the scene of

Cain's murder of Abel and pass by a tent wherein an animatronic Methuselah talks about the cataclysmic event that is about to occur. Entitled "Rejection of God's Word Led to the Catastrophe," the placard explains that "Adam's race became so wicked that God judged the earth with a catastrophic, global Flood, saving only those on the Ark built by Noah (~2348 BC). This global catastrophe resulted in fossils all over the earth." Then comes the Ark construction site, the Voyage of the Ark room, the Flood Geology room, and the Babel room, where visitors encounter the fourth C, "Confusion," on a placard hanging on a pillar. Underneath the title, "Rejection of God's Word Led to Confusion," the text reads, "When Noah's descendants disobeyed God's command to fill the earth, God gave them different languages, forcing them to spread over the earth. The scattering of people explains the formation of different people groups."

Placards for the final three Cs hang from the ceiling along one wall in the waiting area for the Last Adam Theater.[62] The fifth placard is entitled "Christ: The Promise of God's Word" and explains, "The Creator became a man, our relative—a member of the human race. His name was Jesus of Nazareth, who obeyed God in everything, unlike the first man, Adam." The sixth placard is entitled "Cross: The Answer of God's Word"; below the title, the placard says, "The penalty for mankind's disobedience was death. Jesus, the Messiah, died on a cross to pay that penalty. He rose from the dead, providing life for all who trust in Him." The seventh and final placard, "Consummation: The Fulfillment of God's Word," contains the following text: "One day the Creator will remake His creation. He will cast out death and the disobedient, and dwell eternally with all those who trust in Him. Earth will be restored to a perfect place—as it was before sin."

Presented as they are on these placards that appear through the course of the Bible Walkthrough Experience, the Seven Cs tell a linear story that begins with creation and ends with the establishment of God's kingdom on Earth at some unknown point in the future. More than this, the Seven Cs provide a powerful interpretive structure for understanding the meaning of any single story in the Bible. Instead of having to figure out what any particular story means, the visitor needs only to memorize the Seven Cs and to know where among them a story falls to understand its fundamental meaning. In a booklet devoted to the Seven Cs, Ken Ham and Stacia McKeever explain it this way: "Most people look at the Bible as a book that contains many interesting stories and theological teaching. While this is true, the Bible is much more—it's a history book that reveals the major events of history that are foundational to the Bible's important message."[63]

Anyone familiar with American evangelicalism and fundamentalism is sure to hear strong resonances between the Seven Cs of the Creation Museum and dispensational premillennialism, developed in the 1830s by Plymouth Brethren founder John Nelson Darby and popularized by Cyrus Scofield in his 1909 *Scofield Reference Bible*.[64] This is not at all surprising, given that young Earth creationism has great similarities with dispensational premillennialism; in fact, as Mark Noll asserts, "Creationism could, in fact, be called scientific dispensationalism, for creation scientists carry the same attitude toward catastrophe and the sharp break between eras into their science that dispensationalists see in the Scripture." Not surprisingly, a great many young Earth creationists were and are dispensational premillennialists. Among this number are Henry Morris and John Whitcomb, authors of *The Genesis Flood*, who, as Ronald L. Numbers notes, "offered a compelling view of earth history framed by symmetrical catastrophic events and connected by a common [literalist] hermeneutics."[65]

Although AiG appears to remain publicly agnostic on the topic,[66] it is obvious that the Seven Cs of the Creation Museum are very much within the tradition of dispensational premillennialism. At the most foundational level, both understand the Bible to have a referential relationship to reality. What the Bible says is literally true. That being so, the stories the Bible tells of the past and the events it prophesies for the future constitute an accurate record not just of "biblical" history but of all history.[67] In addition, both systems provide a seven-part interpretive grid for understanding that total history. With that grid, followers can figure out the meaning and significance of any story in the Bible and any event in history. That is to say, both systems offer a totalizing history. Nothing happens outside this supernatural narrative.

In keeping with the "seven Ds" of dispensational premillennialism, the Seven Cs of the Creation Museum position the visitor as the subject of knowledge not only of the deep truth of history but of the very will of God. As that subject of knowledge, the visitor knows what God must do about a culture declining into total apostasy and all those sinners making it that way in order to set things right once and for all. That kind of knowledge likely makes for a heady experience for conservative Christians and, indeed, any visitors who feel that contemporary culture

The 7 Ds versus the 7 Cs

7 Dispensations	7 Cs of the Creation Museum
Innocence: Eden to the Fall	Creation
Conscience: Adam to Noah	Corruption (the Fall)
Human Government: Noah to Abraham	Catastrophe (the Flood)
Patriarchal Rule: Abraham to Moses	Confusion (Babel)
Law: Moses to Christ	Christ
Grace: the Church Age	Cross
Millennium: Christ's Reign	Consummation

lost its moorings a long time ago, leaving behind good order, proper morals, and right belief as well as the sort of people, like themselves, who remained faithful to all three. In the end—according to both dispensational premillennialism and young Earth creationism—true "Biblical Christians" will be proven right and all will be made right.

For all their obvious similarities, however, the "seven Ds" and the Seven Cs are not identical (see table).

Note how the Seven Cs differ from the "seven Ds" of dispensational premillennialism in kind. The Seven Cs refer to individual events, bookended by the beginning and end of history: the first four Cs come from the first eleven chapters of Genesis—the great mythical prelude of the Hebrew scriptures (much more on this in chapter 3)—and then there is Christ, his crucifixion, and, sometime in the future, the consummation of history.

In contrast, the dispensations refer to eras of history, each of which has its own distinctive requirements proscribed by God that apply in one era but not in the next. At its core, classical dispensational premillennialism seeks to resolve questions about the relationship between Jews and the church and about the relationship between the Hebrew scriptures and the New Testament. As articulated by John Nelson Darby, there are two peoples of God: the earthly people of God (Israel) and the heavenly people of God (the church). But to borrow from scholar Robert O. Smith, in Darby's system of discontinuous dispensations, "God can only deal with one group at a time." The Jews had their time in the past, when they were governed by the Law, but "when the Jews rejected Jesus," to quote historian Doug Frank on dispensational premillennialism, "God determined to create a 'heavenly' people, a church made of Gentiles who acknowledged Christ as their Savior and who lived not under law but under grace." The Jews nevertheless remain God's people, and after Christians have been caught up in the Rapture, the Jewish remnant that survives the Tribulation will have their earthly kingdom restored.[68]

Some dispensationalists have claimed that dispensational premillennialism marks a rejection of "supersessionism," the notion that Christians replace the Jews in God's plan of salvation. Smith cogently argues that such an argument ignores that in Darby's system Israel remains subordinate to the Church—that is, the earthly realm is subordinate to the heavenly realm—and that for Jews the only hope of eternal salvation is faith in Jesus Christ. That is to say, "Darby viewed Jews not as real persons, but as literary tropes in his world of prophecy interpretation." In dispensational premillennialism, Jews do not exist with their own integrity, but instead serve as bit players in the Christian apocalyptic drama.[69]

This said, there is certainly more attention paid to Jews in the "seven Ds" than in the Seven Cs. None of the Seven Cs has anything to do with ancient Israel. In the museum itself there are a few references to ancient Israel, but not many. There are a number of placards pertaining to the Hebrew prophets, such as Isaiah, but the focus in most of these is on fulfilled prophecies (particularly pertaining to the Messiah) that establish that the Bible really is God's Word. The representation of Moses in the Biblical Authority room includes the following placard (note the focus on the first eleven chapters of Genesis): "Moses gave Israel the first five books of the Bible, known as the Law. The first book, Genesis, tells of creation, man's first sin, Noah's flood, and the birth of the nation Israel." For all the Hebrew text serving as background in the Wonders of Creation room and elsewhere (more on this in chapter 3), it is striking that there is so little in the museum pertaining to the history of ancient Israel, so little pertaining to the last thirty-nine chapters of Genesis and the remaining thirty-seven books of the Hebrew scriptures.

One way to think about this is that in the Creation Museum, as is the case in dispensational premillennialism, Jews have been consigned to playing minor roles in a drama scripted by Christians. In 2007, one month after the Creation Museum opened, Bar-Ilan University (Tel Aviv) professor Noah J. Efron, who was in Cincinnati to give lectures at the famed Hebrew Union College-Jewish Institute of Religion, gathered a group of rabbinical students to visit the Creation Museum. As Efron reports in the preface to his book, *A Chosen Calling: Jews in Science in the Twentieth Century,* he and his students did not experience anti-Semitism at the museum: "Inasmuch as [the museum] portrayed Jews at all, it did so with sensitive respect." But "over the hours [they] spent in the museum," Efron and his "vanload of rabbis" became increasingly uneasy, distressed, and despondent; "by the time" they had "loaded into the van to travel back to the college," their "despondency had blossomed into a sort of aching fear." As Efron puts it, he and the rabbinical students realized that at the Creation Museum—most obvious in the museum's efforts to link evolution and anti-Semitism[70]—"Jews had been co-opted into a battle" against science that they had not asked to join and did not want to fight. That is to say, "Jews had been co-opted into a museum in which we had no place."[71]

It is not just that the Seven Cs and the Creation Museum have virtually nothing to say about ancient Israel and the "dispensation of law" that is so central to the interpretive schema promoted by John Nelson Darby and C. I. Scofield. According to the sixth D of dispensational premillennialism, we now live in the "church age," that is, the "age of Grace." In contrast, with the sixth C as the Cross and the seventh C as the Consummation, the interpretive schema governing the Creation Museum leaps over all of history from the death of Jesus to the future millennium. That is to say, there is no attention to the church[72] or to an "age of Grace" in the Seven Cs; in fact, the word "grace" does not appear on any of the Seven Cs placards or in the Seven Cs booklet put out by AiG.

To be sure, in the museum and in AiG publications one hears many resonances of Darby's notion of the "church age" as marking the ever-growing corruption and apostasy of Christian churches;[73] even more apparent are the allusions to dispensational premillennialist notions such as Rapture, Tribulation, and Last Judgment. This said, the Creation Museum's de-emphasis on the history of ancient Israel and the history of the Christian church—a de-emphasis at odds with dispensational premillennialism, which at its core focuses on the relationship between Jews and the church—points to a deeper innovation that the Seven Cs have made in the Darbyite schema. The Creation Museum has replaced dispensational premillennialism's complicated story of discontinuity (seven different dispensations with different requirements from God for each of the dispensations) with a much simpler story of events along a continuous historical narrative. According to the museum's much simpler story, God created a perfect world, placed Adam in that world, and gave him the Word that he was to obey (Creation). As long as he obeyed that Word, all was well. But when Adam disobeyed that Word, God punished him along with the whole creation by introducing suffering, pain, and death to the world (Corruption).[74] In the years that followed, human beings eventually became so disobedient to God, they became so wicked, that God responded with a global flood that destroyed all life except the obedient Noah and those who were with him on the Ark (Catastrophe). Later, at Babel, Noah's descendants disobeyed God's Word that they were to "fill the Earth," instead staying together to build a great city. God punished them by confusing their languages and scattering them across the globe (Confusion).

This story is continuous, easy to understand, relentlessly repetitive. In effect, the Seven Cs of the Creation Museum simplify dispensational premillenialism, eliminating the different requirements and different stories of the different dispensations, and replacing them with one fundamental dynamic that governs all of history: God sets forth the Word for all humanity to obey; human beings disobey that Word; God, as a righteous and good God, punishes the disobedient (and spares the righteous). Such a story of compelling continuity perhaps helps to explain why, unlike dispensational premillennialism, the Creation Museum pays virtually no attention to ancient Israel (and its particular history), the Christian church (and its particular history), and the relationship of these "two peoples of God." Instead, the overwhelming message at the Creation Museum is that all humanity was and is governed by the same historical logic: command, disobedience, punishment. That is, all humanity was and is part of the same, repetitive, continuous historical narrative.

Of course, the sixth C placard (which hangs just around the corner from the Last Adam Theater) asserts, and the *Last Adam* film reiterates, that Jesus "died on a cross to pay [the] penalty" for "mankind's disobedience," and rising "from the dead" he "provid[es] life for all who trust in him." Thus, Jesus represents a discontinuity, a disruption in the smooth, consistent historical narrative. It is thus striking how little the museum makes of it. That is, it is striking how little "Jesus" is to be found at the Creation Museum. While there are a great many placards throughout the museum with quotes from the Bible, few of these placards contain a quote from Jesus. Visual representations of Jesus in the museum appear to be limited to a white statue that is usually confined to a corner outside the Last Adam Theater (except, it seems, during the Christmas season when it appears in the Main Hall) and images of Jesus as he is being crucified that appear in *The Last Adam* film.

The relative absence of Jesus highlights the essential continuity presented at the Creation Museum: God gives the Word; humans disobey it; God is obliged to punish them. According to this logic, the present is a reiteration of the past. The Creation Museum makes this relationship between the present and the past vivid in the Walkthrough. The Walkthrough begins in the present with a couple of contemporary scientists reflecting on their starting points at the dig site. In the Starting Points room, the present appears again in placards that display photographs of people struggling with suffering, despair, and death. Two rooms later, the present appears again, and quite dramatically, as visitors enter Graffiti Alley and confront newspaper articles and magazine covers about 9–11, school shootings, abortion, gay marriage, school prayer, euthanasia, and so forth. When visitors turn the corner into the Culture in Crisis room, they witness videos representing teens' use of pornography and drugs, teenage pregnancy, failing marriages, and compromising preachers.

At the Creation Museum the present is constructed as a dark place in which sex, violence, and corruption reign, and in which an apostate church is powerless to help. More than this, the museum instructs visitors that the significant events of the past serve as analogies to this dark present. This is made visible by the scenes from the present that appear among the scenes of biblical corruption. Just after visitors pass the two inset cases that show the serpent coming toward the visitor and Eve presenting the fruit to Adam, they turn a corner and enter a short hall with a dramatically different feel. Unlike the lush scene of the Garden, this hall is stark with gray floors and walls. At the end of the hall is a gray, beat-up door that is shut and secured with seven locks. Scratched into its exterior wood surface is the phrase, "The world's not safe anymore."

Next, visitors enter the Cave of Sorrows, which contains the same gray walls and floor. Bright directional lighting shines on a series of large black-and-white images of a mushroom cloud, a child of dark brown skin and distended belly, a white man who is preparing to inject himself with a hypodermic needle, and a white woman screaming from the pain of childbirth. Across from these two walls that come together to form one corner of the room is a third wall consisting of concrete slabs set on angles upon which are projected black-and-white images of Nazi parades, a cemetery filled with identical stone grave markers, an invasion of tanks, and piles of human skulls. From here visitors re-enter biblical scenes, but these, unlike those prior to the Fall, are all about sin and suffering. Thus, the present, which is bleakly depicted and sand-wiched between Adam's sin, on the one hand, and various other scenes of sin from the biblical narrative, on the other, is shown to be essentially the same as those ancient scenes. Just like them, the present follows from Adam's sin and results in violence, pain, and suffering.

From here visitors pass through Methuselah's tent wherein he warns of God's judgment that is coming in response to all that sin. The conclu-sion of the analogy is thus made plain: just as God judged ancient peo-ple for their sins so too will God judge people in the present. Moreover, the punishment that God wrought in the past is a clear sign of what is to come. The scale and intensity of the Flood's destruction reveal the na-ture of God's judgment. Perhaps herein lies the reason that the Creation Museum does not focus on the seventh C—Consummation. Unlike dis-pensational premillennialism, the Creation Museum need not say much about Consummation or the end times since the end is prefigured in the Flood. Put another way, at the Creation Museum, the Flood replaces the end times.

Like the modern natural history museum, the Creation Museum is structured by an underlying narrative that is made visible in its displays and exhibits. Also like the modern natural history museum, the narrative at the Creation Museum is a grand one. It speaks to all history and all peo-ple. No event and no individual can escape it. As a result, like the modern natural history museum, the Creation Museum positions the visitor as the subject of knowledge of all history. Positioned that way, the visitor may un-derstand how to make meaning of any story in the Bible and, indeed, any event from the past or present. Even more, s/he knows what is to come.

That said, the grand narrative of the Creation Museum runs by a logic at odds with that of the modern natural history museum. The modern natural history museum told a story of change over time that was driven by a logic of progress. By contrast, the Creation Museum tells a story of continuity over time that is driven by a logic of repetition. According to the Creation Museum's story, every human being in the present is essentially the same as Adam, the first human who disobeyed God. As that subject, the visitor to the Creation Museum knows s/he will likely disobey God too. Moreover, s/he is also aware that Western cul-ture today, just like God's people from the past, is declining into greater apostasy, sliding rapidly toward God's judgment. Finally, whereas the visitor to the modern natural history museum had a general idea of how evolutionary development would unfold, the visitor to the Creation Museum only need attend to the Flood to know exactly what lies ahead: death and damnation.

The story that drives the Creation Museum is a dark one. But it is not without hope, at least for some. When we look closely at how the Creation Museum employs that third innovation of the modern natural history museum—that is, control of visitors' movement—we see that it does offer visitors a second chance, albeit one of a rather violent sort.

Moving through the Creation Museum

Just as in the modern natural history museum, in the Creation Museum space is carved up, and rooms are designed to move visitors along a unidirectional path, especially once they enter the Walkthrough. Rather than provide as Oppenheimer's museum did, a large open space with many exhibits among which visitors can meander at will, the Walkthrough consists of eighteen rooms arranged in a series. All but one room has one entrance and one exit.[75] Exhibits are arranged inside each room such that the visitor is very likely to move from the entrance past each display in the order that they appear on the way to the exit. The size and/or shape of several of the rooms make it difficult, especially when the museum is crowded, for someone to move along any but a predetermined path. A few rooms, like those devoted to Flood Geology and the Wonders of Creation, are larger and more open, thereby encouraging visitors to move about them more freely. While movement within these rooms may be less linear than in the other rooms, visitors are still likely to pass through them in sequence.

Another way that the Creation Museum encourages visitors to follow a certain path is by making a promise to them. As mentioned above, after visitors move through Graffiti Alley and then Culture in Crisis, wherein they encounter many signs, both public and private, of their decaying culture and the church's impotence in the face of such decline, they come upon two sentences stenciled boldly on a gray wall just before the entrance to the Time Tunnel: "As dark as things appear, *God's Word* gives us *the foundation* to rebuild. The place to start is . . . *at the beginning, six thousand years ago*" (ellipses and emphasis in original). Thus, visitors are promised that the answer for all that darkness and decay will be made known if they will embark on a journey that begins at a particular origin, namely, the first chapter of Genesis, read literally.

As visitors move through subsequent rooms, and especially as they follow the path through the Garden of Eden, that promise is fulfilled.

Indeed, by moving along the path set before them, visitors experience that powerful narrative, described above, according to which the darkness of today is understood as another moment within an ancient and recurring story of God's command, human disobedience, and God's judgment. If visitors were to diverge from the path, they might miss the hidden secret of all history that is revealed by this narrative and that provides the answer to the darkness in the present.[76]

In addition to encouraging visitors to move along a set path, the Creation Museum borrows another strategy of the modern natural history museum that has to do with movement. It will be remembered that the design of modern natural history museums included promenades and other architectural features that encouraged visitors to watch one another. Walking along these promenades enabled more "common" visitors to observe their "betters" so as to learn what "proper" behavior looked like. Moreover, by watching others, visitors also became aware that they too were being watched. That awareness provided strong incentive for visitors to scrutinize their own behavior and, where found lacking, to embark on self-improvement so that they might more closely resemble their "betters."

The Creation Museum has no promenade, although it does have a few larger rooms that could provide opportunities for observing and being observed. Still, other, perhaps even more powerful, encouragements for self-scrutiny can be found in the museum. One such encouragement appears in the Culture in Crisis room. Here, it will be recalled, visitors come upon three sets of windows protruding from what appears to be the exterior wall of a single-family home. One window within each set is a flat screen that plays in a continuous loop a video that shows actors portraying three scenes. In the first, one adolescent boy plays a violent video game while another looks at pornography on the Web. A bag of drugs sits nearby. In the second window, an adolescent girl talks to her friend on the phone about the fact that she is pregnant and wants to "'get this fixed.'" A placard below tells visitors that "she is considering abortion." In the third window, the apparent mother of these adolescents sits in the kitchen with a female friend and, according to the placard below, "complains about her husband, pastor's wife, and others" and shows a "lack of involvement in her children's lives." In the background, a man reclines on a couch while watching TV with a pizza box and what appears to be a bottle of beer at his side.

To see these videos (each seven minutes in length) visitors must approach the windows, since they are relatively small, about the width of an average-size adult body. Also, a black speaker located just below the window frame plays the soundtrack for each video. It is not very loud so that to hear each soundtrack, visitors must approach the speaker. Finally, the words on the placard below each window that explain the scene are also fairly small such that getting close to them is required if one is to read them. In order to see the videos properly, read the placards, and hear the soundtracks, then, visitors typically stand right up next to a black railing that runs along the length of the exterior wall of the house about two feet in front of it and lean toward the windows.

Positioned in this way, visitors become peeping Toms peering into someone else's house. Unbeknownst to the individuals inside, visitors not only watch the private activities of the people inside but also listen to them talk about their secrets. Thus, the visitors are positioned as voyeurs. Moreover, when visitors then look at the fourth window in the room, this one set in the exterior wall of a church, they learn that the individuals they have been watching are fellow Christians.

As it was in the promenade of the modern natural history museum, so it is at the Creation Museum: to watch is also to become aware of being watched. To peer into these private spaces and see what sins are hidden there is also to become aware that within one's own private spaces may be lurking similar secrets. Peeping into these windows, looking where one ought not look, the visitor is invited to wonder what someone else (perhaps God) might see if he or she were to look into the private spaces of the visitor's life. What secrets might be revealed there? What would someone peeping into their windows see the children, mother, father doing? Is their own suburban home subject to the same kind of moral decay as this one?

That the Culture in Crisis room invites this kind of self-scrutiny is also suggested by its striking similarities with Hell Houses. Hell Houses are typically created and sponsored by conservative Christian churches or other Christian organizations. Their interiors are designed to be dark and scary, like a haunted house. And like a haunted house, they are divided into rooms. In each room, amateur actors depict intimate scenes designed to "reveal" the horrific consequences of unrepentant sin. The scenes are typically about alcohol and drug use, premarital sex, abortion, homosexuality, and the like. The purpose of Hell Houses is to scare visitors into scrutinizing their own private thoughts and behaviors toward changing their ways.[77] Just like Hell Houses, the Creation Museum's Culture in Crisis room invites visitors into a voyeuristic relationship with other people's intimate sins and secrets so that visitors may become vigilant in scrutinizing themselves.

In the modern natural history museum, visitors learned to measure themselves against both their "betters" (whom they could observe in the open spaces and promenades of the museum) and the greatest achievement evolutionary history had produced—namely, modern Western man. In this way, visitors became both the subject of the history that unfolded before them and its object. The same dynamic appears to be underway at the Creation Museum. Just as visitors to the Creation Museum are encouraged to become the subjects of the museum's grand narrative such that they may know the hidden truth of all events past and present as well as be able to predict the future, they are also urged to recognize themselves as objects of that same history and, especially, its hidden truth. In short, the judgment that befell Adam and Eve, their descendants, the descendants of Noah, and so forth could befall visitors too. Thus, they are advised to scrutinize themselves with regard to their own disobedience. Do they understand who God is? Do they know what God is like? Are they obeying God's Word? Do they fully understand what God's judgment will mean for those who do not obey?

One place along the Bible Walkthrough Experience that positions visitors powerfully in relationship to these questions is the Voyage of the Ark room. There a miniature diorama depicts a scene in which men, women, and children along with a few animals (two full-grown tigers, a full-grown bear, and a cub) are in various states of misery atop three rock formations poking out of the fast-rising floodwaters. Some people are attempting to climb out of the waters to gain footing on the rock. Others are sprawled upon the rock either face up or face down, unable to sit or stand. Most of those able to stand on the rock have their backs to the visitor and are looking at and waving their arms toward the Ark as it passes by them. A placard nearby explains that "God shut Noah's family into the Ark. This ended any opportunity for people outside the Ark to be saved."

Standing before this diorama, the visitor is positioned in relationship to the Ark in a manner similar to those perishing on the rocks. Like the people on the rocks, the visitor is outside the Ark, looking at it as it passes by, its door shut. As the placard nearby says, the opportunity to be spared this horrifying demise is limited. Once the opportunity passes, it will not come again. Only suffering and death can follow.

This miniature diorama, located in the Voyage of the Ark room, instructs visitors in the stark contrast between the safety and security enjoyed by those inside the Ark who obey God and the catastrophic destruction and suffering delivered upon those who disobey Him. Photograph taken by Susan Trollinger. For more images of and information about the museum, go to creationmuseum.org.

A speech given by Ken Ham at an AiG conference in July 2013 confirms that the Creation Museum means to position the visitor in this manner. In that speech, Ham makes the analogy between the wickedness of the ancient past and the wickedness of today:

> Really, as I read Matthew 24—"as in the days of Noah so shall the coming of the Son of Man be"—in many ways we could look at today and say it's just like the days of Noah. "And God saw the wickedness of man was great in the earth and every intent of the thoughts of his heart was only evil continually." That is certainly true of an increasing part of this culture.

Continuing the analogy, he argues:

> We want to be preachers of righteousness. We want people to understand this event did happen. God did send the animals to Noah. Eight

people did go on board that Ark. You know, the Bible seems to indicate
that when it was fully loaded that Ark stood for seven days before God
shut the door. I don't know what Noah was doing in those seven days.
I like to think that maybe he was out being a preacher of righteousness, standing in front of the door saying, "come through the door to be
saved. There's judgment coming." And nobody else came in [sound of
door slamming shut]. God shut the door [slam]. Let me do that again
[slam]. And then the people outside realized something was wrong and
the Flood did come.[78]

This moment is one of the darkest moments in the museum. Here,
the visitor confronts most vividly what it means to be the subject and
object of this story that is the history of all time. To be its subject is to
know that judgment is coming and that it will be violent, catastrophic,
and global. To be its object is to recognize that within the visitor's own
culture, home, and life is likely to be enough wickedness to call forth
God's judgment at any time. Still, there is hope (of a sort) for the visitor.

As visitors look upon this scene of judgment, they are standing in
the Voyage of the Ark room, a room constructed to look like the interior
of the Ark. When they shift their gaze from this diorama to the rest of
the room, they find themselves in a space distinctive for the warmth of
its décor. The floor is solid wood. The walls are wood paneled. The ceiling is also made of wood as are its hand-hewn beams. All of the wood in
the room is stained a golden color. A wooden table and chairs sit in the
center of the room. A large-piece wooden puzzle of the Ark awaits children who might play with it.

Set into the walls of the room is a series of dioramas. Next to the diorama of the Ark and the people on the rocks is another diorama. This
one depicts a much more pleasant scene wherein the animals are walking two-by-two up a long wooden ramp into the belly of the Ark. Other
smaller miniature dioramas that follow along the perimeter of the room
depict domestic scenes within the Ark.

In one, Noah and his family enjoy a meal in a room that closely resembles the Voyage of the Ark room with its wood walls, wood floor,
and beamed ceiling. Along one wall of the room sit clay jars and baskets
for food storage and preparation. On the other side of the cozy room,
blue and gold tapestries hang behind the family. Oil lamps hang overhead. The family consists of Noah and his wife along with three beau-
tifully dressed young couples. Noah appears to be speaking to the three

couples as he gestures. In each pair, the man and woman sit close to one another. One couple holds hands. All three couples are looking at Noah and appear to be listening intently. A meal is spread before them. Hand-woven rugs lie beneath their feet.

Other miniature dioramas show members of the family feeding and otherwise caring for the animals, who themselves appear to be at ease and comfortable. Elsewhere in the room can be found a grouping of full-size clay pots (like the ones that appear in the miniature diorama depicting the meal shared by Noah and his family), a life-size animal cage, a life-size example of a waste disposal system for the Ark, and an animatronic Noah who sits at an elaborately carved wood desk and patiently answers questions about life on the Ark.

The point is that when visitors observe the miniature people perishing on the rocks, they are positioned as if they are inside the Ark. They are standing within its wood-paneled walls and on its solid wood floors. They are set within the very same space wherein Noah's family enjoys a meal and takes good care of the animals. Another miniature diorama underscores this point. In it, Noah and his family stand inside the Ark, looking out the door. All the animals are inside the Ark with them. And, as a placard explains, God is shutting the door. As the visitor looks at this diorama, s/he is positioned as though s/he is standing behind Noah and his family watching God shut the door on the rest of Creation. On the other side of that door are all the unrighteous, the unrepentant, the people on the rocks. On this side of the door are Noah, his family, the animals, and the visitor.

According to the story told at the Creation Museum, all creation had been warned of the coming judgment. There was a sign, and it was the Ark. As it was being constructed and for seven days after it was completed it was visible to all creation. Moreover, according to the animatronic Methuselah, Noah had been preaching about the coming judgment and inviting people to get on the Ark. But none except Noah's family believed his warnings or heeded God's sign. All of the remaining people on Earth ignored Noah's claim that God's judgment was coming and time was running out. Given that they ignored the call to obey God, the reasoning seems to be, there need be no sympathy for all those men, women, children, and animals on the other side of that door. They got what was coming to them.[79]

In the Voyage of the Ark room, the visitor gets a vivid picture of what the righteous God of judgment is obliged to do to those who disobey; simultaneously, s/he is given the opportunity to imagine in concrete terms (by way of the dioramas and the décor) what it would be like to receive God's favor, escape God's judgment, and reside with the righteous. Of course, the visitor knows that even as the Flood serves as an analogy for the judgment that is to come, it is nevertheless an event located in the past. Thus, the visitor also knows that s/he has not, in fact, escaped God's judgment. Still, knowing well that in the future something like the Ark will emerge again to separate the righteous from the damned, the visitor is strongly encouraged to do whatever it takes to get on board, so to speak. Having made his/her way through much of the Bible Walkthrough already, the visitor has a very good idea what is required of the righteous.

First, the visitor must get history right. S/he must understand that God is obliged to respond to the present state of cultural decay as God has done in the past by bringing forth total catastrophic destruction. Second, s/he must become thoroughly obedient to God's Word. Third, s/he must not be troubled by the fact that all who fail to meet these demands will suffer and die in that judgment. In a nutshell, this seems to be the Creation Museum's answer to all that darkness in contemporary culture that was promised as the visitor exited the Culture in Crisis room and entered the Time Tunnel. These three steps are the Creation Museum's message of hope, in a manner of speaking.[80]

An Answer to the Question, Contemporary or Modern?

The Creation Museum uses state-of-the-art technologies throughout its exhibits. Moreover, in its emphasis on "starting points," it presents not one but two ways to understand history and the visitor's place in it. Does this mean that the Creation Museum is best understood as a contemporary natural history museum that encourages visitors to make their own discoveries, to put technologies to their own uses, or to challenge the so-called wisdom of the experts behind the exhibits?[81]

In *Apostles of Reason: The Crisis of Authority in American Evangelicalism*, Molly Worthen observes that the emphasis on presuppositions ("starting points") is a hallmark of young Earth creationist apologetics: by asserting that "scientific creationism and evolution [are] conflicting 'hypotheses' with different presuppositions," creationists feel they have "the language they [need] to roll back Darwin's dominance." But this should not be read as some sort of commitment to "postmodernism," as some critics have mistakenly concluded about the Creation Museum. Instead, while young Earth creationists—and, Worthen asserts, contemporary evangelicals in general—"demand that presuppositions trump evidence," they then go on to claim "the right kind of evidence as universal fact."[82]

This is precisely what one finds at the Creation Museum. When attention is paid to the "how" of the Creation Museum—that is, to how it displays its objects and constructs its exhibits, organizes those exhibits according to a grand narrative, and positions the visitor in relationship to that grand narrative—it seems that the Creation Museum is not about exploration or putting technology to new uses or contesting the authoritative voice of the museum. Rather, it seems much more to be about instructing the visitor in a totalizing history (or story that purports to account for every event—real or possible) that reveals the hidden truth for all time.[83] Moreover, it positions visitors as both the subjects and objects of that totalizing history and its underlying truth such that they may know what they must think and do in order to end up on the right side of history as well as how they should regard all those who end up on the wrong side.

As with any totalizing history the two options made available at the Creation Museum do not appear to make for much of a choice. One "choice" yields God's favor and escape from judgment. The other brings certain horrific suffering and death. The point is that the Creation Museum does not seem to be about choice or empowering visitors to think carefully through a pair of alternatives and select the one that makes the most sense to them. Rather the Creation Museum is about obedience to that one universal and transhistorical law—God's Word—and about the consequences that follow from disobedience.[84]

Contrary to first appearances, then, there is no love lost at the Creation Museum for Jacques Derrida or anyone else, for that matter, who questions absolute truths or contests totalizing histories. Indeed, as far as the Creation Museum is concerned, when God shuts the door, Jacques Derrida and those of his ilk are sure to be among those perishing on the rocks.

The Strange Object of the Creation Museum

This said, a question remains about the Creation Museum as a museum. What is its object? Traditionally, the kind of object that a museum displays determines what sort of museum it is. Thus, museums that display art are known as art museums. Museums that display artifacts from one or another historical moment are known as history museums. And museums that display "material evidence of the natural and human history of our planet"[85] are known as natural history museums. Given that the Creation Museum appears to be a natural history museum and given its particular argument on behalf of a young Earth, one would expect that its object would be "material evidence" for a young Earth.

To some extent, that is the case in the planetarium, the Six Days Theater, the Wonders of Creation room, the Flood Geology room, Dr. Crawley's Insectorium, and Buddy Davis's Dino Den. In these rooms, visitors watch videos that dramatize a six twenty-four-hour-day creation and a young universe as well as make the case for an intelligent creator. They read placards and study diagrams that explain the effects of a global flood on the Earth's terrain and its creatures. They look upon insects, skeletons, and sculptures of dinosaurs. These rooms are important. Thus, much is said in the next chapter about the exhibits within them and, especially, how they employ science on behalf of young Earth creationism.

But what about the other rooms in the Bible Walkthrough, rooms that the museum's companion book, *Journey through the Creation Museum*, calls "the centerpiece of the Creation Museum"?[86] These are the rooms that are filled with life-size re-creations of scenes from the first eleven chapters of the book of Genesis. And these are the rooms that make certain commentators very uncomfortable. This is so because the exhibits contained within these rooms do not provide obvious material evidence for a young Earth. Instead, they tell stories from the Bible. For visitors familiar with the kinds of objects that natural history museums display and the arguments they make on the basis of those objects, this may seem nonsensical. How can life-size (or any size, for that matter) dioramas depicting stories from the Bible provide material evidence for a young Earth?

For many commentators, at least, they cannot. But in their article on the Creation Museum, "Genesis in Hyperreality: Legitimizing Disingenuous Controversy at the Creation Museum," rhetorical scholars Casey Ryan Kelly and Kristen E. Hoerl argue this is beside the point. According to Kelly and Hoerl, the Creation Museum does not aim to be a real natural history museum. Instead, it only means to look like one, the goal being to trade on the authority of natural history museums to suggest "creationism's scientific veracity without providing visitors with evidence that creationism could withstand scientific scrutiny." But unlike a natural history museum, the Creation Museum consists almost entirely of synthetic objects, with "all of the dioramas . . . comprised of artificially constructed animals, mannequins, and recently manufactured items." According to Kelly and Hoerl, this is crucial: while the objects displayed at natural history museums stand in for all other objects of their kind because they are authentic, the synthetic dioramas at the Creation Museum "[deny] objects' authority to convey knowledge about the past." What the Museum offers is not real but, instead, the "hyperreal"—a synthetic environment that has no material referent— which thus "frees curators from the imperative to prove their objects' authenticity and enables them to destabilize the metonymic relationship between the traditional display object of natural history and the distant past." What makes the exhibits compelling for visitors, Kelly and Hoerl argue, has nothing to do with nature, science, or reality. Instead, for Kelly and Hoerl, what is compelling at the Creation Museum is the "dazzling" manner in which it transforms reality into a three-dimensional hyperreality.[87]

Kelly and Hoerl are right to take seriously the fact that what is on display in the "centerpiece" of the Creation Museum is, for the most part, not authentic artifacts but synthetic objects. Moreover, their argument that by displaying such objects, the Creation Museum undermines the fundamental argument of any natural history museum helps to explain why some commentators report feeling disoriented by the Creation Museum. According to Kelly and Hoerl, the Creation Museum turns the natural history museum on its head.

But is the Creation Museum best understood as constructing a hyperreal—that is, a reality with no material referent? No. While it is true that the rooms depicting scenes from Genesis consist largely, if not entirely, of synthetic objects, it is also crucial to take into account the fact that those synthetic objects appear within dioramas. It will be recalled that dioramas, the most captivating display strategy employed by modern natural history museums, were painstakingly constructed so that they would re-present objects in the most realistic manner possible. Thus, great care was given to re-creating the specimen's natural habitat in vivid detail within the diorama. Great care was also given to constructing the specimen in as realistic a manner as possible. When completed, the diorama was to function as a peephole through which the visitor could catch a glimpse of what some distant place in nature or what some "strange" (at least to them) culture was really like.

The dioramas in the Creation Museum perform a similar function. Just like the dioramas at the Field Museum, the dioramas here were constructed with great attention to detail such that, for instance, tens of thousands of leaves were attached to the Tree of Life by hand. Likewise, the figures that appear within the dioramas were constructed according to the same aesthetic of realism. Moreover, they were constructed and positioned to appear, just like the animals and human figures in a natural history museum, as if they have been caught by the visitor's gaze in a fleeting moment of action. Thus, the visitor seems to stumble upon the moment wherein Adam and Eve look upon one another for the first time.

The dioramas at the Creation Museum, just like the dioramas at any natural history museum, serve as peepholes. They provide windows on a distant world to which the visitor does not otherwise have access. In the case of the traditional natural history museum, the world that was out of reach was some place in nature or some ancient or "primitive" culture made distant by space, time, or alterity. At the Creation Museum, the world that is out of reach for the visitor is the world of Genesis. As visitors move through the "centerpiece" of the Creation Museum, they see in astonishingly vivid detail that distant world of Genesis through the peephole that is the diorama of the Garden of Eden.

The synthetic objects displayed in the course of the Bible Walkthrough Experience do not create a hyperreality for which there is no material referent. Rather, they construct that referent. Or, perhaps better put, they construct a realistic scene for which the visitor, knowing how the peephole works, supplies the material referent. Through the realism of these dioramas, visitors are invited to see the world of Genesis as a real place separated from them only by a few thousand years. These dioramas are not merely offering up dazzling metaphors. Much more powerfully, they are providing the visitor with a representative moment in a re-created Garden that renders material that other distant Garden. That is, with the help of the visitor, the dioramas at the Creation Museum construct the world of Genesis as a real historical and geographical location.

By re-creating scenes from Genesis in the form of the diorama, the Creation Museum provides visitors the opportunity to conjure a material referent for those scenes. In so doing, it enables visitors to experience Genesis not as a collection of ancient and sacred stories but as a historical record. This is important because it means that the grand narrative that undergirds all that appears in the Creation Museum also appears to have a material referent. Those early stories tell of real events, so the logic goes. As Ken Ham puts it, "the Creation Museum is not just about the topic of origins. It is a center that walks people through the Bible, helping them to understand that the history in the Bible is true— and thus the gospel based in that history is true."[88]

Moreover, the fact that that history consists in a series of stories with the same structure—God gives the Word, humans disobey the Word, God punishes them—provides lessons in the present and future. At the Creation Museum, this prophetic narrative is not some literary device. It is what really happened. It will really happen. It is the truth for all history.[89]

One final point. If we follow the metonymic logic at work in the "how" of the Creation Museum one more curious object on display becomes visible. If the scenes depicted in the dioramas are, indeed, peepholes into that distant world of Genesis (which they must be in order to construct that distant world as real) then they must be transparent to it. They must not alter or mediate it in any way. They must simply re-present it. If that is so, then there is no interpretation of Genesis at the Creation Museum. There is only the Word.

In the end, perhaps the object of the Creation Museum is not material evidence for a young Earth as one would expect but, instead, visible proof that it is possible to produce a reading of the Bible that is transparent to the Word of God itself. That is to say, perhaps what the Creation Museum aims to exhibit more than anything else is the literal Word, as a real historical object, as something that actually exists.

2

SCIENCE

On February 4, 2014, CNN streamed a much-anticipated live debate between Ken Ham (CEO of Answers in Genesis [AiG]) and Bill Nye (of *Bill Nye the Science Guy* fame). At the center of the debate was the following question: "Is creation a viable model of origins in today's modern scientific era?" Tom Foreman, of CNN, moderated the more than two-and-a-half-hour debate that was staged in the Creation Museum's fifteen-hundred-seat auditorium, Legacy Hall.[1] The debate consisted of five-minute opening statements, thirty-minute main presentations, rebuttals, counter-rebuttals, and responses to audience questions.

In the months leading up to the debate, mainstream scientists and other commentators criticized Nye's willingness to debate Ham. Jerry A. Coyne, professor of evolutionary biology at the University of Chicago, said that he thought the debate would be "pointless and counterproductive" since Nye "gives [creationists] credibility simply by appearing beside them on the platform." Agreeing, Dan Arel, a freelance writer for *Huffington Post* and *Alternet*, unequivocally stated that "scientists should not debate creationists. Period." Like Coyne, Arel argued that doing so gives credibility to a position that otherwise has none. Writing on behalf of the National Center for Science Education (NCSE), Josh Rosenau issued the same warning and reminded his readers that the NCSE has long recommended against scientists debating creationists.[2] Criticism notwithstanding, the debate went forward.

As to who won, opinions vary. Not surprisingly, evolutionists tended to declare Nye the winner. Science blogger Greg Laden stated by way of a well-worn pun that "Nye didn't really debate Ham. He ate him for breakfast." Even Nye's predebate science critics, like Coyne and Arel, declared Nye the winner.[3] Also, not surprisingly, creationists tended to say Ham won. Conservative Christian (and former member of the Indiana House of Representatives) Don Boys declared on his blog that "Ham won the debate," although he said it was "no grand slam." Other Christian bloggers argued without reservation that Ham won.[4] That said, not everyone lined up in predictable ways. Michael Schulson, for instance, writing in the *Daily Beast*, argued that "if you listened closely, what Ham was saying made absolutely no scientific sense. But debate is a format of impressions, not facts. Ham *sounded* like a reasonable human being, loosely speaking, and that's what mattered" (emphasis in original).[5]

The conclusion few contested was that the debate was an important media event. Although numbers varied, all indications were that the online audience for the debate was huge. The *New York Times* reported that five hundred thousand people watched the debate live. MSNBC said eight hundred thousand were registered to watch, and the *Christian Science Monitor* put the number at more than three million. AiG claims that more than 5 million viewers watched it within the first twenty-four hours. YouTube numbers indicate that as of October 2014, another 3.6 million viewers watched the "official" online recording.[6] In addition to online spectators, seventy media organizations attended the live debate along with the lucky audience members who managed to purchase a ticket before they were sold out online in just two minutes.[7]

Referred to by some (including AiG) as Scopes II, the debate had historical significance, but certainly not because it settled the creation/ evolution question.[8] To be sure, it did no better at settling that debate than did the Scopes trial.[9] Importantly, the historical significance of the Ham/Nye debate finds its origins in the understanding of the Scopes trial that emerged among conservative Christians some decades after the actual trial.[10] By the 1980s, conservative Christians had resolved never again to suffer the humiliation Bryan had brought upon them. Never again would they look like backwater fools who knew nothing of science, who equivocated on the age of the Earth and, even worse, wavered on the literal meaning of Genesis. On a national stage located within the Creation Museum and witnessed by millions of Americans, Ken Ham put that ghost to rest.[11] How did he do it?

Clearly, he did not do it by coming away the undisputed winner in either substance or style. Instead, he did it by providing an unrelenting performance of what might be called the "not-afraid-of-science Christian apologist." In the course of the debate, Ham argued that only a literal reading of Genesis that flies in the face of mainstream geology and evolutionary science will save America from complete moral decline. And as he made that case, he spoke the word "science" 105 times. On our count, that was 2.4 more times than Bill Nye "the *Science* Guy" used the word. Perhaps to make sure no one missed the point, near the end of his thirty-minute presentation, he enthusiastically exclaimed, "I love science!"[12]

Ham did not merely speak the word "science." He made arguments about it and with it. In an effort to counter one of Nye's main arguments that belief in young Earth creationism threatens scientific and innovative progress, Ham argued that there are plenty of hard-core scientists and inventors who are young Earth creationists. To support his claim, he played videos of scientists and inventors with impressive credentials who testified to their belief in young Earth creationism. In another of his main arguments, he claimed that science, properly done, "confirms" a literal interpretation of the Genesis account of origins.[13] To back up that claim he argued, among other things, that the Bible says we are all one race and that biogenetics has proven the Bible to be right. Finally, more than a third of his remarks in the debate were dedicated to instructing mainstream scientists and Christians alike in what appears to be for Ham an absolutely crucial distinction between "historical science," which is speculative, and "observational science," which is objective, repeatable, and verifiable.[14]

Ken Ham is not obliged to mobilize science on behalf of his reading of Genesis. He does not have to argue that a literal interpretation of Genesis 1–11, according to which the Earth is no more than ten thousand years old, can be "confirmed" by scientific observations. Indeed, he could skip the scientific arguments altogether and simply say that a proper Christian should accept a young Earth as a matter of faith regardless of what the geological column or genome mapping might indicate to the contrary. If the Bible really is the uncompromised Word of God, what difference should it make to a faithful Christian that mainstream science contradicts it? One could even argue that seeking to ground a faith-based argument in science undercuts the very faith that science is being used to buttress.

Ken Ham has not chosen that argumentative path. That is likely because AiG is an apologetics ministry that seeks to be persuasive. And mobilizing science on behalf of young Earth creationism affords at least three important rhetorical advantages. First, in a culture that largely understands science to be *the* discourse of truth—that is, the methods and practices that yield real facts and reliable conclusions—an argument for a young Earth that can claim scientific support is likely to be much more persuasive to believers and unbelievers alike.[15] Second, and closely related, if one is going to make the claim that the account of origins in Genesis is accurate history in a culture that believes in science, one is going to be expected to provide evidence of the events in that history. That is, if there really was a global flood, then it should have left behind some evidence of its effects. Embracing rather than ignoring or contesting this logic makes the argument for special creation much more persuasive, assuming such evidence can be provided. A third rhetorical advantage is specific to the evangelical/fundamentalist subculture. By embracing science, Ham is able to perform (in the debate and elsewhere) the not-afraid-of-science Christian apologist that conservative Christians in America have been seeking. If by the end of the twentieth century conservative Christians were humiliated by the way William Jennings Bryan sacrificed a literal reading of Genesis at the altar of modern science, at the end of the Ham-Nye debate, conservative Christians likely felt triumphant for having witnessed a young Earth creationist put science in the service of the inerrant Word. The conservative Christian need no longer be troubled by science. Science now bows to the inerrant word. Or, at least, that is the idea.

"Science" at the Creation Museum

In his opening remarks at the debate, Ken Ham pointed out (as he often does) that "science" comes from the Latin word *scientia* and that it means "state of knowing" or "knowledge."[16] He then went on to argue that there is much confusion about science in contemporary culture because people do not understand the difference between observational science (also called experimental or operational science) and historical science (also called origins science).

According to Ham, "operational [or observational] science deals with knowledge that is gained by observation and repeated testing in our present world."[17] Important in this definition is the notion that observational science mobilizes the practices of the scientific method to produce knowledge about phenomena in the present. Ham continues: "Whether scientists are Christian or non-Christian, they can all perform the same operational science."[18] Because observational science involves applying the scientific method to phenomena in the present and because its methods are repeatable, it is objective. Anyone who follows correct scientific procedure, no matter his or her religious perspective, should arrive at reliable results.

By contrast, Ham argues, "Historical science deals with history—the past." It seeks to answer the question, "What happened in the past?" According to Ham, an event or process that occurred in the past is not available to observation in the present and thus cannot be understood by way of observational science. As Ham puts it: "We do not have access to the past. We only have the present. All the fossils, all the living animals and plants, our planet, the universe—everything exists in the present. We cannot directly test the past using the scientific method (which involves repeating things and watching them happen) since all the evidence we have is in the present."[19] Faced with a fossil and wondering how it came to be, Ham reasons, the scientist cannot hope to come to any reliable answers by way of observational science. Instead, the scientist must use historical science. Historical science involves applying a set of assumptions or a belief system, like evolution or special creation, to questions about the past.[20]

Ham offers the following example to clarify the difference between observational science and historical science: "If we are talking about the fact that we can see and study apes today, and we can see and study people, that is operational science. But if somebody claims that millions of years ago ape-like creatures turned into people, what is that? That is historical science. What if we said that God made Adam from dust, what is that? That is historical science, too."[21] In short, when scientific methods are applied to evidence in the present, and conclusions are drawn about that evidence that pertain to the present, then one is doing observational science. When inferences are made about the past by way of evidence that exists in the present, then historical science is underway. Moreover, historical science, unlike observational science, always involves the application of a set of beliefs and assumptions about the past. Thus, whereas observational science is objective, historical science is not.

Although observational science and historical science are distinct, they are related. Observational science cannot answer questions about the past, but historical science can. Historical science does so by constructing models that explain how things might have happened in the past. Observational science then supports the construction of models by providing evidence to "confirm" them.[22] Ham and his colleagues at AiG do not use the language of "proof" in this connection. Observational science can never *prove* whether a model that aims to account for the origins of the Earth is true. That is impossible because, again, no scientist living in the present is able to witness something from the past.

For Ham, this distinction between observational science and historical science is crucial because it corrects a problem in people's understanding about evolution and special creation.[23] Ham argues, "Instead of perceiving the real issue, [people] have been deceived into believing that evolution/millions of years is science and that the Bible's account of origins is religion. But this is not so."[24] Evolution is not real science, according to Ham, because it makes claims about phenomena and processes that cannot be observed since they are in the past and because it makes inferences about those things in the past based on a set of presuppositions. It is not objective. Evolution is historical science. Likewise, special creation or the biblical account of origins is historical science.[25] By making this distinction, Ham removes evolution from the category of real science.[26]

Although Ham's language of historical science and observational science is an innovation, the overall argument is not.[27] Young Earth creationists have argued that evolution is not real science on the grounds that it makes inferences based on things that cannot be observed and that it is presuppositional—that its inferences always depend on a set of assumptions that may not be true—since the early twentieth century. Likewise, young Earth creationists, especially since the emergence of "creation science" in the early 1970s, have argued that, given this, evolution and special creation must be understood to have the same epistemological status.[28] Since evolution and special creation are both models, evolution can claim no special scientific status as fact.[29]

While the Creation Museum rarely, if ever, employs the language of observational versus historical science, this distinction repeatedly appears. For instance, in the Dig Site room, the first room in the Bible Walkthrough Experience, visitors see a series of cases that display objects that look like pieces of evidence that a scientist might study, such as an artificial sapphire, a cast of a fossil, and an artist's reconstruction of Trilobite tracks. To the right of each case is a placard. Each placard follows the same form. At the top is a statement in large font: "The evidence is in the *Present*" (emphasis in original). Below that statement is an identification of the evidence, such as "*Archaeopteryx Fossils*" (italics

This life-size diorama, the first in the Bible Walkthrough Experience, features two paleontologists—one who believes in an old Earth and one who believes in a young Earth—that marks the beginning of the museum's argument that mainstream science is no more credible than creation science since it too is grounded in certain untestable presuppositions. Photograph taken by Susan Trollinger. For more images of and information about the museum, go to creationmuseum.org.

in original). Below that, in smaller font, is a description of the object: "This is a cast of an *Archaeopteryx lithographica* specimen from the Solnhofen Limestone (Jurassic System), Germany" (italics in original). Below that is another heading that says, "But what happened in the *Past?*" (emphasis in original). Below that several questions are listed: "When did the animal live? What did the animal look like? (*For example, was it a bird?*) How did the animal behave? (*For example, could it fly?*) How was the animal related to other animals? (*For example, is it related to dinosaurs?*)" (all emphasis in original). Opposite this series of cases and placards is the centerpiece of the Dig Site room—namely, a diorama featuring what the placard in front of it calls an "artist's reconstruction" of a paleontology dig site complete with two paleontologists working to liberate the skeleton of a *Utahraptor* from what looks like rocky ground. The accompanying placard follows the same form as those just discussed, identifying the skeleton and then asking a series of questions about when it lived, how it died, where it lived, and how its body was preserved.[30]

Without saying so, these displays and placards instruct the visitor that there is a big difference between observational science, which studies objects that exist in the present, and historical science, which posits conjectures about those objects in the past. Given that the ostensive claim of the Creation Museum is that a proper understanding—that is, a literal interpretation—of *Genesis* reveals the truth that the universe was created fewer than ten thousand years ago, given that the Creation Museum looks to be a natural history/science museum that defends that claim, given that Ken Ham and his colleagues at AiG "love science," and given that observational science (not historical science) is the only real science, one would expect to see plenty of observational science on display at the Creation Museum in support of the claim for special creation and a young Earth. Is this, in fact, what we find?

Writing to the AiG website, one visitor to the Creation Museum complained: "My parents and I recently visited the Creation Museum . . . and were greatly disappointed. We did not feel that enough scientific evidence was given to support creationism. We think a museum should be more about evidence than faith. There seems to be plenty of evidence out there—it is just not present at all."[31] Jason Lisle (PhD in astrophysics, director of the planetarium at the Creation Museum, and author and designer of *Created Cosmos*—the film typically shown in the planetarium) responded: "Of course, there is plenty of science that confirms biblical creation. . . . Much of this science is presented in the Creation Museum. Some of this science in the museum is very apparent (such as the information presented in the planetarium, or the Flood . . . geology room). But much of the science is 'behind the scenes,' and you may not have noticed it."[32]

Since a visit to the Creation Museum does not include access to the science that resides "behind the scenes," and since Lisle says the places to look for the science that "confirms" a special creation are the Stargazer's Planetarium and the Flood Geology room, it makes sense to take a close look at those spaces. And since elsewhere AiG says that "much of the scientific content of the museum is presented in [the] videos" displayed on monitors in the Wonders of Creation room,[33] it makes sense to take a close look at those, too.

The analysis that follows takes a close look at the science presented in these spaces. Its focus is not so much on determining the quality or accuracy of the information presented. In keeping with the previous chapter, which analyzed the museum on its own terms as a natural science museum, this chapter seeks to analyze the mobilization of science at the Creation Museum on its own terms. Thus, this chapter focuses on whether the Creation Museum delivers on the promise implied in Lisle's response—namely, that the Creation Museum presents "plenty of science" that "confirms biblical creation."

Breaking that question down, three subquestions emerge. First, according to its own definition of observational science, does the Creation Museum present real science? That is to say, is the science presented here observable in the present, and are those observations repeatable? Second, is there, as Lisle implies, plenty of evidence? Since opinions may differ on what constitutes "a lot," the analysis will endeavor to provide a clear idea of how much observational science is on display in these rooms so that the reader may decide. Finally, does the science that appears in these rooms "confirm" a special creation?

Notably, Ham warns against mobilizing the notion of "proof" when it comes to questions about origins. In the Ham/Nye debate, for instance, he said, "I've been emphasizing all night, you cannot ever prove, using the scientific method in the present, you can't prove the age of the earth. So you can never prove it's old."[34] Ham insists that processes that occurred in the past cannot be observed in the present, and, therefore, any claims made about them cannot be proven. At best, such claims can be "confirmed." Accepting that qualification, the question for the analysis that follows is not whether the Creation Museum *proves* a special creation on the basis of scientific evidence but, rather, whether the Creation Museum presents a *strong case* based in real (observational) science that the universe was created by the God of the Bible in six twenty-four-hour days less than ten thousand years ago.

The analysis that follows evaluates evidence that appears in the Creation Museum against the criteria associated with "observational science" and "historical science." This is not because the distinction between observational science and historical science makes sense. It may or it may not make sense. Instead, this distinction is applied in this analysis because it plays a crucial role in the argument of the Creation Museum (and AiG). Indeed, this distinction between observational science (or real science), which can be repeated, on the one hand, and historical science, which cannot because it occurred in the past, on the other, is what enables Ham and his colleagues to remove evolution from the category of real science and put it into a category of something akin to religion. Given that argument, it is fair to ask whether the science displayed at the Creation Museum is, according to Ham's and AiG's criteria, "real science."[35]

The Stargazer's Planetarium

Created Cosmos, one of two films screened on the 30-foot dome of the Stargazer's Planetarium every day, is a 23-minute trip from Earth to deep space and back.[36] Along the way, the film provides information about the Earth, the solar system, the Milky Way, and distant stars. For instance, viewers learn that the Earth is about 8,000 miles in diameter, that the Apollo astronauts traveled approximately 240,000 miles from Earth, that the distance between the Sun and the next nearest star system (Alpha Centauri) is equivalent to 4,278 solar systems stacked end to end, that Sirius (the brightest star in our night sky) is 50 trillion miles away, that Betelgeuse (a red super giant star) is 600 times the diameter of our Sun, that from the location of the nearest stars in Orion's Belt our solar system is invisible to the naked eye, that a nebula (cloud of hydrogen and helium gas) is very colorful and sometimes hot, and that the globular cluster called M4 is 40 million billion miles from Earth and consists of about 100,000 stars densely packed together.[37]

Does the planetarium present plenty of observational science? The answer seems to be yes. The size of planets, their distance from the solar system, the characteristics of nebulae, and so forth fit the definition of observational science. These facts (assuming they are accurate) are observable, and their measurement can be repeated. Moreover, the fact that the film devotes a full seventeen minutes (or 77 percent of the film) to the presentation of this observational science indicates that there is plenty of science here.

But does *Created Cosmos* make a strong case that the observational science presented therein confirms a biblical creation? In that regard, the film makes three basic arguments on behalf of a biblical creation. The first argument is that the immensity, beauty, and/or diversity of the universe, which scientific observation makes visible, testify to an infinite God who is their creator. For example, after describing nebulae at some length, the film reflects on their characteristics and concludes, "These amazing creations can rightly be called the artwork of God." In a more developed version of this argument, the opening statement of the film says, "As we learn more about the universe, we are continually amazed at the astonishing diversity and beauty we find. Though marred by the curse, the universe still exhibits the handiwork of the Lord. By learning more about the intricacies of the celestial realm, we gain an infinitesimal glimpse into the infinite mind of God. One extraordinary aspect of creation is the incredible range of sizes and distances we observe."[38] Following this statement, the film offers five minutes and thirty-five seconds of uninterrupted information about, for instance, the sizes of planets and their distances from the Sun.

Of course, this kind of argument is a variation of long-standing arguments that reason that some characteristic of nature—especially its beauty and complexity—cannot be explained by chance and, therefore, must be taken to indicate an intelligent designer.[39] To be sure, this kind of argument is compelling to some people. Still, questions remain about its articulation here. Why do immensity, diversity, and beauty serve as signs for God? Do they reflect God? Are they characteristics that are favored by God? Moreover, if immensity, diversity, and beauty indicate God, would smallness, homogeneity, and/or ugliness counterindicate God? If the universe displayed smallness, homogeneity, and/or ugliness, which it undoubtedly does somewhere, would they indicate that God does not exist? Even if these characteristics do support the notion of a divine designer, what indicates that this divine designer is the God of the Bible? Finally, even if that designer is the God of the Bible, what evidence suggests that these nebulae and planets and such were created in six twenty-four-hour days fewer than ten thousand years ago?

Another kind of argument that the film makes takes a completely different tack. When speaking of worlds deep in space, the film says, "Since current technology is not able to observe these worlds directly, we can only speculate what they look like. But we can be certain that their richness declares the majesty of their maker." In arguments like this one, rather than reason that some observable characteristic of the universe serves as a sign that the God of the Bible created them, the argument presents no observable phenomena and assumes that the God of the Bible is their creator. Importantly, this argument cannot be taken to confirm a biblical creation based on observational science since no observational science is presented.

A third kind of argument made in the film says that observational science provides evidence that contests the notion of an old universe.

Regarding blue stars, for instance, the film says, "Blue stars, like Al-nilam, are very luminous. They expend their fuel quickly and cannot last billions of years. So, blue stars remind us that the universe is much younger than is generally claimed. . . . [More than this,] star formation is riddled with theoretical problems and has never been observed."[40] First, it should be noted that even if some observational science challenges the notion of an old universe, taken by itself, it would not necessarily confirm a biblical creation. Second, assuming that it is true that blue stars "cannot last billions of years," the crucial question here is whether it is right also to assume that Alnilam was created when the universe was created. If one need not assume that, then the fact that blue stars cannot last billions of years poses no problem for an old universe. Finally, the film makes the more general claim that "star formation is riddled with theoretical problems and has never been observed," without ever identifying those "theoretical problems" or explaining how those problems would undermine the notion of an old universe.

To be sure, the planetarium presents plenty of observational science. But it is hard to see how all of that "real science" confirms that the God of the Bible created the universe in six twenty-four-hour days fewer than ten thousand years ago. Yes, the film offers observational science on behalf of a supernatural designer. But does it provide a good reason to infer that that designer is the God of the Bible? Yes, the film offers observational science that is supposed to challenge the notion of an old universe. But does that data really contest an old universe? Yes, the film makes claims that grow out of the assumption of biblical creation, but where is the observational science to support those claims?

The Wonders of Creation Room

The Wonders of Creation room shows fifteen videos in continuous loop on about the same number of small flat-screen monitors mounted high on the walls that form the perimeter of the room. On average, these videos last about two minutes and twenty seconds and feature a combination of photographic images, video clips, computer-generated images (still and moving), diagrams, and biblical text. All but one of the videos have a voiceover that provides information about the topic at hand.[41] For instance, the "Solar System" video tells viewers about the size of each planet, its location in the solar system, what it is made of, its temperature at the surface, how many moons it has, and so forth. The pres-

Photographs and video loops displayed on numerous flat-screen monitors in the Wonders of Creation room impress upon the visitor the argument that the beauty, precision, and complexity so clearly displayed in the universe and on Earth testify to the creation of all things by the God of the Bible. Photograph taken by Susan Trollinger. For more images of and information about the museum, go to creationmuseum.org.

entation of descriptive information takes up one minute and fifty-one seconds of the total two minutes and thirty-two seconds of the video (or 73 percent of the time in the video).[42] Similarly, the "Stars" video, which is the shortest video in the Wonders of Creation room, dedicates fifty-two of the total seventy-five seconds of the video (or 68 percent of its time) to the presentation of descriptive information such as what stars are made of, how much energy the Sun gives off, how many stars are in the Milky Way, and so forth.[43]

In short, and in keeping with *Created Cosmos*, the videos in the Wonders of Creation room appear to feature plenty of observational science. As regards the arguments made by the videos, while there are a few that (also in keeping with *Created Cosmos*) suggest that the universe cannot be as old as mainstream science claims, the most frequent argument is made on behalf of divine design. Indeed, thirteen of the fifteen videos on display argue that something observable in nature suggests a divine designer. These arguments take something that is observable in the present—the Earth's location in relation to the Sun, the complexity of the structure of bird wings, or the fact that DNA resembles a language—and argue that it serves as a sign for divine design. For instance, the "Designed for Flight" video argues on the basis of what feathers look like and how bird muscles work: "Every feature of birds from the muscles in the chest to the feathers on the wings is well *designed* for flight" (emphasis added).[44] Likewise, the "DNA" video reasons that since DNA consists of four nucleotides that appear in various sequences to produce different genetic information, "DNA seems to represent a language—the language of life. An unseen author, the creator of heaven and Earth has left a testimony of his existence in the DNA of every living thing."[45] Other videos argue that observed complexity; similarity among the chemicals found within organs, organ systems, and DNA; resemblances among species within a scientific family; or the fact that all organisms require cooperative relationships to survive point to an intelligent designer.[46]

While most of the videos in the Wonders of Creation room make explicit arguments that the Creation was made by a divine designer, one does not. Clocking in at four minutes and thirty-one seconds (nearly twice the average length of the other videos) the "Creator Clearly Seen" video is unique among the fifteen in that it has no voiceover and, thus, does not present any oral scientific information or argument for a divine designer. Likewise, no diagrams or computer-generated images of scientific models appear. Instead, this video displays a stream of still and moving images that very quickly change from one to another in accordance with the rhythms of two classical music pieces.

Everything that appears in these images—whether a body of water, a flower, an insect, a bird, a dog, or a person—is beautiful. Water is clean; grass is lush; smiles are bright; snow is white; skin is clear. There are no imperfections, no decay, no violence, no pollution, and no destruction. The argument that appears to be implied in this video is that a world filled with so much beauty could not be an accident and, therefore, must be the creation of a divine designer.

Again, the argument that the beauty, complexity, and apparent purposefulness that is observable in nature indicates a divine designer is persuasive to many people. Still, questions remain. Why is it that beauty, complexity, similarity, apparent purposefulness of design, and the cooperative nature of certain relationships confirm a divine designer? What is it exactly about beauty, similarity, complexity, or cooperativeness that indicates an intelligent designer? Would their opposites—ugliness, simplicity, difference, or competitiveness—contest the notion of a divine designer? To be sure, all of these qualities are also observable in nature. Do these observable characteristics in nature contest the notion of a divine designer? Now, one could argue that these opposite qualities in nature do not contest a divine designer but, instead, are symptomatic of the Fall. The argument then would seem to be that ugliness, simplicity, difference, and competitiveness are all the result of sin. But that would be a troubling argument to make. Difference is the result of the Fall? Simplicity is a sign of sin? Moreover, who determines what is beautiful or ugly, complex or simple, similar or different, cooperative or competitive? Is the assumption that these qualities are universally recognized by all peoples? What if they are not? How would that impact the implied argument here? The videos are silent on these important questions.

Unlike *Created Cosmos*, the videos in the Wonders of Creation room are not content merely to suggest *some* kind of supernatural designer. They take the additional argumentative step of saying that the divine designer of the universe is the God of the Bible. Typically, they do this by displaying a verse from the Bible at the end of the video, while the voice-over makes some suggestion that the verse speaks to whatever aspect in nature has been featured in the video. Thus, the observational science in the video appears to confirm what the Bible says.

Sometimes the connection between the observational science and the biblical text seems fairly direct, whether or not it is convincing. For instance, the "Eyes" video closes with the following scripture, which appears against the backdrop of a digital animation of an eyeball: "The Lord has made . . . the seeing eye. Proverbs 20:12" (ellipsis in original).[47] Similarly, the "Designed for Flight" video closes with the following verse superimposed in Hebrew script, which (oddly) appears upside down: "God created every winged bird after its kind, and God saw that it was good. Genesis 1:21."[48]

Given that these verses appear out of context in the videos, it is difficult for the visitor to determine whether, in fact, the texts really do reiterate the observational science in the videos. But other connections between the ultimate claim of the video and the closing scripture are even less direct. For instance, the "Communities" video, which argues that the cooperative relationships seen in nature are the creation of the God of the Bible, closes with the following scripture: "One Spirit . . . one

Lord . . . one God. Ephesians 4:4–6." Viewers are left scratching their
heads for the connection here. This is not at all surprising given that
this heavily excised text, which was reduced from forty-one words to a
mere six, is cryptic. Aiming, perhaps, to fill in the blanks, the voiceover
says, "Cooperative relationships, cycles of provision, communities of organisms, a complex interwoven web of life where every organism has a
place and a function in the larger whole. The value of the individual and
the necessity of community—these are beautiful manifestations of a triune creator God. Three persons in one."[49] Assuming the visitor accepts
the suggestion that this scripture is talking about "a triune God," how do
cooperative relationships within nature confirm such a God? Moreover,
would observational science that takes note of competitive relationships
in nature contest a triune God?

Even less clear is the relationship between the central claim of the
"Solar System" and its closing scripture. The claim is that the "beauty"
of the planets and the "clockwork precision" of their orbits confirm
that the God of the Bible is the creator of the solar system. The closing
scripture says, "The heavens declare the glory of God. Psalm 19:1."[50]
Given that this verse comes from Psalms, a book of poems, one is in-
clined to read it figuratively. Thus, one would probably not take this
verse to be a historical account of the origins of our solar system. Is the
suggestion here that one should imagine that the heavens *literally* declared God's glory? If so, how would such a vague declaration support
the central claim of the video? Finally, how does all this support the video's claim that the God of the Bible created this solar system in a single
twenty-four-hour day fewer than ten thousand years ago?

AiG is right. The videos in the Wonders of Creation room provide
plenty of "real science" measured against Ham's criteria. But to what
end? Much of the observational science that appears in these videos
is used in support of the argument that a divine designer created the
universe. All the beauty, complexity, similarity, and apparent purposefulness of design that is observable in our world and beyond cannot be
the result of mere chance, so the argument goes; there must have been
some kind of divine designer who created it all. Assuming for the moment that the reasoning behind that argument makes sense, does it also
make sense, based on bits of Bible verses often only tangentially related
to the evidence provided, that the God of the Bible is that designer? As
with the film in the planetarium, such claims seem to have little or no
connection with observational science.[51] But perhaps the rooms devoted

to flood geology, the science that has been at the heart of young Earth creationism for more than a century, make a strong scientific case for a biblical creation.

Flood Geology

Four rooms in the lower level of the museum, including the Flood Geology room, the Natural Selection room, the Facing the Allosaurus room, and the Post-Flood room, present the museum's case that flood geology explains the appearance of an old Earth. Three of these rooms exhibit placards large and small, flat-screen monitors, miniature encased dioramas, and other inset lighted cases. The Facing the Allosaurus room— renovated in 2014 to accommodate this exhibit—features a life-size skeleton of an Allosaurus dinosaur standing on a bed of white gravel behind a glass partition. This skeleton, consisting of actual bone and bone casts, is thirty feet long and ten feet high. On either end of the skeleton, also behind the glass, sit two flat-screen monitors on stands showing videos about the dinosaur along with a few small placards. Elsewhere in the room there is a glass case that holds, according to the accompanying placard, "the actual fossilized skull of Ebenezer, the *Allosaurus*."

Taken together, these rooms (which we call the flood geology section of the museum) exhibit some thirty-eight placards. Given that placards dominate these rooms visually by their number and often large size, and given that they appear to present plenty of information on behalf of a biblical creation, the analysis that follows focuses on the placards. In addition to placards, these rooms also show seven videos in continuous loop on flat screens.[52] These videos are similar in length to the videos in the Wonders of Creation room, averaging two minutes and fifty-one seconds. They explain the models depicted on the placards that accompany them. In addition, three computer-generated animations are on display. One has no text and appears to be a simulation of a model designed to explain the Flood's role in plate tectonics. The other two appear inside the glass partition with the Allosaurus. One shows the dig site where the skeleton was found, and the other provides descriptive information about the skeleton.

Flood geology has served as the scientific keystone of young Earth creationism since the early twentieth century when George McCready Price, a Seventh Day Adventist, published a textbook on the topic—*The New Geology*. Then in 1961, it was made widely popular among evangelicals and fundamentalists when Henry M. Morris and John C. Whit-comb put forward what was largely a restatement of Price's argument in *The Genesis Flood*. The centrality and popularity of flood geology is due to the fact that it holds out the promise that the fossil record and other presumptive evidences of an old Earth can be shown scientifically to support a young Earth instead.[53] That being so, one might expect the vast majority of placards in the Flood Geology room to feature ob-servational science since observational science is the only true science according to Ken Ham and his colleagues at AiG. Surprisingly, fifteen of the thirty-eight placards on display in the Flood Geology room (40 percent) are identified by white print in the top right corner as "mod-els." These models include the "residual catastrophism model," the "Bretz Columbia River flood model," the "catastrophic plate tectonics model," the "Vardiman hypercane model," and the "biogeographic raft-ing model," just to name a few.

According to AiG, models (including big ones like creationism and evolution) are historical science and not "real science." Roger Patterson, who serves on AiG's Educational Resource team and writes on a range of topics, including evolution, polygamy, and Easter, explains:

> Operational science deals with testing and verifying ideas in the pres-ent and leads to the production of useful products like computers, cars, and satellites. Historical (origins) science involves interpreting evidence from the past and includes the *models* of evolution and special creation. Recognizing that everyone has presuppositions that shape the way they interpret the evidence is an important step in realizing that historical science is not equal to operational science (emphasis added).[54]

Since models are interpretations of the past rather than observations made and repeatable in the present, not one of these fifteen placards, by AiG's criteria, qualifies as "real science."

That said, the fact that a placard self-identifies as a model does not necessarily mean it displays no observational science. Given that, a close content analysis was conducted of all thirty-eight placards in the flood geology section, the purpose being to ascertain whether, in fact, these rooms deliver on the Creation Museum's promise to provide "real sci-ence" that confirms a biblical creation.

It makes sense to start by asking how many of the thirty-eight plac-ards look like they display scientific information according to a com-monsense notion of science—that is, information that appears to have

been culled from careful observation, experimentation, and/or measurement. It turns out that twenty-six of the thirty-eight (68 percent)
pass that test by displaying information such as the length of a dinosaur
skeleton, the age of rock layers, the amount of ash deposited by a vol-
canic eruption, the depth of ice in the middle of Greenland, and so forth.

This means that 32 percent (nearly a third) of the placards in the
flood geology section display nothing that even resembles scientific evidence, never mind evidence that can meet AiG's much more stringent
criteria for observational science. Instead, these twelve placards provide
introductions or summaries for a certain series of placards, speculate
on processes that might have occurred during or after the Flood, make
moral arguments for the destruction of all land-based life that did not
get on the Ark, display a picture or map, or define terms such as "natural
selection" and "evolution."

So, what about the twenty-six placards that appear to present scientific evidence? It turns out that ten of these twenty-six (39 percent)
do not display "real science" according to AiG's definition. Often that is
because the information presented is based on observations that took
place in the past and cannot be repeated. For instance, two placards in
the Flood Geology room make the crucial argument that the kinds of
geological changes that evolutionists say take billions of years can actually be accomplished in a very short period of time. This argument relies
on the example of the Mount Saint Helens eruption, which occurred on
May 18, 1980, and its aftermath. In support of the argument, these placards identify processes set in motion by the eruption. They refer to "mud
flows," the "lava dome," the "ash cloud," "pyroclastic deposits," and "bark
rubb[ing] off floating logs." These processes, which took place decades
ago, are neither observable in the present, nor repeatable.[55] Thus, the
statements made about these processes, such as that "mudflows cut canyons out of solid rock in just a few years," must be regarded, according
to Ham's instruction, as statements about the past and therefore as historical, not observational science.[56]

Another placard among these ten, which hangs in the Post-Flood
room, explains the "residual catastrophism model" according to which
the Earth's "crust settles down after the Flood" as "supervolcanoes" and
"superquakes" gradually subsided. Listed on that placard are five vol-
canic eruptions along with the amount of ash they produced and a map
suggesting the area covered by the ash. The five eruptions named there
are the Yellowstone eruption that produced the Huckleberry Ridge Tuff

(600 cubic miles of ash), the Yellowstone eruption that produced the Lava Creek Tuff (240 cubic miles of ash), the Long Valley eruption that produced the Bishop Tuff (150 cubic miles of ash), the Crater Lake eruption (seventeen cubic miles of ash), and the Mount Saint Helens eruption (.25 cubic miles of ash).

No dates appear on this placard to indicate when these eruptions occurred and were observed. Other placards indicate that Mount Saint Helens erupted in 1980, but what about the other four eruptions? According to the United States Geological Survey, the most recent of the four (at Crater Lake) happened about seventy-seven hundred years ago and the most remote (the first Yellowstone eruption) about 2.1 million years ago.[57] Of course, according to the story that is told in the Creation Museum, none of these eruptions could have occurred more than about forty-three hundred years ago since, as the model makes clear, all five were residual effects of the Flood. Although it would be helpful if the placard provided AiG's dates for these eruptions, the important point here is that no scientist living today could have witnessed four of these eruptions or measured the amount of ash deposited by them. And although a scientist living today may have witnessed the Mount Saint Helens eruption, any observations they may have made at the time cannot be repeated.

A third example provides a good sense of how information that appears scientific on these ten placards fails to meet AiG's criteria for "real science." The placard called "Canyons Erode," which appears in the Post-Flood room, explains the "breached dam model," according to which "lakes filled up behind the Kaibab Upwarp" after the Flood, then "broke through, carved Grand Canyon, stripped slopes clean, and drained away," leaving "1000 cubic miles of sediment . . . in the Salton Sea trough." This placard is important because it makes an argument that instead of taking millions of years to form, "canyons were carved rapidly after the Flood."

A crucial piece of evidence for this argument on the placard is "the small difference in age between lake deposits upstream and surge deposits down stream." This information is key because a small difference in age could indicate that these processes occurred in quick succession, thus diminishing the amount of time required to produce canyons. Although "the small difference" identified here is not specified, and although the magnitude of the difference is crucial to the argument, still one can imagine a geologist performing repeatable experiments to determine the age of the two deposits. Thus, although vague, this crucial evidence looks like it passes AiG's test for "real science."

But does it? According to Ham, dating methods are not "real science." As he put it in his debate with Bill Nye,

> My point is all these dating methods actually give all sorts of different dates. In fact, different dating methods on the same rock we can show give all sorts of different dates. You see, there are lots of assumptions in regard to radioactive dating. Number one, for instance, the amounts of the parent and daughter isotope at the beginning, when the rock formed—you have to know them, but you weren't there; see, that's historical science.[58]

Since no scientist actually observed the formation of either the "lake deposits upstream" or the "surge deposits downstream," since no one today can observe their formation, and since any attempt to discern their ages using dating methods amounts to historical science, this evidence must be regarded as not "real science."

The point here is not that the evidence on these placards is, in fact, bogus or unreliable. Maybe it is, and maybe it is not. The point is also not that it is unreasonable to draw inferences from evidence that comes from the past. The point here is that while the information on these ten placards looks like science—thus contributing to the visitor's impression that the flood geology section displays plenty of science—it does not meet the criteria for "real science" that AiG uses to dethrone evolution as sham science.

If these ten placards, which feature historical science rather than observational science, are removed from the list of twenty-six placards that appear to provide scientific information when measured against a commonsense notion of scientific evidence, just sixteen (42 percent of the total number of placards) remain that display what AiG calls "real science." To what purpose are they put?

One purpose to which this observational science is put is simply to provide information. Three of the sixteen placards that display observational science do that. All three appear in the Facing the Allosaurus room. One, titled "Why the Long Face?" describes Ebenezer's skull, noting (among other characteristics) that it is "34 inches long and 22 inches high, and is 97 percent complete," and that "there are 53 curved, saw-edged, saber-like teeth, up to 4.5 inches long including the roots." A second placard, titled "What Was Found?" provides "details from the dig," including the fact that "this animal was found buried lying on its left side," that "the skeleton was oriented with the head at the western end and the tail pointed southeast," that "the remains of the animal's spine . . . were found lying in a curved alignment," and that "due to erosion of the hillside he was found in, most of Ebenezer's limb bones and mid-section were not recovered."

Neither placard makes any connection between the observational science presented and a claim for a biblical creation. Then there is a third placard, titled "Who Am I?" which explains that "Allosaurus" means "'other lizard,'" that Allosaurus "belongs to the suborder of dinosaurs that contains theropods," that Ebenezer was likely a "formidable, carnivorous dinosaur," and that a "carnivore is something that eats mostly meat." The placard also notes that this skeleton was named Ebenezer because "some of the people involved in excavating this dinosaur . . . saw him as a reminder of God's judgment of the world and how He preserved mankind and animal kinds on Noah's Ark." Although this and one other biblical reference appear on the placard, it does not make an argument that any of the information provided about the dinosaur skeleton supports a claim for a biblical creation or a global flood. The information about the dinosaur simply appears alongside statements about a biblical story that includes the name "Ebenezer."

When these three placards, which display observational science in a purely informational way, are removed from the list of sixteen placards that display observational science, just thirteen (about a third) remain that might make a connection between observational science and a claim for a biblical creation. This is surprising given that flood geology is the scientific keystone of the young Earth creationist argument, and given AiG's own claim that this is one of the areas in the museum that displays "plenty" of science that confirms a biblical creation.

A Particular Kind of Science at the Creation Museum

Importantly, AiG's distinction between observational science and historical science, so crucial for demoting evolution from the status of real science, is steeped in a tradition long familiar to evangelicals and fundamentalists. According to that tradition, proper science is a thoroughly empirical enterprise grounded in careful and precise observation. All presuppositions or preconceptions are to be put aside so that observations may be true to the object under study. Not even hypotheses are appropriate in this scientific approach. To bring any preconceptions whatsoever to the practice of observation only distorts the observation and, thus, renders it unreliable. When properly done, science begins with numerous disciplined observations, and only on the basis of those observations does one venture an inference about what might be the case more generally.

Understanding science in this way dates back to the nineteenth century when, as Mark Noll points out, evangelicals believed "that strict induction from verified individual facts to more general laws offered the best way to understand the data of any subject." Likewise, argues Noll, creationists insisted upon "no speculation without direct empirical proof, no deductions from speculative principles, no science without extensive empirical evidence."[59] According to Ronald L. Numbers, the point of taking this strict view of science has been to delegitimate evolution: "Even when promoting their beliefs as science, creationists sometimes denied similar legitimacy to evolution. They did so by limiting science to Baconian fact-gathering or, more typically, by appealing to dictionary definitions of science as factual knowledge."[60]

According to this view, "real science" is not just about whether observations are repeatable in the present but even more about whether observations serve as the grounds for inferences. According to this tradition within evangelicalism and fundamentalism, the only proper ground for a conclusion is observation, indeed multiple observations. Given the Creation Museum's indebtedness to this view, as evidenced by its emphasis on the distinction between observational science and historical science, one would expect a corresponding emphasis on a scientific method in which observations made free of preconceptions serve as the grounds for cautious inferences. Is this the kind of scientific method that is on display in the flood geology section?

When the remaining thirteen placards are examined closely, it turns out that only seven (18 percent) of the total thirty-eight placards, both display evidence that appears to pass the test for observational science and use that evidence as grounds for some kind of claim that could support a larger case for a biblical creation.[61] Of those seven, four reason from observational science to a claim not directly on behalf of a biblical creation but against evolution instead. All four of these placards are found in the Natural Selection room, and all four argue that natural selection is not a process that leads to evolution.

Three of these four placards discuss specific examples of natural selection: blind cave fish, blind mice, and antibiotic-resistant bacteria. On each of the three placards, observational science is presented in the form of observable examples: cave fish that live in very dark waters and that are albino and blind, mutant mice whose DNA produces a defective protein and that exhibit deficiencies such as blindness, and certain bacteria that produce mutant forms of a protein that antibiotics cannot affect. On the basis of this evidence, each placard claims that the mutations that produce these changes involve the loss of genetic information (to produce color, eyes, sight, or a certain kind of protein) rather than any gain. Since evolution requires the production of new genetic information to create new species, the reasoning goes, these three examples support natural selection, not evolution.

The fourth placard that grounds a critique of evolution in observational science also emphasizes the differences between natural selection and evolution. But here, rather than take up a specific genetic change within a particular species, the focus is on the sorts of changes that might appear in the context of a "kind." For young Earth creationists, the category of the "kind" is crucial. It is bigger than a species and, according to AiG and other young Earth creationists, is akin to the scientific "family." Because a kind encompasses many species and, with the help of natural selection and a rich genetic pool, can produce many more species, the notion of a kind both reduces the number of animals that had to be on the ark and can help to explain how lots of species were produced by the relatively small number of kinds that survived the Flood.[62]

The two "kinds" discussed on this placard, which is titled "What about Speciation within Kinds?" are finches and dogs. One placard argues, "Although often viewed as an icon of evolution, Darwin's finches serve as a perfect model of variation within a created kind." This is so since, "baraminologists (scientists who study created kinds) have determined that 'kind' is likely equivalent to the 'family' level (in relation to common classification terminology)." Thus, "all of Darwin's finches belong to the same family, Emberizidae, and therefore likely belong to the same created kind. . . ." Armed with this distinction between "kind" and "species," the placard argues that "natural selection . . . has caused fluctuations in the populations of finches and possibly led to finch speciation but never a finch evolving into another kind of organism."

The placard makes a similar argument regarding dogs. A beginning premise of the placard is that "it has been determined genetically that all dogs (wolves, coyotes, foxes, jackals, dingoes, and domestic dogs) are related to each other." Given this and the fact that "God said, 'Let the earth bring forth the living creature according to its kind'" . . . all dogs, whether wild or domestic belong to the same family—Canidae. Thus, all dogs likely belong to the same created kind." As in the case of finches, so too with dogs: "Natural and artificial selection have acted upon variation within the populations of dogs leading to the wide variety of dogs we see today." The placard continues, "Dogs have never evolved into another kind of organism. Instead what we observe is the selection of existing information . . . not the addition of new genetic information as required for molecules-to-man evolution."

But is it true that all of these examples involve only the loss of genetic information? Even if that is true, can these examples stand in for all instances of genetic mutations or alterations? That is, do genetic changes in other instances ever produce new genetic information? Also, is it true that all evolutionary processes must involve the production of new information? Could a mutation that causes a loss of genetic information participate in larger evolutionary processes? If that were a possibility, then these three examples could just as well be taken to support a claim for evolution.

More than this, simply contesting evolution does not make a case for biblical creation. So, it is important that only three placards in the flood geology section make an argument based in observational science on behalf of a biblical creation. While none of these three placards provides evidence for a six-day, twenty-four-hour-a-day, creation, all three do, at least at first glance, seem to reason from observational science to a claim in support of a global flood.

One placard, which appears in the Flood room and is titled "The Flood Buries the Earth," presents the "Paleozoic-Mesozoic flood model." This placard displays plenty of observational science[63] from the Grand Canyon for an argument on behalf of a global flood. Evidence includes the fact that "thousands of feet above sea level, the Redwall Limestone has abundant fossils, including billions of nautiloids," that "the Coconino Sandstone covers 300,000 square miles of area in western North America," that "the Redwall Limestone is continuous with limestones across the United States, including the rocks where Mammoth Cave is found in Kentucky," and that "the Tapeats Sandstone is continuous with sheet sandstones covering most of North America." While no specific claims are associated with any of this evidence, visitors could conclude that such information supports the placard's subtitle, "The Flood Covered All the Continents in Sediment."

Other evidence on the placard is offered to support specific claims about processes that may have (note the repetition of the word "suggest") occurred during the Flood. For instance, the fact that "Coconino Sandstone is several hundred feet thick . . . suggest[s] rapid, thick deposition." Likewise, a "six-foot layer containing billions of nautiloids suggests rapid deposition," "abundant, well-preserved fossils suggest rapid, thick deposition," "thin layers also suggest rapid deposition," and "tight bending of the Tapeats Sandstone through the Kaibab Formation suggests one mile of sediment was deposited quickly and was still soft when bent." How it was that thick sandstone was rapidly deposited by the Flood, how billions of nautiloids were rapidly deposited by the Flood, how "thin layers" of something (which is not identified by the placard) was rapidly deposited by the Flood, or how the Flood bent the Tapeats Sandstone is left entirely to the imagination of the visitor.

In a video titled "Rock Layers" that is displayed next to this placard, Andrew Snelling (who earned a PhD in geology from the University of Sydney and serves as director of research at AiG) and the video's voiceover help the visitor out as they discuss some of the evidence that appears on the placard. Surprisingly, the video does not argue directly that evidence observed in the Grand Canyon indicates a global flood. Instead, the video argues that processes observable in the present are incapable of producing the evidence displayed on the placard, like fossils, rock layers, or sand carried across continents. Snelling asks, for instance, "What river or what process today is carrying material from the Appalachians to the Southwest?" With no answer given, the viewer is invited to conclude that there is no such river or process in the present. An alternative answer is offered by the voiceover: "a global flood could carry sand and mud enormous distances." Likewise, Snelling observes, "In the Grand Canyon area we've got limestone bed—the Red Wall Limestone—the same limestone can be found over here in the Eastern United States." The voiceover says, "A global flood could form thick layers across entire continents."[64]

Thus, the global flood serves as a possible answer for "mysteries" that, according to the video, cannot be explained by processes observable in the present.[65] Of course, the repeated use of the word "suggests" gives away the fact that the claims made in this video and implied in the placards nearby are speculative. But if speculating is appropriate, despite the fact that the global flood is not observable in the present, why not consider other possibilities? Are there any other geological processes that might have taken place in the past that could explain this evidence? If so, would it not behoove a geologist to consider multiple possibilities in order to arrive at the best possible explanation?

The second of the three placards that appear to ground a claim for a global flood in observational science is titled "Backyard Scale," and, like the previous one, it is in the Flood Geology room. This placard presents observational science about river floods: "River floods commonly erode and deposit less than 1/1000 cubic miles of sand and mud in minutes to hours." No claim appears on the placard, but the placard appears in a series of six other placards that, together, suggest a larger argument. The first placard of the series claims, "In our own *backyards* we can see the layers rain produces on a small scale" (emphasis in original). The five placards that follow talk about ever-increasing scales: from "Backyard Scale" to "County Scale" to "State Scale" to "Regional Scale" to "Global Scale." Taken together, one of the arguments that this series of placards makes is that visible layers (small to big) indicate catastrophes (small to big). In that context, the "Backyard Scale" placard functions as evidence for the claim that a small catastrophe in the present can serve as a mini-analogy for a global one of the distant past.

But can it? The data displayed on this placard is about how much erosion a river flood produces in a short period of time. But the claim that the data are supposed to support is that "in our own *backyards* we can see the layers rain produces on a small scale" (emphasis in original).[66] Is a river flood to be understood as a backyard event? The image of a river flood that appears on the placard, which is not identified in any way, looks more like it would take out a backyard than be considered a "backyard event." In the end, this placard does not display any evidence (observational or otherwise) in support of the claim associated with it.

Despite the fact that the observational science on this card seems, at best, ill chosen, the argument itself reappears in somewhat different form in a video shown in the Post-Flood room. The video, titled "Canyons," talks about the same scales that appear in this series of placards and presents them in the same order. In the course of the video, the voiceover says, "In soft mud or sand in your own backyard you can see the power of heavy rains on a small scale. A rainstorm can create miniature canyons in only minutes. Though these canyons are very small, cut into mud, they share many of the same characteristics as the world's great canyons."[67] In support of this claim, the video shows images of rivulets cutting tiny "canyons" into wet soil. Thus, unlike the "Backyard Scale" placard, the video does appear to present observational science appropriate to the claim that small "catastrophes" can give clues about big ones.

But with the evidence scaled down to the backyard, does the reasoning make sense? In what sense does the process of a rivulet cutting a miniature "canyon" in one's backyard give clues about a global flood cutting the Grand Canyon? Are the hydrodynamics of a rivulet anything like those of a global flood? Does this analogy work?

Perhaps one of the most baffling aspects of all this analogical "evidence" regarding the Flood is that the Creation Museum itself makes the case that such analogies are out of order. The very first placard that visitors encounter in the flood geology section, titled "The Key: God's Word," says, "The present is not the key to the past," thus repudiating a central principle of geology dating back to Charles Lyell. According to this placard, one cannot understand the past by means of evidence in the present, especially when it comes to the global flood since, "Noah's Flood and times following involved more violent catastrophism than anything known in the present."

If the argument is that the global flood was so huge and so catastrophic that there can be no analogies in the present, then neither the "Backyard Scale" placard nor the "Canyons" video can provide any support whatsoever for a global flood or a biblical creation. More than that, if no process in the present can serve even as a mini-analogy for a global flood, then it does not matter how fast a rivulet can cut a minicanyon in one's backyard. Indeed, it does not matter how fast any process observable today can do anything. According to this logic, even if Mount Saint Helens erupted today and cut a brand new canyon in an hour, it could provide no support for the argument that a global flood could have radically transformed the Earth's surface in a short period of time. If this is the case, then a crucial claim of the Creation Museum—namely, that "real science" (observations made and repeatable in the present) can provide support on behalf of a global flood that produced fossils, canyons, and rock layers in a matter of years—is, at best, seriously compromised.[68]

The final placard among the three that appear to ground a claim for a biblical creation in observational science, titled "What Happened to Ebenezer?" appears in the Facing the Allosaurus room. This is an important placard because it is the only one in this exhibit that goes beyond simply presenting information about the skeleton to making some kind of argument about the skeleton's relationship to the global flood. The placard is divided into three sections: the "Pre-Flood World," "During the Flood," and "Entombed by the Flood." The observational science displayed on the placard appears in the first section where it says that "the fossil record contains many more varieties of fossilized plants and animals than we see among the plants and animals living today." Whether or not this difference is actually borne out by the fossil record, this evidence would seem to count as observational science.

The claim that this observational science is used to support is, "Before Noah's Flood, the world bore little resemblance to present-day Colorado where Ebenezer was found," and further that "Ebenezer probably lived on a vast supercontinent with a tremendous assortment of plant and animal life." The placard makes no connection between the fossil record, Ebenezer's habitat before the Flood, and the Flood itself. Visitors are left to come up with the link. But, if they have been paying close attention to claims made elsewhere in the flood geology section, they might take the placard to be saying that the fossil record, which indicates that there was once greater diversity of plant and animal life than there is now, suggests that the catastrophic effects of a global flood reduced that diversity. How the fossil record provides any evidence for a "supercontinent" is not made clear here.[69]

Even less clear is the connection between the fossil record, which is the only observational science on this placard, and the placard's main argument, which begins to take form in the second section. This section merits quoting in full:

> So, what were the final moments like for this dinosaur? During Noah's Flood, the fountains of the great deep were broken up. As rain fell, water levels rose; and elevated land areas were progressively covered. Violent waves surged increasingly inland over the continental fragments where dinosaurs and other land creatures lived in the pre-Flood world. Fleeing the encroaching danger could only be a temporary reprieve as the Flood waters continued to rise and surge. The accumulating sediments quickly covered and fossilized many footprints that dinosaurs—maybe even Ebenezer—left behind.

The third section continues in the same way as the second as it describes Ebenezer's demise: "Overwhelmed by the rising Flood waters, Ebenezer eventually drowned as his body was swept away in a sediment debris flow consisting of a mixture of sand and pebbles." In the end, "his body was quickly buried lying on his left side. Entombed, his bones were rapidly fossilized."

No observational science appears in either of these two sections. Even the fact that the dinosaur was buried "lying on his left side" cannot be considered observational science since, now that the skeleton has been removed from the soil, it cannot be observed today.

Given that this placard appears in the room wherein a truly impressive skeleton of a real dinosaur is on display, and given that this placard tells a story that seeks to link that skeleton to the Flood, it is surprising that the placard makes no mention of the skeleton itself. Not one piece of physical evidence from the skeleton is mobilized in any way by this placard. No inferences whatsoever are drawn from the skeleton about Ebenezer on this placard. It is as if nothing about Ebenezer is relevant for the story that is being told about him. Thus, the visitor is left wondering what value, if any, Ebenezer holds for a case on behalf of a global flood and a biblical creation.

Oddly, the placard argues that Ebenezer has great value since his demise is prototypical: "Ebenezer is only one example of what occurred on a global scale to billions of creatures. Except for those on Noah's Ark, all of the dinosaurs, other land creatures, birds, and humans died in the global catastrophe." But how can Ebenezer-the-skeleton (not Ebenezer-the-dinosaur that was once living and is now inaccessible) serve as an "example" when nothing about Ebenezer-the-skeleton is apparently relevant? More to the point, how can Ebenezer-the-skeleton indicate anything about what happened to other creatures in the global flood when Ebenezer-the-skeleton apparently can say nothing about what happened to himself?

In the end, Ebenezer-the-skeleton appears to make no contribution to an understanding either of his demise or any other creature's. The real evidence for the argument being made on this placard has nothing to do with Ebenezer. It turns out that the real evidence—that is to say, the evidence that does provide support for the claim that all creatures not on the Ark suffered this horrible fate—is the story of the Flood that appears in Genesis 7:21–23 and that is rewritten on this placard as a series of physical events involving "the fountains of the deep," "violent waves," and "accumulating sediments." It is on the basis of this extrapolation from Genesis, not Ebenezer-the-skeleton, that the story of Ebenezer's demise is spun. According to this placard, Ebenezer-the-skeleton is irrelevant for figuring out how he died, much less for how all the other creatures died. Since an extrapolation from Genesis serves as the evidence here, and since that story already tells what happened to all of the land creatures that did not make it on the Ark, no example in the form of a skeleton, or anything else, is needed.

The Creation Museum emphasizes the point that "real science" demands repeatable observations in the present. But when it comes to presenting arguments that the global flood created the geological formations we see today in a very short period of time, observational science plays no significant role at the museum. Observational science does not appear to serve as the ground for claims about a global flood or a biblical creation.[70]

If observational science does not serve as the basis from which inferences are made on behalf of the Flood or a biblical creation, then what is its function? So far, we have seen that it serves as the criterion by which AiG renders evolution akin to a religion rather than "real science." We have also seen that it serves as the evidence that the Creation Museum displays "plenty" of science. But does it make any direct contribution to a case for a biblical creation?

Ken Ham pointed in the direction of an answer to this question when, at the end of his opening remarks during his debate with Bill Nye, he stated that "creation is the only viable model of historical science, *confirmed by observational science*, in today's modern scientific era" (emphasis added).[71] Later, in his thirty-minute prepared remarks, he used the word "confirm" fifteen more times in statements about observational science "confirming" this or that aspect of a biblical creation. For Ham, the function of observational science is to "confirm" a biblical creation. But what does that mean?

In *The Lie*, Ham describes a peculiar scientific method that mobilizes observational science in support of young Earth creationist apologetics:

> We can take what the Bible says about history and see if the evidence of the present does fit. If we take the book of Genesis, which is a detailed account of our origins, we can see what it says concerning how the world was created and what subsequently happened. We can decide what we would expect to find if the Bible is true (this is our worldview, or model, built on the creation account). Then we can look at the world to see if

what we observe confirms the account in God's Word (and it does, over and over again).[72]

For Ham, observational science "confirms" a biblical creation when it provides evidence for a model that has been constructed by young Earth creationists to support a biblical account of origins. Returning to an example provided by Ham and referred to above clarifies the scientific method he advocates: "Genesis tells us that because of man's wickedness, God judged the world with a worldwide Flood. If this is true, what sort of evidence would we find? We could expect that we would find billions of dead things (fossils) buried in rock layers, laid down by water and catastrophic processes over most of the earth. This is exactly what we observe. Observational science *confirms* the Bible's historical science" (emphasis added).[73] Based on this description, Ham's scientific method appears to have four steps: (1) assume that what the Bible says about the creation and Flood is historically accurate; (2) construct a model based on the biblical account of creation to explain some event, process, or phenomena; (3) look for evidence observable in the present that can confirm the model and, thus, the biblical account; and (4) if observations support the model, then a biblical creation can be said to have been confirmed by science.

Notably, in both of Ham's descriptions of this scientific method, he mobilizes a conditional statement at a critical point. In the first description he says, "We can decide what we would expect to find *if the Bible is true*," and in the second description, he says, "*If this* [aspect of the biblical account] *is true*, what sort of evidence would we find?" (emphasis added). In both cases, Ham suggests by way of the conditional statement the possibility that the biblical account may or *may not* be true. Given this language, one might think that in Ham's scientific method it is possible either to confirm *or deny* the biblical account. Indeed, given this language one might imagine that Ham's scientific method is like the scientific method used in mainstream science: begin with a hypothesis, conduct observations or experiments, and either confirm *or deny* the hypothesis.

But this is not the case. As Ham put it in the debate: "We build models based upon the Bible. And those models are always subject to change. The fact of Noah's Flood is not subject to change, but the model of how the flood occurred is subject to change."[74] AiG's statement of faith makes the point even more clearly: "By definition, no apparent, perceived or claimed evidence in any field, including history and chronology, can be valid if it contradicts the scriptural record."[75] Observational science may confirm the model and, thus, a biblical creation. Observational science may never contest a biblical creation. Thus, our phrasing of the last step in Ham's model must be revised: if observational science proves consistent with the model, then it can be said to confirm a biblical creation, and if observational science appears inconsistent with the model, then either the observational science or the model must be changed until both confirm a biblical creation. Under no circumstances may observational science lead a young Earth creation scientist to raise questions about the truth of a twenty-four-hour, six-day creation or a global flood.[76] Given Ham's absolute conviction that the biblical creation can never be challenged no matter what observations are made, one can only wonder why he uses the conditional "if" when he describes his method.

With clarity about Ken Ham's peculiar scientific method, which moves from an uncontestable assumption to a model, to some observation, it is worth taking a second look at all thirty-eight placards.[77] It turns out that of the thirty-eight placards in the flood geology section, fourteen placards, or just over a third of the total number of placards, reason according to Ham's odd scientific method. All fourteen assume that Noah's Flood occurred about forty-three hundred years ago and was global. All but three identify a model that explains one or more effects of this global flood.[78] And all suggest that evidence (whether that evidence can pass muster as observational science or not) is consistent with the model describing the effects of a global flood and, thus, confirms a biblical creation.

Strikingly, of those fourteen placards that reason according to Ham's scientific method, only eight display observational science. A typical placard among these eight appears in the Post-Flood room. The visitor who begins at the top of the placard, where the boldface title appears, and works his or her way down experiences the steps in Ham's scientific method. The title at top left assumes the global flood: "The Flood Recedes." Below it, and in smaller font, a subtitle begins to work out the model: "Continents were scoured by retreating Flood waters." Beneath that, more explanation of the "regressional erosion model" is provided: "Flood waters dropped thousands of feet of mud and sand onto the world's continents. Retreating Flood waters swept some of that pile of sediment away." Then, below the explanation of the model appear three photographs of the "Kaibab Upwarp," the "Grand Staircase," and the "Red Butte and Cedar Mountain." These photographs function as the observational science on the placard since these geological formations can be observed today. Finally, statements below each of the photographs suggest how these geological formations, which are visible today, "confirm" the model and, thus, the Flood.

Below the photograph of the Kaibab Upwarp, the placard says that "warping in the late Flood times left a fold in the crust know as the Kaibab Upwarp. At that time there was probably another mile of sediment over this area." How one would know that "warping in late Flood times left a fold in the crust," or what "warping" is, or why that process would have "left a fold in the crust," or in what sense the Kaibab Upwarp is a "fold" is not explained. Moreover, statements under this photograph and the photograph of the "Grand Staircase" about the amount of sediment the Flood moved around or deposited may sound like observational science that confirms the model. However, since the post-flood processes that left behind all that sediment, according to the Creation Museum, occurred more than 4,000 years ago, they certainly cannot be observed or measured now. Thus, the visitor is left to wonder how AiG scientists determined those amounts.

The visitor who considers this placard—even the visitor who accepts Ham's curious scientific method—is obliged to fill in a lot of blanks as causal connections are barely hinted at and little observational science is provided (save three photographs). But if thoughtful visitors must fill in a lot of blanks on "The Flood Recedes" placard, even studious visitors will be at a loss when they consider "Rafting." This placard, which also appears in the Post-Flood room, seeks to explain how plants and animals were distributed across the continents after the Flood. Given that the Ark contained all of the living plants and land animals when the floodwaters receded, an explanation as to how those plants and animals were disseminated so as to repopulate the Earth in a relatively short period of time is crucial. The time frame is tight since the Flood occurred only about four thousand years ago, and since in its aftermath 938 "kinds" had to mutate by way of natural selection into the roughly 8 million species that populate the Earth today.[79]

Like the previous placard, this one assumes that the global flood occurred and then puts forth a "biogeographical rafting model" to explain processes that occurred afterward: "When the Flood destroyed the world's forests, it must have left billions of trees floating for centuries on the ocean. These log mats served as ready-made rafts for animals to cross oceans." Then comes the observational science that "confirms" the model and, thereby, the biblical account of the global flood. Below this appear three maps of the continents. On the first map, certain coastal areas are shaded in yellow, coral, lavender, and mint. Areas shaded the same color indicate coastal areas populated by a particular species. For instance, an area along the southeast African coast and another in southern Asia are shaded coral. According to the map, observational science shows that both of these areas are home to "mammal *Rhinocerotidae*." A pair of coral-colored arrows that, together, form an oval over the Indian Ocean appear to indicate a migration pattern. The suggestion here seems to be that at least four rhinoceros (two male, two female) boarded the log rafts produced by the global flood and, following "the paths of ocean currents," found themselves on one of these two coastal areas, where they reproduced. How those rhinoceros managed to get on a log mat and, even more, survive the voyage across the Indian Ocean is not suggested. Nor is any explanation given for why a rhinoceros would get on a log mat in the Indian Ocean in the first place.

The second map shows the same image of the continents as the first map and features the Ark, in yellow, situated on the land between the Black Sea and the Caspian Sea, where it presumably came to rest when the floodwaters receded. Yellow arrows extend from the Ark into central and south Asia, northern Africa, and Europe. No explanation is offered of these arrows. However, since the placard is all about migration patterns, it seems reasonable to infer that they represent the routes taken by animals (and perhaps also by plants somehow) from the Ark across and into other continents. Other arrows extend from red dots in places like present-day Iran, Mozambique, and Panama into southern Asia, sub-Saharan Africa, and Mexico and South America, respectively. The only text associated with this map says that these are "places of high diversity (probable landing sites)."[80] Blue arrows on the oceans seem to suggest patterns of migration that animals (and plants?) may have taken to and from those landing sites. Again, little observational science is provided to support the claims implied by the arrows. If visitors are to conclude from this map that observational science confirms the "biogeographical rafting model," they will be able to do so not because they have actually encountered observational science, but rather because they trust that "real science" lies behind the red dots.

The third map shows the same image of the continents. Like the other two maps, it displays a migration pattern, though not one seen on either of the other two maps. This migration pattern is for a particular species, the *Geochelone* (or giant) tortoise. Above the map appears the phrase, "distribution of *Geochelone* tortoises." Three small areas on the map are highlighted: the Amazon basin in western Brazil, the Galápagos Islands off the western coast of South America, and the Seychelles Islands off the eastern coast of Africa.[81] Projected from each of the three highlighted areas is a photographic image of a tortoise. Five blue arrows that appear on the water masses between South America and the Seychelles indicate the migration route of the tortoises. All of the arrows point in the same direction—that is, from South America toward the Seychelles. Appearing to follow ocean currents, the first arrow begins off the west coast of South America and ends just off the southern coast of the Galápagos Islands. The second arrow points straight west from the Galápagos Islands across the Pacific Ocean. The third weaves its way between New Guinea and Indonesia, to the north, and the northern coast of Australia, to the south. A fourth appears to follow currents across the Indian Ocean. The last makes a sharp turn northward between the east African coast and the Seychelles.

The claim that this map seems to make is that the location of *Geochelone* tortoises today in the Amazon basin and the Galápagos and Seychelles Islands confirms the "biogeographical rafting model" according to which land animals migrated from the Ark to distant locations by way of log mats and ocean currents. Thus, so the reasoning would seem to go, observational science confirms the biblical account of the Flood and a biblical creation.

But closer examination of the model raises a number of questions. The observational science on this map suggests that giant tortoises currently inhabit the Amazon basin and the Galápagos and Seychelles Islands. Justin Gerlach (a zoologist with particular expertise in *Geochelone* tortoises, senior member of the faculty at Robinson College at the University of Cambridge, and scientific coordinator of the Nature Protection Trust of Seychelles) says that whereas giant tortoises can be found on the Galápagos and the Seychelles, they are not to be found in the Amazon basin.[82] If that is so, then they cannot serve as observational science and, thus, cannot confirm the model. Moreover, the supposed observational science on the map indicates that these three groups of tortoises share an ancestor—namely, the two tortoises that survived the Flood on the Ark. The fact that mainstream scientists who study these tortoises disagree about whether they have a common ancestry raises questions about that assumption.[83] If they do not share an ancestor then, again, they cannot confirm the model or the account of the Flood in Genesis.

Putting aside, for the moment, the issue of whether the evidence and assumptions in this argument are accurate, what about the migration model presented here? According to the model, the first leg of the voyage would require two giant tortoises (one male, one female) to travel over land from the Amazon basin to the western coast of South America. Assuming they took the most direct route possible, they would have had to travel more than 700 miles over land. The top speed of a *Geochelone* tortoise is .2 miles per hour.[84] Assuming these tortoises moved at top speed and never stopped to rest or eat, this leg of the trip would have taken at least 145 days (or nearly five months) to complete. And that would just be the beginning.

The second leg of the trip would also be challenging but perhaps possible for the tortoises given strong ocean currents that flow from the western coast of South America to the Galápagos Islands.[85] To pull off this leg of the trip, the tortoises would (according to the Creation Museum's model) need to board log mats and travel a distance of more than 650 miles. That would likely take a week at ocean current speeds. No small feat. Once on the Galápagos, the two tortoises would either have to be joined by another pair from South America or would have to produce a pair of descendants. Either way, two tortoises would need to remain on the Galápagos to produce the population that is there now, and two more tortoises would have to embark on the third, much longer, leg of the voyage.

To complete the third leg of the trip proposed by the model, two tortoises would have to board log mats once again. This time, however, instead of riding strong and direct currents for 650 miles, these tortoises would have to find their way across the full breadth of the Pacific Ocean between the Galápagos and the southern shores of New Guinea. Assuming they took the most direct route possible across the Pacific, they would have to travel more than 8,400 miles to arrive in New Guinea. From there, they would have to steer their way through the waters between New Guinea, the northern coast of Australia, and the southern coast of Indonesia. Once in the Indian Ocean, they would have to drift all the way across the Indian Ocean. The fourth leg of the trip from New Guinea to the Seychelles would add more than 6,200 miles to the voyage across the Pacific. All together, to get from the Galápagos to the

Seychelles, they would (according to the best case scenario) cross approximately 14,600 miles of sea. Notably, the majority of that trip would be in the middle of the Pacific Ocean.

According to Gerlach, to imagine two tortoises completing the last two legs of the trip (from the Galápagos to New Guinea and from New Guinea to the Seychelles) is simply not reasonable. Although tortoises are known for their ability to disperse via water, they could not survive the (at minimum) three months it would take them to cross the Pacific Ocean plus the (at minimum) additional two months that it would take for them to cross the Indian Ocean. Of course, all calculations here assume both ideal traveling conditions lasting months on end and the tortoises' ability to navigate a straight line for thousands of miles across the Pacific Ocean on a log mat despite currents. This is not a reasonable expectation. As Gerlach points out, the tortoises "would have to switch from one current system to another and then wait for the seasons to change. They would be stuck in the middle of the Indian Ocean for months or years."[86]

It may be that one can observe big tortoises that appear to resemble one another in the Amazon basin and the Galápagos and Seychelles Islands. It may be that one can draw arrows on a map from point A to point B to point C. But does it really make sense to imagine that a couple of tortoises made this trek from the Amazon basin to the west coast of South America, then to the Galápagos, and then across the Pacific Ocean and the Indian Ocean? Indeed, in what sense does the so-called observational science presented in conjunction with the "biogeographical rafting model" confirm anything, much less a global flood?

The Creation Museum claims to put forth plenty of "real science" on behalf of a biblical creation. In the course of this analysis, we have scrutinized this claim by looking closely at those places identified by AiG as displaying the science of young Earth creationism. In sum, what do we find?

The planetarium and the Wonders of Creation room do present plenty of observational science, but little of it is mobilized in support of a biblical creation or a global flood. Instead, it is used to support the much more modest claim that the complexity, beauty, and apparent purposefulness that can be observed in nature suggest a divine designer who could be the God of the Bible.

As expected, the flood geology section is the space wherein "real science" is mobilized specifically on behalf of a biblical creation. But does it provide plenty of science to that end? A close analysis of the thirty-eight placards on display indicates that while twenty-six appear at first glance to display some kind of science, only sixteen pass AiG's test for "real science." Of those sixteen, three merely contain information and make no connection whatsoever to a biblical creation. Thus, just thirteen placards (about a third) on display in the Flood Geology room both display observational science and make some argument in support of a biblical creation.

Of those thirteen, only seven reason in the tradition of inductive science from observations to conclusions, and most of those (four) focus on critiquing evolution rather than advancing an argument for a biblical creation. Of the remaining three, one merely suggests that a global flood can explain "mysteries" observable in the present; another makes an analogy between observations in the present and a global flood, but that analogy is undercut by another placard nearby that says no such analogies are possible; and the third, on closer examination, assumes that the global flood occurred based on the biblical account and then infers from that assumption what a particular dinosaur's demise must have been like. In the end, only two placards in the Flood Geology room that offer arguments on behalf of a biblical creation reason in the traditional scientific way—that is from observations to conclusions—and mobilize scientific evidence that passes muster as observational science. Put another way, just 5 percent of all of the placards in the Flood Geology room reason from "real" scientific evidence to a global flood. Whatever one's definition of "plenty of science," to borrow AiG's Jason Lisle's statement about the amount of science that appears in the Creation Museum, 5 percent seems unlikely to rise to it.

If we accept Ken Ham's curious scientific method, we find that fourteen (just over a third) of the placards follow that pattern of reasoning from uncontestable assumption to model to observation. If we apply AiG's criteria for "real science," only eight of those fourteen placards pass the test. Thus, even when Ham's particular kind of science is the measure, just one in five placards can be said to "confirm" a biblical creation. To call this "plenty" hardly seems accurate either.

None of this says anything about whether or not the arguments made on any of these placards are reasonable. Do the arguments in the flood geology section make sense? Our close analysis of many of the arguments on these placards suggests that the answer is no because they depend on unwarranted assumptions (such as that *Geochelone* tortoises can be found in the Amazon basin or that they could float on a log from the Galápagos to the Seychelles Islands) or because they leave many questions unanswered (such as, in what sense a tiny rivulet caused by a rain shower is analogous to a global flood) or because placards in one spot contest the reasoning of placards in another (such as the placard that says no analogies can be made with the Flood that flatly contests the reasoning of several other placards that draw such analogies). Not to put too fine a point on it, the viewer who takes a critical view of the reasoning on these placards is likely to be bewildered by the science in the Creation Museum.

Bewildering or not, a crucial underlying assumption at the Creation Museum is that the scientific claims that are made throughout the entire museum align with what the Bible says is true about our world and beyond it. In short, the implied claim of the Creation Museum is that its science is the Bible's science. Is that, in fact, the case?

The Bible's Science

In February 1616, Cardinal Robert Bellarmine issued a private warning to Galileo on behalf of the Inquisition. In March of that same year, the Catholic Church issued a decree banning Copernicus's book, *On the Revolutions of the Heavenly Spheres*. In June 1633, Galileo was found guilty of heresy at a trial of the Inquisition and sentenced to house arrest for the rest of his life.[87] Why did the Catholic Church and the Inquisition take these grave actions? They did so for one simple and crucial reason: Copernicus's and Galileo's scientific writing—which suggested that the Earth moved and that the Sun was stationary—contradicted the obvious truth of the Bible and thus had to be silenced so as to protect the faithful from false teaching.[88]

So what was the biblical cosmology that Copernicus and Galileo were contradicting? Put simply, it was the cosmology of ancient Near Eastern cultures that had been understood and verified by human perception and experience for centuries. That cosmology consisted of a three-tiered universe with the Earth in the middle, the heavens above, and the "netherworld" below. The Earth in the middle was a fixed, disk-shaped landmass surrounded by a sea. On that landmass were mountains that held up the sky, which was a solid structure in the form of a tent or a dome that held back the cosmic waters. Now and again, rain would fall from those cosmic waters through openings in the sky. Beyond the celestial waters were the heavens where a deity or deities (depending on one's religion) dwelt. The Sun moved across the sky each day and passed into the netherworld, which was beneath the Earth, at night. The Moon moved in a similar fashion. The Sun, Moon, and stars were embedded in tracks on the surface of the sky that enabled them to move across the sky.[89]

Given that the biblical writers were steeped in that ancient cosmology, it is not surprising to find it reflected in both the Old and New Testaments. Thus, for instance, Genesis says that God created a "dome" (NRSV) or "firmament" (NKJV) to separate "the waters from the waters" (NRSV 1:6–7)[90] and, in the Flood story, tells of the "windows" in the heavens that opened to release the waters held back by that firmament or dome (7:11). Genesis also echoes the notion that the Sun, the Moon, and the stars were embedded on the surface of the dome or sky when it says, "God made the two great lights—the greater light to rule the day and the lesser light to rule the night—and the stars. God set them in the dome of the sky to give light upon the earth" (1:16–17). Psalm 19:6 affirms the readily observable idea that the Sun crossed the sky each day when it says, "Its rising *is* from one end of heaven, And its circuit to the other end" (NKJV, emphasis in original).[91] The book of Isaiah reiterates the ancient notion that the Earth was a flat disk and that God resides above it when the text says that God "sits above the circle of the earth" (40:22). Importantly, the Hebrew word translated here as "circle" refers not to a sphere but to a two-dimensional shape.[92] Finally, in Proverbs, the notion that the Earth is a flat circle surrounded by water appears in the verse describing God's creation of the Earth: "He drew a circle on the face of the deep" (8:27). In the New Testament, the apostle Paul refers to a three-tiered cosmos when he says that "at the name of Jesus every knee should bend, in heaven and on earth and under the earth" (Philippians 2:10). Here Paul is underscoring that *all* will bow by naming the three regions that Near Eastern people understood to constitute the cosmos.

When twenty-first-century readers of the Bible put aside their modern cosmology and read the Bible as it is actually written, they can see that it assumes an ancient cosmos. In that ancient cosmos, the Earth is flat, circular, and immovable and is surrounded by a sea. Enclosing both the Earth and the sea is a fixed dome or firmament with stars embedded in it. The Sun crosses that dome each day. Above the dome are heavenly waters and a heavenly realm beyond that.[93]

The Creation Museum's Science

In the course of his debate with Bill Nye, Ken Ham asserted, "By Creation, we mean here at Answers in Genesis and the Creation Museum, we mean the account based on the Bible. Yes, I take Genesis as literal history, as Jesus did. And here at the Creation Museum we walk people through that history."[94] For Ham, the Creation Museum, and young Earth creationism in general, the whole point of arguing that the universe was created in six twenty-four-hour days fewer than ten thousand years ago is to support a literal reading of the Bible. Given this, and given that a literal reading of the Bible yields a three-tiered universe, one would expect to see a three-tiered cosmos on display at the Creation Museum.

In the Starting Points room, the Earth appears at least ten times in association with God's Word (versus in association with Man's Reason). Each time it appears in a two-dimensional representation of a sphere floating in space. In *Six Days*, which is the film shown in the Six Days Theater, the Earth appears five times as a three-dimensional sphere floating and rotating in space. In one image, one large land mass and one smaller land mass can be seen. Neither appears as a flat circle but, instead, as an irregular organic shape.[95] In the Wonders of Creation room appear two-dimensional images of the Milky Way and its spiral arms, the Earth as a sphere floating in space, and the solar system with the Sun at the center and orbital rings suggesting the movement of the planets around it. Videos like "The Solar System" add many more images of the Earth, the other planets in the solar system, and the Sun at its center. Similarly, the animation of Adam's creation shows the Earth floating in space with the Moon nearby and the Sun in the distance. The Voyage of the Ark room features a video displayed on a composite of four large flat screens that shows a three-dimensional spherical Earth floating in space as floodwaters quickly consume the surface of the Earth. In the flood geology section three placards feature seven images of the Earth in either two-dimensional or three-dimensional representations of a sphere floating in space. In addition, five maps of the continents in the Post-Flood room show them in their present-day shapes with oceans and seas separating them, rather than in a single flat circle. Finally, in the *Created Cosmos* video featured in the planetarium, the Earth is shown numerous times as a sphere floating in space. Its landmasses appear in their contemporary shapes with oceans and seas among them. The solar system appears at least three times. In each case the Sun is a sphere at the center with spherical planets orbiting it. Distant stars float in space rather than sit fixed on the surface of a dome. A distant solar system is depicted with planets orbiting it as well. The Milky Way, along with many other galaxies, is also shown as a three-dimensional form floating in space.

In all, the Earth appears more than twenty times on placards in the Creation Museum. Each time, it appears as a sphere floating in space. Many more representations of the Earth as a sphere floating in space appear in videos shown in the Wonders of Creation room, the flood geology section, and the planetarium. Other images on placards and in videos depict the solar system with the Sun at its center and the Milky Way floating in space. Images and maps of the continents show these landmasses not as one single flat circle surrounded by water but instead in their contemporary shapes amidst oceans and seas on the surface of a spherical Earth. Put simply, at the Creation Museum, a modern universe is clearly on display.

In the Starting Points room, one of the first placards visitors see is titled "Same Universe." At the center of the placard is an image of the Milky Way. Superimposed over the left side of the image are four images that depict, in order, the "Big Bang" in the form of what appears to be matter expanding; the beginnings of the solar system with the sun and two planets; the Earth, Sun, and Moon; and the Earth with water, a landmass, and clouds. These four images and their labels that refer, respectively, to the "Evolution of galaxies," the "Evolution of our solar system," the "Evolution of earth and moon," and the "Evolution of continents" appear under the heading "Human Reason." Below the fourth image appears the phrase, "Billions of Years Ago." Superimposed over the right side of the image of the Milky Way are four more images that depict, in order, a blue sphere; the Earth with water and clouds; the Earth with landmass, water, and clouds; and the Earth with a landmass, water, clouds, Sun, and Moon. Next to each of these images are labels such as "Day 1 Let there be light," "Day 2 Let there be a firmament," and so forth. Above this series of images appears the title "God's Word" and below the fourth image is a label "Thousands of Years Ago."

The point of the placard is a simple one repeated again and again in the museum—namely, that evolutionists and young Earth creationists start with the same evidence, in this case the "Same Universe," but draw different conclusions about that evidence on the basis of their "Starting Points." Indeed, they do. Both start with a modern universe complete with spherical Earth rotating on its axis and orbiting the Sun along with the other planets in a solar system that is located on the edge of the gigantic Milky Way, which is but one galaxy among many in an enormous universe.

But how can that be so? If the Creation Museum promises, as Ham declares that it does, a literal interpretation of the Bible, how is it that visitors do not see a three-tiered universe and a flat, stationary Earth? The answer, quite simply, is that when it comes to representing the universe at the Creation Museum, a modern universe is preferred over the ancient cosmology indicated by a literal reading of the Bible.

Why is this so? Why do AiG and the Creation Museum impose a modern universe on the Bible, which clearly reflects in its language and phrasing an ancient cosmology? While we cannot speculate on intentions in this regard, it does seem obvious that if the Creation Museum were to display an ancient cosmology, it would be jarring to visitors who assume a modern universe. Indeed, to display an ancient cosmology would cement the idea, lamented by many fundamentalists decades after the Scopes trial, that accepting the notion of a young Earth means embracing a view of the universe that mainstream science disproved hundreds of years ago.

Moreover, imposing a modern universe on the Bible has certain rhetorical advantages—namely, it enables the visitor to have the experience of modern-day science seeming to confirm the Bible. If the cosmology of the Bible appears to be the same as modern-day science, then confirmation by scientific evidence can appear to be underway. Indeed, it is likely the case that in order for the science at the Creation Museum to "confirm" the Bible at all, the Bible's cosmology must be made to match a modern universe. The physical rules of an ancient cosmology are quite different from those of a modern universe. Given that all the many presuppositions and rules of modern science reflect the structure of a modern universe, how could modern science confirm an ancient cosmology with a completely different structure and set of relationships among its heavenly bodies? For modern science to confirm a young Earth, the Bible's cosmology must be made modern.

Commentators have argued that the Creation Museum seeks to dethrone science as a discourse of truth in order to elevate the status of young Earth creationism and a literal interpretation of the Bible.[96] This is certainly true. While that demotion of mainstream science is especially visible in the Starting Points room and the flood geology section, it undergirds much that is seen throughout the museum. That said, when it comes to cosmology something else is clearly underway at the Creation Museum. Rather than dethrone science, science appears to be privileged to the point of imposing a modern universe onto the text of the Bible, in the process, as John Walton has observed, "making the text say something that it never said." [97] Indeed, when it comes to how the universe is made to appear to visitors, it seems that modern science rather than a literal Bible rules at the Creation Museum.

3

BIBLE

Writing in the second quarter of the fourth century BCE, just about a half century after papyrus arrived in Greece, Plato has Socrates comment in the *Phaedrus* on the dangers of a new technology of writing that was transforming his world from an oral to a written culture. Socrates says of written words,

> You'd think they were speaking as if they had some understanding, but if you question anything that has been said because you want to learn more, it continues to signify just that very same thing forever. When it has once been written down, every discourse rolls about everywhere, reaching indiscriminately those with understanding no less than those who have no business with it, and it doesn't know to whom it should speak and to whom it should not. And when it is faulted and attacked unfairly, it always needs its father's support; alone, it can neither defend itself nor come to its own support.[1]

For Plato's Socrates, writing is a kind of sham speech. It resembles oral speech—it seems to have understanding, to be able to communicate, to defend itself—but because the author is absent it can do none of these. Worse yet, unlike words spoken to an intended listener, written words may go anywhere and address anyone with no regard for whether the reader will get their meaning or completely misunderstand it.[2] The chief problem with writing, in other words, is that as a technology of dissemination, it invites unintended readers and, thus, enables mistaken readings. In short, the problem with writing is reading.

Matters only got worse with the arrival of another technology of communication: the book. And once words were separated by spaces, and conventions of word order, punctuation, and the like were established, both writing and reading became easier. More books were written, and more people wanted to read them. Nicholas Carr suggests that this technological change effected a profound transformation in human culture:

> For centuries, the technology of writing had reflected, and reinforced, the intellectual ethic of the oral culture in which it arose. The writing and reading of tablets, scrolls, and early codices had stressed the communal development and propagation of knowledge. Individual creativity had remained subordinate to the needs of the group. . . . Now, writing began to take on, and to disseminate, a new intellectual ethic: the ethic of the book. The development of knowledge became an increasingly private act, with each reader creating, in his own mind, a personal synthesis of the ideas and information passed down through the writings of other thinkers.[3]

In the presence of the book and the privacy of his or her own thoughts, the reader had final say over the meaning of the text. With no one present to control another's reading, many different interpretations were possible.

The dissemination of written texts (along with the potential for multiplying meanings) accelerated exponentially in 1440 with the invention of the printing press. And the stakes regarding the proliferation of interpretations went through the roof when, in 1455, Gutenberg saw fit to print and, thereby, disseminate more widely than ever before the very word of God. When Martin Luther emerged on the scene in the early sixteenth century, all hell broke loose. We know it as the Reformation.

For Luther, and for Protestants generally, religious authority ultimately rests in the Bible, an idea traditionally summed up by the Latin phrase *sola scriptura*. This became "dangerous," to borrow from Alister E. McGrath's *Christianity's Dangerous Idea*, when it was combined with Luther's assertion of the "priesthood of all believers," which includes within it the notion that "all Christians have the right to interpret the Bible for themselves," thus "bypass[ing] the idea that a centralized authority had the right to interpret the Bible." Whereas once the medieval Church "had declared itself to be above criticism on biblical grounds," Luther's translation of the Bible into German empowered Christians to develop their own understandings of the text and, if they wanted, to challenge the doctrines and practices of the Church on the basis of those alternate interpretations.[4]

While early on Luther "clearly believed that the Bible was sufficiently clear for ordinary Christians to be able to read and understand it," it did not take him long to figure out that he could not control how other Christians understood the text they were reading. So he produced various forms of catechetical materials to provide assistance to readers; in the wake of the 1525 Peasants' War, he stressed the "importance of authorized religious leaders, such as himself, and institutions in the interpretation of the Bible."[5] These efforts notwithstanding, the text, to borrow from Plato, rolled about indiscriminately, to be read in a whole variety of ways. In the course of the Reformation and through the centuries that followed, earnest Christians have come up with all manner of biblical interpretations. They have included but are by no means limited to the following: the radical idea that Jesus's work on the cross renders us all already reconciled to God, to neighbor, and even to enemy such that violence is ruled out of order; the apocalyptic idea that at the end of history 144,000 "spirit-anointed" believers will be called into heaven to rule with Christ; and the baffling idea that at the heart of the Gospel is a weight loss program according to which shedding pounds and keeping them off is a matter of faith.[6]

Given the complexity of this sacred text and the intensity with which Protestants have sought to glean its truths, it is not surprising that Luther's "dangerous idea" has yielded an endless variety of theologies and practices, and thus countless splits, schisms, and sects, with no end in sight. And while each of these groups claims (whether explicitly or implicitly) that it has the true word of God, none has been able to control the proliferation of its meaning. But this has not stopped efforts to arrest the flow of interpretations, to freeze for all time the One True Interpretation. Enter young Earth creationism, and the Creation Museum.

The Inerrant Young Earth Creationist Word

On the face of it the Creation Museum's raison d'être would seem to be the marshaling of scientific evidence to discredit evolution and the notion of an old Earth on behalf of the claim that the God of Christianity created the whole universe approximately six thousand years ago in six twenty-four-hour days. As noted in the previous chapter, the museum offers up its own "observational and historical science" as an alternative to the ways in which mainstream science explains the origins of the solar system and the emergence and development of life on Earth. But while elaborate models and a flood of evidentiary bits are provided in behalf of a "young Earth," they are secondary to the museum's primary assertion, which is that we know that the Earth is but six thousand years old because this is what the Word of God tells us. To be more specific, the effort to discredit thoroughly the science of evolution and the notion of an old Earth is very much on behalf of the fundamental claim that the Bible is inerrant, that is, factually accurate in all that it has to say, and without error.

The AiG Statement of Faith contains a succinct summary of these ideas in the very first proposition in its list of "basics": "The Bible is divinely inspired and inerrant throughout. Its assertions are factually true in all the original autographs. It is the supreme authority in everything it teaches. Its authority is not limited to spiritual, religious, or redemptive themes but includes its assertions in such fields as history and science." AiG expends great energy elaborating on its understanding of the Bible. Readers of *Answers* magazine are repeatedly told that the Bible does not just contain the truth; the Bible *is* the truth: "Every *word* is 'inspired (God-breathed), not just the thoughts behind them ('verbal' inspiration). And all of the words are inspired, not just the ones we like ('plenary' inspiration)" (emphasis in original). Because God spoke every word, there are neither errors nor contradictions; if we mistakenly think we see an error or contradiction, we simply have to look more closely at the text and "clear up our own ignorance." Because God spoke every word, we do not need scientists to tell us if it is true: "If the Bible says something happened, we are to believe it, up front, right away, without hesitation. . . . Its truth does not wait for verification from us."[7]

According to AiG, not only is the Bible the errorless, consistent final authority on all matters on which it speaks, but it is also perspicacious, so clear that everyone everywhere and at all times can understand what the text says and what it means. According to Tim Chaffey, all that is required is for readers to take a historical-grammatical approach to the Bible, that is, note the "historical and cultural settings in which the book was written," attend to "standard rules of grammar," and ascertain "what the author intended [the text] to mean." In the words of AiG speaker Brian Edwards, these commonsensical "principles are easy to follow, and they are within the reach of everyone who prayerfully and carefully uses them." Follow these principles, and one will end up with the plain (literal) sense of the Bible. Henry Morris, co-author of *The Genesis Flood*, put it this way in *The Henry Morris Study Bible*:

> A literal interpretation is not an interpretation at all, for it takes the words at face value, assuming that the Holy Spirit was able to say exactly what He meant to say, using the thoughts and abilities of the human writer whose words He inspired. Any kind of allegorical or figurative interpretation of those words (unless directly indicated as required in the context) assumes that the interpreter knows better than the Holy Spirit what He should be saying, and such an attitude is presumptuous, if not blasphemous.[8]

The perspicuity of the Bible includes the first eleven chapters of Genesis. While some scholars try to render these chapters as "myth," a bad word in the AiG lexicon, Ken Ham and his comrades argue that a commonsensical reading makes it obvious that the only way to read it is as literal history. In Ham's words, "There is only one view of Genesis—it means what it says. It is written as history and is meant to be taken as history." According to AiG contributor Terry Mortenson, "Jesus took Genesis 1–11 as straightforward, reliable history, [and] several of His statements show that He believed in a literal six-day creation a few thousand years ago and the global Flood at the time of Noah." Not to believe that the first few chapters of Genesis is a literally accurate and errorless historical account raises questions about God's trustworthiness and undermines biblical authority; as the editors of *Answers* responded to a reader who wondered why the magazine places such emphasis on a six-day creation, "If we question His first words, what does that do to our respect for the Author and the rest of what He says about His plan of salvation?" Ken Ham repeatedly asserts that "the history in Genesis 1–11 is foundational for all Christian doctrine, including the gospel itself." As an *Answers* contributor pointed out in a blistering response to theistic evolutionists, for "Christ's work on the Cross" to make any sense it is necessary to believe in a literal Eden, a literal Adam and Eve, and a literal Fall from grace that requires the atoning death of Jesus Christ. In short, Christians can no more allow dissent from Genesis as a literal historical account than they can allow for dissent from Christ's literal Virgin Birth and literal Resurrection.[9]

Given the clarity of God's Word, given how obvious it is that Genesis is a literally true historical account, it is reasonable to ask how false readings of Genesis came to be. The simple answer is sin: people will-fully reject what they know in their hearts to be true, that is, the literal accounts of Creation and the Flood. More specifically, sin is manifested in the desire to modify our understanding of God's Word in order to ac-commodate human reason and the reigning views of science. The pro-cess of compromising the Bible began in the seventeenth century with Galileo. Responding to critics who argued that his heliocentric universe was at odds with the Bible's teachings that "the earth was the center of the universe and the sun moved around it," he "suggested that the Bible might use figurative language and so biblical interpretation should 'ac-commodate' new scientific discoveries." This notion that "science can change interpretation of the Bible, but the Bible cannot change scien-tific interpretations" became increasingly popular, but "it took several centuries for the destructive implications" to become clear.[10]

By the nineteenth century scholars and church leaders were adjust-ing their understanding of the Bible to fit with old Earth geology and Darwinian evolution. This readiness to bow to regnant science even in-fected the early fundamentalists, many of whom modified their under-standing of God's Word to accommodate evolution and an old Earth. But as an *Answers* contributor points out, such modifications are at odds with *sola scriptura*: "It is not millions of years plus the Bible. It is not the 66 books of the Bible plus the book of nature . . . It is not the 'Bible plus' but the Bible alone." While Ken Ham grants that it is possi-ble to believe in an old Earth and still be a Christian, he asserts that once you "add a concept (millions of years) from outside Scripture into God's Word," you have opened "the door of compromise." It is thus "a salvation issue indirectly," as "Christians who compromise on millions of years can encourage *others* toward unbelief concerning God's Word and the gospel." The result of such compromise has not only been the spread of unbelief, but the secularization and corruption of Western culture, which was once securely grounded in biblical principles.[11]

It is this inerrant authoritative young Earth creationist Word of God that is the primary focus of the Creation Museum. That this is the case is made very clear at the beginning of the Bible Walkthrough Experience, in the Starting Points room. This entire room is dedicated to the notion that the perspective through which we view the world is determined by our "starting point." There are only two possibilities from which to choose: human reason or God's Word. While reason is acceptable, as it is God's gift to human beings, it is only acceptable if it knows its place and does not contest God's Word. Unfortunately, history reveals that over the millennia sinful humans have used their reason against the Bible, the result being various attempts (as described in a large display in the Biblical Authority room) to "Question," "Destroy," "Discredit," "Criticize," "Poison," "Replace," and "Attack" God's Word, the latter referring to the effort to replace the six-day creation with "millions of years." But those who keep reason in its place, whose starting point is the Bible, do not question, criticize, or replace. Instead, they seek to emulate the prophets and apostles who are displayed on the other side of the room, men who proclaimed and obeyed God's Word. At the Creation Museum it could not be clearer that obedience is the only appropriate response to the inerrant and authoritative Word of God.

Not Reading the Word at the Creation Museum

For all of this emphasis on the Bible as the final authority, it is striking that there are no Bibles in the museum for visitors to read. We located four Bibles, but all are inaccessible.[12] One is displayed at the very beginning of the Bible Walkthrough. It serves as a prop in the Dig Site exhibit; it is open to Genesis 1 (with verses 24–31 highlighted in yellow), reinforcing the point that one of the two archaeologists is a (young Earth creationist) Christian. The other three Bibles are locked up and under or behind glass. One appears in the room with Luther and the printing press; it is open, but the text is in Greek. A second encased Bible is to be found at the end of the Bible Walkthrough: it belonged to Ken Ham's father, and its presentation here serves as a memorial to his father's love of the Word. Finally, and most strikingly, a Bible appears in a floor-to-ceiling glass case in a series of cases featuring fossils and skulls. Reinforcing the notion that humans are special creatures who can communicate with God, this particular Bible is in the hands of a full-size human skeleton that stands next to an ape skeleton.

To be fair, the Creation Museum does not claim to deal with the entire Bible, or even a substantial portion of the Bible. Its focus is on Genesis 1–11. This said, there is no exhibit, no wall display, no video presentation that simply provides visitors with the first eleven chapters of Genesis. It seems reasonable to suggest, in light of the repeated emphasis on the perspicuity of God's Word—so clear that anyone can read and understand that Genesis is a literally true historical narrative—that the Creation Museum should include at least one prominent display of

A human skeleton displayed next to a chimpanzee skeleton in a glass case just off the Main Hall in the Creation Museum holds one of the few Bibles that appear among the exhibits in the museum. Photograph taken by Susan Trollinger. For more images of and information about the museum, go to creationmuseum.org.

the complete text from the first eleven chapters of Genesis. There could even be benches placed in front of the display, as is done in the Wonders of Creation room, so that visitors could sit at their leisure and read the Word of God.

Rather than a display of the complete text, visitors find bits of Genesis in the form of individual verses and snippets from verses on various placards, murals, columns, and screens throughout the museum. But before looking more closely at how the museum presents biblical text, it is crucial to establish the visual and aural context. We begin by returning to the Biblical Authority room, with its prophets and apostles who proclaimed and obeyed the Word and its display recounting humanity's foolish attempts to challenge it. At the center of this room is a flat screen mounted on a wall, which plays a continuous video loop that instructs visitors how to be proper Christians. In the course of the video, individuals (both male and female and of various ages and skin colors) appear by themselves against a dark backdrop. Lit by a very bright and glowing light, each individual speaks a single verse or portion of a verse. They articulate the verse slowly, deliberately, and calmly. They look directly and intently at the camera. After they have spoken, the text appears below their image and then both the individual and the text fade to black.

No context is given for these biblical excerpts except the identification of chapter and verse. The viewer has no idea what is going on in the surrounding text. That said, the verses are grouped into nine categories: one, perfect, righteous, eternal, unchanging, true, good, beautiful, and powerful. One simply sees the name of the category on the screen and then one individual after another speaking a particular verse. Thus, for instance, in the category of "Power" a young boy says, "Great is our Lord, and of great power." Or for "True" an adult Caucasian man says, "Your word is truth." Or for "Unchanging" a young white woman says, "I am the Lord, I change not." Excerpted from the biblical context, the verses seem to reiterate the nine abstract characteristics attributed here to the God of the Bible.

Standing before this flat screen, the visitor is invited to observe and mimic the proper relationship that the Christian ought to have with the Word. Like the individuals in the video, the task is to receive the Word. Viewers are not to endeavor to figure out what it means given its biblical context, never mind its historical or cultural context. Nor do viewers need to discuss it, to see if others understand it in the same way. On the contrary, when it comes to encountering the Word, the video strongly suggests that individuals need nothing (neither other text nor other people) to understand it. The Word is transparent to meaning. Rather than read it, much less analyze it, readers are to revere it in all its awesome power and simplicity.

Visitors depart the Biblical Authority room and enter into the Biblical Relevance room with a display of Luther, a large placard documenting the decline of the church and Western civilization, and another video playing, this one with photos and narration regarding the Scopes trial. Then visitors head into Graffiti Alley, with its sounds of the city and its collage of newspaper and magazine fragments highlighting the corruption of contemporary culture. Then it's on to another noisy street, but this time the noise comes from three separate videos of one American family sinking into its own particular morass of middle-class corruption (gossip, violent video games, drugs, pornography, abortion), while across the street one can hear their mainline Protestant minister preaching a sermon in which he exhorts the congregants to surrender the Gospel to the demands of science, particularly mainstream geology.

It is easy to be overwhelmed by the sounds and sights of a culture that has abandoned the Bible and is thus in desperate crisis. But there is an answer. To access it visitors only need to pass through the Time Tunnel (a dark hallway with small white lights all around) and into the back of the Six Days Theater. The opposite wall is curved. Upon the length and height of it appears a short animation that depicts, presumably scene-by-scene, exactly how the Creation unfolded according to Genesis 1. The animation is punctuated with text announcing each day: "The First Day," "The Second Day," and so forth. Behind these words appears what looks like Hebrew script. As the animation unfolds, a voice speaks fragments from Genesis. Here, of course, the same logic obtains between the Word and meaning as was seen in the Biblical Authority room. There, a verse was presented as directly reflecting a characteristic of God. Here, the verse appears to mirror physical phenomena as they are "re-created" through the technologies of digital animation.

When the film fades to black, visitors exit the dark theater and enter a large, brightly lit room with a brown ceramic tile floor. This is the Wonders of Creation room, the perimeter of which consists of white walls and columns. At eye level on the walls are various framed and illuminated photographs and digital renderings of plants, animals, Earth, other planets, the solar system, and the double helix. Above these large images are smaller flat screens upon which videos constantly play. As noted in chapter 2, and according to the official Creation Museum book, these videos "scientifically confirm" the "first verse of the Bible . . . 'In the beginning God created.'"[13] These videos include images of all sorts of things like fish, birds, wishbones, plants, and airplanes. For each video there is a male voiceover (reproduced as text below the image) that speaks Bible verses and talks about what the Bible does and does not say about natural phenomena. Backless benches, such as you would expect to see in an art gallery, are positioned at various points inside the perimeter, so visitors can sit as they take it all in.

At the center of the room are large (nearly floor to ceiling) placards

(with embedded flat screens) as well as fat columns. On both the placards and the columns there is a great deal of text that is in both English
and Hebrew. Scattered throughout the room, including above and
below the video screens, are digitized fragments of Genesis 1 (in a font
that simulates handwriting) on what is made to appear as jagged pieces
of parchment; perhaps the point is to simulate a fragile, ancient (although not too ancient, given the text is in English) manuscript that had
been retrieved from an archaeological dig. All of this looks very much
like wallpaper. So does the Hebrew script which serves as backdrop to
verses on video screens and is presented upside down on at least one
video display. There is even Hebrew script on screens displaying New
Testament verses (e.g., Colossians 1:16) that were, of course, originally
written in Greek. But the crucial point of this room is to be found at its
center, where one finds an edited version of Romans 1:20 divided up
onto three floor-to-ceiling half-columns (rounded on one side and flat
on the other, with flat screens embedded in them): "For the invisible
things of God from the Creation of the world are Clearly Seen" (I) "Even
His eternal Power and Godhead" (II) "So that men are Without Excuse"
(III).[14] In this room, with its ever-repeating "15 Amazing Science Videos
on the 6 Days of Creation," visitors are clearly expected to conclude that
they have no excuse for rejecting the claims that God created the universe in six twenty-four-hour days.

As museumgoers move to the far side of the room and turn a corner, they are confronted with a very large flat screen upon which plays
an animation of the creation of Adam. Seeming to rise above viewers,
he is blessed with a perfect body, stylish hair, and a nicely trimmed
beard. From here visitors pass between two white pillars and enter into
the lush re-creation of the Garden of Eden. As seen in chapter 1, this
three-dimensional environment presents a lot to take in. In addition
to the room-size dioramas, there is the ubiquitous male voiceover, here
unceasingly speaking excerpts from the first few chapters of Genesis,
which are spoken without introduction, context, or commentary. Moreover, in front of each scene from the Bible is a series of three or more
placards. The placard in the center of them repeats a portion of the
scripture that we hear overhead, in the voiceover; the placards on either
side of that excerpted text present the Creation Museum's instructions
for understanding these biblical fragments.

Every time we have visited the Creation Museum, we have come out
of it feeling a certain sensory exhaustion. To make one's way through

the Bible Walkthrough Experience is to be subject to textual, visual, and aural overload. Text is everywhere—on placards, walls, columns, and flat screens. It is excerpted and fragmented. Visitors' attention is drawn here then there. They try to read this and are interrupted by that. All the while images are moving and changing. Videos are looping. A voice is talking. An animatronic dinosaur lunges for visitors' heads.

In his book, *The Shallows: What the Internet Is Doing to Our Brains*, Nicholas Carr presents the results of scientific studies that make the case that the new technologies associated with the Internet are reconfiguring the human brain. According to Carr, for centuries humans learned from the technology of print how to read line after line and page after page of text. People learned how to read in the quiet of their own thoughts, to think about what they were reading, even to get lost in it. Readers would imagine the people talking in the texts, the scenes within which they appeared and acted. Readers would anticipate the line of argument being developed over many pages. In short, people inhabited those lines and lines of text and the spaces they described and the arguments that they made. This kind of reading Carr calls "deep reading."[15] According to Carr, deep reading has been crucial for the development of human society because it encourages logical, critical, and creative thinking. And the Internet, Carr dramatically argues, is bringing deep reading to an end.

The structure of the Internet and its pages is not linear. It is a web that is all about making as many connections as possible. Within the space of the Internet readers are invited to hop, skip, and jump around. Even when they stay in one place, embedded videos, moving banners, and pop-up windows demand attention and interrupt or disrupt engagement with any particular text. Unfortunately, Carr argues, human brains are adapting to the logic of the web with all its fragmentation and interruption. The result is that readers are being transformed into skimmers, on average spending only twenty seconds on a web page. They skim the first few lines and then glance to the bottom of the page. Easily distracted by hyperlinks, it does not take them long before they are pages away from what they originally had been reading. It is not surprising that they have great difficulty remembering what they have read.

Although the Internet may be the technology that best deals in the logics of fragmentation and interruption, other technologies are quickly following suit. Cell phones, televisions, Kindles, and even the printed page (in, for instance, the form of magazine layout) are being reconfigured to mirror the logic of the Internet. Everywhere people are invited to glance here, notice that, skip to this, and jump to that. Rarely are people invited to slow down, consider, or deliberate.

The speed and volume of information pose real challenges. While the human brain has great capacity to retain long-term memories and develop connections among them, it takes time to move new information to long-term memory. According to Carr, the rate at which people are able to read printed text on a page is perfectly suited to this somewhat slow process through which the human brain moves information from short-term to long-term memory. That is why people have the ability to remember well what they have read in a book. By contrast, when people are bombarded with too much information too fast, and especially when that information comes in bits and fragments, their brains cannot make sense of it and cannot remember it.

Even if Carr is too apocalyptic in his discussion of how the Internet is altering the human brain for the worse, his comments are illuminating when it comes to the Creation Museum. Of course, the Creation Museum is not the Internet, but it deploys technologies of communication organized according to a similar logic. The museum consists of too much text: text on placards, murals, columns, and screens. Moreover, throughout the museum text appears in great varieties of font and size. It appears as background; it appears in the foreground; certain words are big; certain words are barely legible. Visitors are thus encouraged to skim and ignore the context. And text is not just put before the eyes. It also fills the ears by way of the constant voiceover. Sometimes the dialogue comes from the Bible, and sometimes it provides information about birds, frogs, plants, rising floodwaters, or geological transformations, and on and on it goes. And as visitors try to take all of this in, they are further distracted by flat screens playing endless video loops. Like pop-up windows, the screens are everywhere, sometimes as many as twenty of them in just one room.

It is not a stretch to say that the presentation of information in the Bible Walkthrough Experience at the Creation Museum echoes a web page; just like a web page, the space within the museum demands that one make the inevitably futile effort to absorb an overload of disparate informational bits. In this regard the Starting Points room is illuminating. As noted above, in this room visitors are told that they must choose God's Word and forever subordinate reason to it. Interestingly, the technologies deployed in here do much to make that so. As visitors move through these spaces (in much the same way as they move through a web site), as they encounter all these technologies, as their brains are overloaded, as they are obliged to skim, reason is indeed subordinated to a word. Whether it is the Word of God or the word of someone else is an open question. But, by passing through this cacophony of texts, sounds, and images, visitors are little able to engage it thoughtfully.

Displaying the Word at the Creation Museum

This said, there is much biblical text on display at the Creation Museum. The best way to see how the Creation Museum presents and uses the Bible is to examine closely the three stories that receive the most attention in the museum: the Tower of Babel, Noah's Ark and the Flood, and the Garden of Eden. We begin with the Babel room, where there is a prominently displayed placard with the following text:

> It came to pass that they found a plain in the land of Shinar, and they dwelt there.
>
> And they said, "Come, let us build ourselves a city and a tower, whose top may reach to heaven; and come, let us make ourselves a name, lest we be scattered abroad upon the face of the whole earth."
>
> And the Lord said, "Behold, the people is one, and they all have one language; and this they begin to do. Let us go down, and there confuse their language, that they may not understand one another's speech."
>
> So the Lord scattered them abroad from there upon the face of all the earth: and they left off building the city. (Genesis 11:1–9)

Most museumgoers would naturally assume that the text is, as labeled, Genesis 11:1–9. But a comparison with the Bible reveals that the curators have eliminated verses 1, 3, 5, 6b, and 9.[16] More than this, the museum has substituted (here and elsewhere) "Lord" for "LORD," the latter being the standard means (used by the King James Version, the New Revised Standard Version, and so forth) by which to indicate that the word being translated here is YHWH, the covenantal name revealed by the God of Israel to Moses. This relaxed approach to displaying what the museum understands to be the inerrant Bible is also found in the exhibit's centerpiece, a video entitled "Babel" that includes computer-generated animations and printed text, the source of which is not identified, but which is read aloud by an unseen but solemn narrator. Some of the text comes from Genesis 11:1–9. Some verses not included on the aforementioned placard are here, but they have been edited. For example, the section of verse 2 omitted from the placard ("as they journeyed from the east") is rewritten in the video as "they moved down from the mountains of Ararat"; given that the Ark came to rest "on the mountains of Ararat" (Genesis 8:4), this editorial decision underscores the museum's claim that it was Noah's descendants who settled in Babel. Then there is verse 5. While according to the KJV "the LORD came down to see the city and the tower, which the children of men builded," in the video it is reworked to say that, "God, the Creator of the heavens and the earth, looked down upon them and saw the city." Perhaps this revision is designed to ensure there is no confusion about God's omniscience and omnipotence, as God would not have had to come to Earth to see what was going on in Babel.[17]

The last part of the video departs from the biblical text altogether in connecting the God-established "language barriers" with the development of human diversity: "Each family group became isolated and developed distinct physical traits and cultures." But Babel is used here to explain more than human dispersion and diversity. On the wall adjoining the video display one finds a large signboard with a picture of the Tower, portrayed as a ziggurat-like structure with a large staircase leading into the sky, two pillars topped with burning torches on either side of the steps. A hooded figure (apparently a pagan priest) is standing near the top, with one hand holding a crook, and the other hand reaching upward. The heading at the top of the signboard reads, "The Entrance of Human Religion." While it may not be obvious from the biblical text that Babel marks the origin of "all human religions," the signboard instructs visitors that at Babel "humans rejected God's plan, worshipping the creation rather than the Creator, following their own way rather than God's way." All subsequent religions have followed Babel's "example" by "inventing myths to replace God's account of creation and Noah's Flood."

The latter is the focus of six rooms at the museum: Ark Construction, Voyage of the Ark, Flood Geology, Natural Selection, Facing the Allosaurus, and Post-Flood. As noted in chapter 1, in the first room visitors encounter a replica of the Ark being built as well as an animatronic Noah giving instructions to the construction workers. An accompanying placard helpfully suggests, "After centuries of righteous living, Noah may have been wealthy enough to hire shipbuilders" (how he found shipbuilders far from the coast is left unaddressed). More than this, these "paid workers may have been his audience," as "Noah, 'a preacher of righteousness,' warned them of the coming judgment." The animatronic Noah tries to persuade a nearby worker to join him on the impending voyage: "Judgment is coming my friend, but if you come along, you will be safe in the Ark." But the workers positioned further from their boss mumble about their "religious fanatic" employer: "What a fool! Why does he believe the word of his God?" Noah's workers were not the only ones warned of impending disaster. According to another signboard, "the Ark itself would have been like a billboard, a bold statement of Noah's faith, forewarning of coming judgment. A large ship, built far inland, would likely have been known the world over. In fact, the Bible claims that Noah, by preparing the Ark for the entire world to see, 'condemned the world.'" In the end, "God spared not the old world, but saved Noah."

It is striking how much of this is not found in the Genesis account of the Ark and the Flood. There is no reference in Genesis to Noah preaching the impending judgment, naysayers mocking Noah, Noah trying to persuade others to join his family on the Ark,[18] or anything akin to the Ark serving as a "billboard" of judgment for the entire world. So where is all this coming from? Much of it comes from the New Testament, in particular, sections of one verse from Hebrews and five verses from 2 Peter. The highlighted portions below are what appear on the museum placards: "By faith Noah, being warned of God of things not seen as yet, moved with fear, prepared an ark to the saving of his house; by which he **condemned the world**, and became heir of the righteousness which is by faith" (Hebrews 11:7, KJV); "And [**God**] **spared not the old world, but saved Noah** the eighth person, **a preacher of righteousness**, bringing in the flood upon the world of the ungodly" (2 Peter 2:5, KJV); and, "Knowing this first, that **there shall come in the last days scoffers**, walking after their own lusts, And **saying**, 'Where is the promise of his coming? for since the fathers fell asleep, **all things continue as they were from the beginning** of the creation. **For** this **they [are] willingly** are **ignorant** of, **that by the word of God the heavens were of old, and** the earth standing out of the water and in the water: Whereby **the world that** then was [**was then**], **being overflowed with water, perished**" (2 Peter 3:3–6, KJV).[19]

The museum puts these New Testament verses to good use, although even with these verses it is not obvious that, according to the Bible, Noah attempted to secure additional passengers for the Ark or that people across the globe knew about the Ark and understood its purpose. Moreover, there is no reference on the placards as to how these verses fit within Hebrews and 2 Peter. The museum's practice of providing biblical text without context is even more obvious in the Flood Geology section. Filling an entire wall at the entrance to this section is the following: "Ask now the beasts and the birds of the air and they shall tell you/Speak to the earth and it shall teach you/Who knows not in all these that the hand of the Lord has wrought this Job 12:7–9."[20] There is no question that the museum is encouraging visitors to understand these verses, particularly "this" at the end of verse 9, as referring to the global flood. Such a reading seems creative, at best, when one looks at this passage in its context. At this point in the book, Job has undergone a series of horrific afflictions; understanding himself to be blameless, he wants answers as to why God has done this to him. One of his friends responds by saying that God's wisdom is inaccessible to humans and beyond critique, and thus there is no question Job has sinned and is responsible for the ills he has endured. In Job 12:7–9 (the passage on the placard) Job sarcastically retorts that even the animals understand that God is in control. Job then proceeds with a savage parody (12:13–25) of God's so-called "wisdom," giving numerous examples of God's capricious and wanton use of power to wreak havoc in both nature and society. One of Job's examples comes in verse 15, when he mentions that on occasion God "withholdeth the waters," thus producing drought, and on occasion "he sendeth them [the waters] out," thus "overturn[ing] the earth." As Carol Newsom notes in her commentary on the book of Job, what we have here is an "unmotivated, destructive manipulation of water," a striking contrast with other biblical passages where "water is dried up and released specifically as punishment and blessing."[21]

The bitter sarcasm, the savage parody, the drought and other God-produced catastrophes, even the fact that the reference to God sending waters to overturn the Earth comes six verses after the passage on the placard: none of this information is provided to museumgoers. Unencumbered by context, visitors could easily conclude that Job 12:7–9 refers to God's just inundation of a wicked Earth.

Of course, the museum gives much attention to the Flood account itself. In the six rooms devoted to this topic there are numerous snippets from Genesis 6–9. They are often found on placards with explanatory titles such as "Judgment of the Whole World," "Noah Loads the Ark," "God Shuts the Door," "The Waters Rise," "The Ark Is Alone," and "God Renews His Promises." On occasion these signboards also contain further explanation as to how a particular passage should be read, such as this statement accompanying Genesis 7:19–23: "This passage of the Bible is not referring to some local event, but rather a Global Flood." As elsewhere in the museum, visitors are not informed when and where— as is frequently the case—verses and portions of verses have been removed from the passages on the placards.[22]

That there are deletions at all seems rather odd. Given how much space and text the Creation Museum devotes to the Ark and the Flood, there would seem to be very good reason to expect that the museum would provide a display with the full text from Genesis 6–9:17. But to do so would make visible one of the strangest stories in the Bible:

> When people began to multiply on the face of the ground, and daughters were born to them, the sons of God saw that they were fair; and they took wives for themselves of all that they chose. Then the LORD said, "My spirit shall not abide in mortals forever, for they are flesh; their days shall be one hundred twenty years." The Nephilim were on the earth in those days—and also afterward—when the sons of God went in to the daughters of humans, who bore children to them. These were the heroes that were of old, warriors of renown.
>
> The LORD saw that the wickedness of humankind was great in the earth, and that every inclination of the thoughts of their hearts was only evil continually. And the LORD was sorry that he had made humankind on the earth, and it grieved him to his heart. So the LORD said, "I will blot out from the earth the human beings I have created—people together with animals and creeping things and birds of the air, for I am sorry that I have made them." (Genesis 6:1–7, NRSV)

A 2013 *Answers* article claims that, in contrast with *The Epic of Gilgamesh*, the Genesis account of the Flood is "reasonable," "plausible," and "word-for-word history." Leaving aside the question of whether the "breaking open of the fountains of the great deep," the inundation of the entire planet, and eight people and thousands of animals surviving this global cataclysm in a large boat fits into the category of "reasonable," it is hard to imagine how Genesis 6:1–7 qualifies as a "plausible" historical narrative. Some of the weirdness in this passage revolves around the question of how we should understand "the sons of God." While some have argued that these are best understood as either royalty or as godly men, evangelical scholar Gordon Wenham notes that "the earliest Jewish exegesis . . . and the earliest Christian writers" along with "most modern commentators" understand these sons of God as "nonhuman, godlike beings such as angels, demons, or spirits."[23] These divine beings were attracted by the beauty of human women and had sexual relations with them. While Wenham suggests that "those who believe that the creator could unite himself to human nature in the Virgin's womb will not find this story intrinsically beyond belief," its strangeness (even to those who believe in the Virgin Birth) could explain why there is no reference to this story in the Creation Museum. More important, this story flows right into the Flood narrative. From the museum's perspective there would be real benefits in not having visitors muse on the possibility that it was the matter of god-men impregnating human women that prompted a disgusted God to inundate the Earth, especially given that it could raise questions in the minds of some museumgoers about the degree to which humankind bore responsibility for its own annihilation.[24]

A full display of Genesis 6–9:17 could raise other questions. Attentive visitors might not only note the story's exceedingly uneven narrative—to give just one example, Noah and his family enter the Ark twice (7:7 and 7:13)—but they would also encounter some odd inconsistencies. In 6:19, God commands Noah that of "every living thing, of all flesh, you shall bring two of every kind into the ark," while in 7:2 the Lord commands Noah to "take with you seven pairs of all clean animals, the male and its mate; and a pair of the animals that are not clean, the male and its mate." For another, in 7:17 "the flood continued forty days on the earth"; however, if one examines the flood's beginning in 7:11 ("in the six hundredth year of Noah's life, in the second month, on the seventeenth day of the month") and its ending in 8: 13 ("in the six hundred first year, in the first month, the first day of the month, the waters were dried up from the earth"), the flood lasts approximately a year. And at one point in the text, Noah sends forth a raven to see if the waters had abated. This act is immediately followed in the text by Noah sending forth a dove for the same purpose (8:6–12).

Some scholars have worked hard to reconcile these differences, making the case that, for example, the admonition in Genesis 7:2 to bring onto the Ark seven pairs of each kind of "clean animal" was simply an elaboration on the commandment in 6:19 to bring "two of every sort."[25] Another explanation is that the Genesis account of the Flood is a wonderful case study in behalf of the documentary hypothesis that posits that the first five books of the Bible, the Torah, are really a combination of sources, with redactors who brought them together to create what we now have as the Pentateuch. In the standard version of the documentary hypothesis the sources are designated J (for the document's consistent reference to the deity as Yahweh/Jehovah), E (for the document's consistent reference to the deity as God, or in Hebrew, Elohim), P (for the document's focus on priestly matters), and D (for the document only found in Deuteronomy).[26] In his lucidly written brief in behalf of the documentary hypothesis, *Who Wrote the Bible?*, Richard Elliott Friedman argues that the biblical account of the Flood is actually a combination of J and P, with both stories in keeping with the J and P stories elsewhere in the Torah. Once the two accounts are separated into two stories, the inconsistencies make sense: P has one pair of each kind of animal, while J has seven pairs of clean animals and one pair of unclean animals; P pictures the flood as lasting a little over a year, while J says forty days and forty nights; P has Noah sending a raven, and J has Noah sending a dove; P's God (Elohim) is a "transcendent controller of the universe," while J's God (YHWH) is anthropomorphic, grieved, personally closes the ark, and smells Noah's postflood sacrifice.[27]

In his commentary on Genesis, Walter Brueggemann flatly states that "it is beyond dispute that [the Genesis account of Noah and the Flood] conflates two strands of tradition, commonly designated J and P." But while Donald Akenson cheerfully notes that the "contradictions" in the "two different versions of the Flood" are "merely epidermal," at the Creation Museum the presence of contrasting stories cannot be countenanced, given that the Flood account, in particular, and Genesis 1–11, in general, are understood as an inerrant and thus perfectly smooth historical narrative. According to Ken Ham and his colleagues, the documentary hypothesis is "a thoroughly debunked idea put forth by anti-supernatural liberal critics."[28] Instead of Genesis being the compilation of sources by an editor or editors hundreds of years after the events described, a God-inspired Moses—"very likely . . . working with written documents"—wrote Genesis and the rest of the Pentateuch. Jesus said so in John 5:46: "For if you believed Moses, you would believe me; for he wrote about Me" (John 5:46). If Jesus was wrong here—that is, if he lied or erred—then he is "a sinner and less than God," and thus "could not die on the Cross for our sins."[29]

The museum's need to massage the text into a consistent historical narrative is particularly pressing when it comes to the two Creation accounts contained in Genesis 1:1–2:4a and Genesis 2:4b–25. A careful perusal of these stories reveals a transcendent God (Elohim) and a six-

In this life-size diorama, Adam and Eve serve as a model to the visitor for right relationship to God as they live blissfully amidst the beauty of the Garden of Eden and in obedience to God's simple commands, some of which the placard nearby delineates. Photograph taken by Susan Trollinger. For more images of and information about the museum, go to creationmuseum.org.

day creation in the first account and an immanent God (YHWH) and implied one-day creation in the second. In addition, there are radically different orders of creation. In the first account, the sequence is light, firmament, dry land, plants, lights in the sky, sea and sky creatures, land animals, humans. In the second account, the sequence is man, garden with trees and river, land animals and birds, woman. Peter Enns argues that "to stitch" these two accounts "into a seamless whole would dismiss the particular and distinct points of view that the authors were so deliberate in placing there."[30]

But at the Creation Museum the notion that the two creation accounts are one is simply assumed. When visitors begin their walk along the path through the Garden of Eden, the museum would have them understand not that they are entering a second narrative, but, instead, simply continuing their journey through one consistent creation account. Along their path there are numerous placards with verses and snippets of verses from Genesis 2–3. As elsewhere in the museum, there are rarely indications where text has been removed.[31] Most placards include labels summarizing the verses being quoted, such as "Adam Names the Animals," "God Forms Eve from Adam's Side," "The Tree of Life." Often there is commentary. On the large placard entitled "Curse" (with "Serpent," "Eve," and "Adam") visitors are reminded of a central argument of the museum: "According to God's Word, thorns came after Adam's sin, about six thousand years ago, not millions of years ago. Since we have discovered thorns in the fossil record, along with dinosaurs and other plants and animals, they all must have lived at the same time as humans." On occasion, the commentary serves as a theological gloss on a verse whose meaning may not be obvious to the museumgoer. For example, regarding Genesis 3:15—"The Lord God said to the serpent, 'Because you have done this, I will put enmity between your seed and the woman's seed; it will bruise your head and you shall bruise his heel'"—the accompanying commentary elaborates: "God promised a descendant—a seed of Eve—who would pay in full our sin account and destroy forever the effects of the curse."

As is the case in the flood geology section, along the Garden of Eden pathway there is creative use of biblical passages from beyond Genesis. There are quotes from Jesus, but the only identifying information is "Jesus Christ." Chapter and verse are not provided. Like the "God Says" billboards alongside many interstate highways, such an approach gives the impression that since these words are coming directly from the Almighty, no further information is needed.

Particularly interesting is the series of signboards with passages from the books of Isaiah and Revelation that are used to describe the Garden of Eden before Adam's sin: "No conflict" ("The wolf shall dwell with the lamb, and the leopard shall lie down with the kid; and the calf and the young lion and the fatling together; and a little child shall lead them," Isaiah 11:6); "No venom" ("And the sucking child shall play on the hole of the asp, and the weaned child shall put his hand on the viper's den," Isaiah 11:8); "No struggle for survival" ("They shall not hurt nor destroy in all my holy mountain," Isaiah 11:9); "No burdensome work" ("They shall not labor in vain," Isaiah 65:23); "No poisons" ("'They shall not hurt nor destroy in all my holy mountain,' says the Lord," Isaiah 65:25); "No suffering" ("God shall wipe away all tears from their eyes, and there shall be neither sorrow nor crying," Revelation 21:4); "No disease" ("There shall be no more pain," Revelation 21:4).[32]

Of course, the repeated use of the word "shall" indicates that these verses are describing a millennial future. On each of the placards there is thus a parenthetical statement—"promise of a future time, similar to Eden"—tucked in after the verse snippet. The extraordinarily tight analogy between the biblical past and biblical future is simply assumed here, despite the fact that there is no textual evidence to substantiate such a claim, and despite the fact that Isaiah 65:23 and 25 and Revelation 21:4 are preceded by passages that emphasize that the future described here is explained as something completely new:[33] "For, behold, I create new heavens and a new earth: and the former shall not be remembered, nor come into mind" (Isaiah 65:17, KJV); "And I saw a new heaven and a new earth: for the first heaven and the first earth were passed away; and there was no more sea" (Revelation 21:1, KJV). Whatever theological argument there may be for the symmetry of Edenic and millennial states, such symmetry is not grounded in the biblical text.

We conclude this examination of how the Creation Museum presents and makes use of the Bible by returning to Genesis 1 and the most important biblical words in the museum. First, the word "day." For Ken Ham, for the Creation Museum, for all young Earth creationists, the word "day" in Genesis 1 must refer to a twenty-four-hour day, the result being, of course, a six-day creation. This point is repeated again and again in the museum, including a video loop in the Dragon Hall Bookstore in which Ken Ham explains at great length that the Hebrew word *yom*—the word that has been translated as "day"—can only be legitimately understood to mean a twenty-four-hour day. If one is going to take the original text seriously, if one is going to read the Bible literally, then "day" must mean "day." In his *Study Bible* Henry Morris expands upon the point: "It is undeniable that God intends us to know that the days of creation week were of the same duration as any natural solar day. The word *yom* in the Old Testament almost always is used in this natural way, and is never used to mean any other definite time period than a literal day."[34]

In his book, *The Lost World of Genesis One*, Wheaton College (IL) professor of Old Testament John Walton agrees that "the best reading of the Hebrew text" of Genesis 1 is clearly that "these are seven twenty-four-hour days"; Walton goes on to observe (and again in agreement with Ham and Morris) that old Earth creationists made the mistake of "stretch[ing] the meaning of *yom*" to mean "long era" because they saw no other "way to reconcile seven twenty-four-hour days of material creation with the evidence from science that the earth and universe are very old." But then Walton goes on to suggest that such definitional stretching is unnecessary, given that, *contra* the young Earth creationists, "the seven days and Genesis 1 as a whole have nothing to contribute to the discussion of the age of the earth." According to Walton, the best way to understand Genesis 1 is that in these seven twenty-four-hour days "the cosmos is being given its functions as God's temple, where he has taken up his residence and from where he runs the cosmos." In such an understanding of the cosmos the Israelites were very similar to other ancient peoples, who were also primarily concerned with the world's functions and who made it function. Central to Walton's argument is that *bara,* the Hebrew word translated as "created" in Genesis 1:1, is best understood as a matter of "assigning functions," as opposed to material creation.[35]

Walton asserts that he is reading Genesis 1:1 literally, making the point that a truly literal reading requires a knowledge of Hebrew and ancient Israelite culture: "Someone who claims a 'literal' reading based on their thinking about the English word 'create' may not be reading the text literally at all, because the English word is of little significance in the discussion." Walton's argument that what is "created" in Genesis 1 is the "bringing [of] functionality to a nonfunctional condition rather than bringing material substance to a situation in which matter was absent" is strengthened by the second verse of the chapter: "And the earth was without form, and void; and darkness was upon the face of the deep. And the Spirit of God moved upon the face of the waters" (KJV). But at the Creation Museum the suggestion that there was existing matter at the beginning of God's work of creation is ignored, and there is very little comment regarding the proper translation of *bara* (in striking contrast with all the attention paid to *yom.*) Instead, throughout the museum it is flatly asserted that what we have in Genesis is the account of creation out of nothing, that is, *creatio ex nihilo.* As noted on a signboard entitled "Scripture Alone" (and quoting from the Westminster Confession): "It pleased God the Father, Son, and Holy Ghost . . . in the beginning to create, or make out of nothing, the world, and all things therein, whether visible or invisible, in the space of six days." Henry Morris celebrates the point in his *Study Bible*: "The concept of the special creation of the universe of space and time itself is found nowhere in all religion or philosophy, ancient or modern, except here in Genesis 1:1."[36]

But as Peter Enns observes, most biblical scholars translate Genesis 1:1 in keeping with how it is translated in the New Revised Standard Version—"In the beginning **when** God created the heavens and the earth" [emphasis added]—as "this clause then introduces v. 2, which depicts the prior chaotic state." Not only does Genesis 1:1 fail to express the notion of *creatio ex nihilo*, but there are no passages in the Bible that can be properly understood as making the case for this idea. In fact, as theologian Gerhard May points out, it is not until the latter half of the second century, when theologians—reasoning beyond the strictures of the biblical text—developed the doctrine of *creatio ex nihilo*. None of this matters at the Creation Museum, where the very real possibility that *creatio ex nihilo* is not a literal reading of Genesis is simply not broached.[37]

Freezing the Interpretation

In a 2011 *Answers* article, Terry Mortenson, AiG speaker and writer, decries what he sees as a "dangerous trend" in contemporary evangelicalism, in which a "growing number of evangelical scholars and leaders . . . are functioning as 'popes.'" While "they don't directly declare themselves to be infallible and supremely authoritative," their "writings or lectures" are treated as such by many evangelical laypersons. As "one not-so-subtle example" of this trend, Mortenson points to John Walton and *The Lost World of Genesis One*. Walton's argument that "God did not create anything in Genesis 1" but instead "gave pre-existing things a new function" rests on esoteric knowledge of Hebrew and Near Eastern literature inaccessible to most laypersons, the implication being that "you can't understand the Bible on your own, so let the scholarly experts explain it to you." The result is that, as Mortenson explains, Walton and other "supposed experts" are "stealing the Bible from the people in the pew." This was precisely what Martin Luther was opposing when he "raised his voice against unquestioned rule by religious experts." Mortenson concludes by calling on good Christians to join with Luther and the apostle Paul in holding to "the written Word of God, not any man or group of men" as the "final authority for determining truth."[38]

Mortenson's argument that Walton and his fellow "evangelical popes" are substituting their ideas and their authority for the Bible echo Ken Ham's blog diatribes against *The Lost World of Genesis One*. "Regardless of what John Walton claims," Ham declares, "there is no doubt in my mind he had developed his ideas about Genesis to accommodate the supposed millions of years of earth history." While such "bizarre and elitist ideas" may satisfy Walton's academic audience, a commonsensical reading of Genesis makes clear that God created the universe in six twenty-four-hour days. In a post entitled *"The Genesis Flood*—The Battle Still Rages!*,"* Ham proclaims that we do not need Walton and the other "'Protestant popes' popping up all over the Christian world" to tell us what the Bible means. It means what it says, and everyone can understand what it means: "If the infinite God, who created language, cannot move people to write His 'God-breathed' words so all people (regardless of culture) can understand them, then there is something dreadfully wrong."[39]

What is interesting about Ham's claim that the Bible is perspicuous, clear enough for anyone to read and understand without the assistance of experts, is that the Creation Museum is a $27 million edifice that goes to great lengths to instruct visitors on how they should understand Genesis 1–11. As we have seen, throughout the museum there is a veritable barrage of explanatory text—on placards, on video screens, in the omnipresent voiceovers—indoctrinating visitors in the correct way to read this passage, that snippet, or (in the case of "day") one particular word. In short, the Creation Museum leaves nothing to chance in ensuring that museumgoers depart the premises fully inculcated in the proper understanding of the biblical text.

In this regard, it would seem that, for all the emphasis on the clarity of scripture, the Creation Museum lacks faith in the folks who walk through their doors to be able to read Genesis 1–11 and come up with the "correct" interpretation of the text. But to be fair to the museum, there is good reason for this lack of faith. As noted earlier, the Reformation and its emphasis on *sola scriptura* and "the priesthood of all believers" has resulted in an incredible and ever-increasing diversity of biblical interpretations. And the notion of biblical inerrancy, which has formed conservative Protestant understandings of the Bible for the past 150 years, has not solved this problem. One can maintain that the Bible is "God-breathed," errorless, and true on all matters upon which it speaks. One can proclaim with great conviction that it is consistent and perspicuous. One can promise to read it "literally." But the question of what the text means remains, and the disagreements over what the text means remain. As Christian Smith observes in *The Bible Made Impossible*, his fascinating study of how all this has worked out in practice within American evangelicalism:

> The very same Bible—which biblicists insist is perspicuous and harmonious—gives rise to divergent understandings among intelligent,

sincere, committed readers about what it says about most topics of interest. Knowledge of "biblical" teachings, in short, is characterized by *pervasive interpretive pluralism* . . . In a crucial sense it simply does not matter whether the Bible is everything that biblicists claim theoretically concerning its authority, infallibility, inner consistency, perspicuity, and so on, since in actual functioning the Bible produces a pluralism of interpretations. (Smith's emphasis)[40]

Here is a great conundrum. Biblical inerrancy does not change the reality that written texts are unstable, "roll[ing] around everywhere" (in the words of Plato), falling into the hands of various readers who come up with multitudes of wide-ranging interpretations. Christian Smith makes abundantly clear that biblical inerrancy does not resolve the fundamental tension inherent to *sola scriptura*. As Kathleen Boone points out in her study of fundamentalist discourse, *The Bible Tells Them So*, "the sole authority of the text is subverted by the very nature of texts." That is to say, how does one "control interpretation, while claiming at the same time that nothing but the text itself is authoritative?" The only answer, according to Boone, is to depend on an external authority, a magisterium, as it were, to declare that this is indeed what the text means. But for conservative Protestants holding to the inerrant Word of God as the sole authority, this interpretive magisterium must be unseen. As Boone puts it, "the distinction between text and interpretation" must be effaced, an "effacement especially apparent in literalistic reading when it is claimed that the interpreter does nothing more than expound the 'plain sense' of the text."[41]

This is precisely what one finds at the Creation Museum. One would not know from the museum that "pervasive interpretive pluralism," regarding Genesis 1–11 and more, is endemic to evangelicalism and fundamentalism. Instead, everything the museum has to say about Genesis 1–11—the Creation, the Fall, Noah and the Flood, Babel, and more—is simply presented as the "plain sense of the text." That is to say, according to the museum it is simply relaying to museumgoers what the inerrant Word of God clearly says about the creation of the Earth and its early history. The museum is serving as magisterium, but its role is hidden. There is no magisterial interpreter in sight. Like the constant voiceover in the walkthrough, the magisterial interpreter is to be obeyed but not acknowledged. The act of interpretation has been completely effaced. Young Earth creationism is not an interpretation. It is the Truth.

Ken Ham makes this clear in a heated blog battle with biblical scholar Peter Enns, whose 2012 book, *The Evolution of Adam: What the Bible Does and Doesn't Say about Human Origins,* makes the case against understanding Adam, Eve, and the Fall as literal history. Responding to Ham's attacks, which included both a harsh review of *The Evolution of Adam* on the AiG webpage as well as a critique of Enns at a homeschool convention where both were speaking, Enns suggested that Ham's "engagement of his *Christian* opponents be more shaped by his acknowledgment of their shared Christian *bond*" (emphases in original). Ham responded by decrying Enns's "agree to disagree" plea: "What Enns is doing is destructive to Christianity because he is deliberately undermining the clear teaching of the Word of God . . . [and while] Enns wants me to consider his view of Genesis a possible valid one . . . I would never say such a thing as I will never knowingly compromise God's Word!" Ham charged that Enns "reinterprets God's Word to fit man's fallible words instead of judging man's words by God's inerrant Word," a reinterpretation clearly motivated by "clear academic pride" for which the author of *The Evolution of Adam* "need[s] to humble himself as a child and repent . . . before a holy God."[42]

Holding fast to "the clear teaching of the Word of God," Ken Ham and the Creation Museum cannot acknowledge they are presenting an interpretation, nor can they consider the possibility that other interpretations—including other conservative Protestant interpretations—of Genesis might be correct. They cannot seriously consider, for example, the possibility that Genesis 1–11 is not a literal historical narrative, that Genesis contains two creation stories and two flood stories, that their interpretation of "create" is not a literal translation of "*bara*," or that the notion of *creatio ex nihilo* is not sustained by the Bible alone. If there is any chance that any of these possibilities is true, then their truth is not Truth, they are the ones with the "fallible words," and the whole AiG enterprise—including the Creation Museum—is washed away.

The unwavering commitment to a young Earth creationist interpretation of Genesis 1–11 helps explain the museum's apparently perplexing treatment of biblical text: the overwhelming flood of visual and aural distractions that makes it almost impossible to read; the absence of a display with the eleven chapters of Genesis; the inconsistent use of translations and the creative editing; the lack of ellipses indicating where text has been removed from a passage; the failure to provide relevant context for the passages that are displayed. All of this seems oddly

loose, given the Creation Museum's stated commitment to biblical inerrancy and the very words of the Bible as "God-breathed." But to find this
perplexing is to miss the point of the museum. The museum is not primarily interested in the careful presentation of the inerrant biblical text.
James Barr has observed that, while fundamentalists profess to defend
biblical authority, they are much more committed to imposing a particular, conservative interpretation upon it. This is certainly the case at
the Creation Museum: much more than presenting the Bible itself, the
museum is most interested in making the case for one very particular
interpretation of the inerrant biblical text—the young Earth creationist
interpretation of the inerrant biblical text—as the Truth.[43]

This effort to put an end to pervasive interpretive pluralism, to
freeze for all time how we should understand the Bible, has precedents.
Take, for example, the *Scofield Reference Bible*. C. I. Scofield was a theologically self-educated Congregationalist (later Presbyterian) minister
from Dallas who in the late nineteenth century became a leading light
in the Bible and Prophecy movement, which sought to promote biblical
inerrancy and dispensational premillennialism among American Protestants. Over time Scofield—whose 1888 book, *Rightly Dividing the
Word of Truth*, soon became essential reading for dispensationalists—
became convinced that the Bible needed to be bolstered with notes that
ensured readers rightly understood God's Word. So in 1903 he resigned
his ministry in order to develop his own edition of the KJV. Issued in
1909 by Oxford University Press (a second edition came out in 1917), the
Scofield Reference Bible quickly became a publishing phenomenon. By
1930, one million copies had been sold. By 1945, the number had risen
to two million.

In *Bible: The Story of the King James Version, 1611–2011*, Gordon
Campbell notes that "Scofield's edition marked a new phase in the history of the KJV." Using both extensive commentary and an elaborate
system of cross-referencing with other verses in the Bible, "Scofield
act[ed] as a guiding hand, at every point helping the reader to understand what the passage means." In this regard, the *Scofield Reference
Bible* is most famous for its aggressive promotion of dispensational
premillennialism. The amount of interpretive instruction he provided
for the books of Daniel and Revelation (on occasion over half the page
consists of his notes) was simply overwhelming, as Scofield went all out
to instruct readers on how to understand (actually, to decode according
to an elaborate dispensationalist formula) these very strange books.[44]

But Scofield also had a great deal to say about Genesis 1–11. Take, for example, Genesis 1:1–2, in which Scofield makes the case for the "gap theory" of creation. For these two verses Scofield added four notations with nine cross-referenced verses from elsewhere in the Bible. He also inserted an explanatory heading for each verse: *"The original creation"* for verse 1 and *"Earth made waste and empty by judgment,"* the latter with its own biblical cross-reference (Jeremiah 4:23–26). Then there are the interpretive footnotes. Regarding Genesis 1:1, Scofield inserts a note regarding the Hebrew word *Elohim* that is translated "God" (which includes the claim that "the Trinity is latent in *Elohim*"); he also attaches a note to the word "created," in which he points out that there are three distinct *"creative* acts" in Genesis 1, the first of which (in Genesis 1:1) involves the creation of the heaven and the earth, and takes place in "the dateless past, and gives scope for all the geologic ages." Then, regarding Genesis 1:2, Scofield attaches an extended note in which he explains that "the earth was without form, and void" because of a cataclysmic "divine judgment," perhaps related to a "previous testing and fall of angels," a suggestion for which he finds evidence in the books of Ezekiel and Isaiah.[45]

As with the Creation Museum, it is hard to square all of Scofield's overbearing commentary with the traditional Protestant idea of individual believers reading the perspicuous Word of God for themselves. But as is also the case with the Creation Museum, the commentary is presented as simply being the plain sense of the scripture and not as interpretation (a point underscored by the virtual absence of references to other biblical interpreters in Scofield's notes). This was certainly how his comments were understood by American fundamentalists, for whom the *Scofield Reference Bible* was the Bible of choice from the movement's beginning in 1919. As Dewey Beegle observed in his 1973 book, *Scripture, Tradition, and Infallibility,* "the influence of the Scofield notes became so dominant within evangelical circles that many adherents considered the commentary as *the true interpretation* of the Bible, thus implicitly granting the commentary equal authority with the biblical text" (emphasis Beegle's).[46]

This was a remarkable achievement. It is no stretch to say that C. I. Scofield was wildly successful in his ambitious effort to set in stone what the Bible means, at least for conservative Protestants in America. And as regards his gloss in behalf of the "gap theory" of creation, Scofield was writing in the early twentieth century, at a time when almost all conservative Protestants were old Earth creationists, and thus what he had to say regarding Genesis 1:1–2 seems to have generated virtually no controversy. But times have changed, young Earth creationism has eclipsed old Earth creationism in conservative Protestantism, and Scofield's statement that God's "first creative act refers to the dateless past, and gives scope for all the geologic ages" has consigned the creator of the "fundamentalist Bible" to an ignoble fate at the Creation Museum.

To see what has happened to Scofield we need to go to the Biblical Relevance room, which serves as the pivot between the Biblical Authority room and Graffiti Alley. At first glance this room seems to be a celebration of the Reformation, given that it contains a model of Gutenberg's printing press (credited with making the Bible "the biggest seller of all time"); a signboard celebrating the Reformers, "who called the church back to the authority of Scripture Alone"; and the iconic tableau of Martin Luther nailing his ninety-five theses on the door of the All Saints Church in Wittenberg in 1517. But then one notices that the quotes on the "Scripture Alone" signboard—including John Calvin's "the duration of the world, now declining to its ultimate end, has not yet attained six thousand years"—are there to establish that the Reformers were young Earth creationists. One also notices that the Luther figure is not actually posting the ninety-five theses, but, instead, the following:

> If I profess with the loudest voice and clearest exposition every portion
> of the Word of God except precisely that little point which the world
> and the devil are at that moment attacking, I am not confessing Christ,
> however boldly I may be professing him. Where the battle rages there
> the loyalty of the soldier is proved; and to be steady on all the battle front
> besides is mere flight and disgrace if he flinches at that point.

While the museum credits this call to culture war to "Martin Luther, correspondence," it turns out that the quote does not actually come from Luther, but instead from a character named Fritz in a nineteenth-century historical novel by Elizabeth Rundle Charles.[47]

Most important for our purposes is the strange mural that appears on the wall next to the printing press. Entitled "Scripture Questioned: The Church Compromises God's Word," the mural consists of a timeline of the past five centuries of Western history, a timeline that is shaped into an arc that gradually and then rapidly descends. Underneath the descending timeline is the heading "Human Reason" and a list that includes Rene Descartes (who "revolutionized philosophy, making man the measure of all things, not the Bible"), Charles Lyell, Charles Darwin, and others. Much more prominent and clearly more important (given the mural's title) is what is found above the timeline. At the very top of this arc stands Martin Luther and the statement "Scripture Alone." Curiously, or not so curiously, the timeline begins to descend literally right under Luther's feet; [48] as the arc moves downward, from "Scripture Alone" to "Scripture Questioned," one finds Francis Bacon, who "argued that the Bible is not needed to understand the world," as well as Galileo (more on him below). As the descending timeline makes the transition from "Scripture Questioned" to "Scripture Abandoned" one finds nineteenth-century evangelicals Thomas Chalmers and Hugh Miller who developed, respectively, the gap theory (there were millions of years between the first and second days of creation) and the day-age theory (each Genesis day of creation represented an immense period of time).[49] Then it is on to the full-fledged abandonment of scripture in the late nineteenth and early twentieth centuries. Directly below the heading "Scripture Abandoned," theologically literate visitors will perhaps be surprised to discover conservative Presbyterian theologians Charles Hodge and B. B. Warfield. While Hodge and Warfield were perhaps the most sophisticated and influential proponents of biblical inerrancy, here they are portrayed as moving history on an accelerating downward trajectory for "argu[ing] that 'day' can mean millions of years in Genesis 1" (Hodge) and for "accept[ing] the possibility that God directed the evolution of life (theistic evolution)" (Warfield).

And then, further down the descending timeline, one finds the *Scofield Reference Bible* (1909), here because it "popularized long periods of time between Genesis 1:1 and 1:2 (gap theory)." Above this caption is what appears to be a photograph from the *Reference Bible*, with Genesis 1:1–2 and Scofield's notes, with red lines under the egregiously offending passages from Scofield's commentary. Notwithstanding the efforts at creating the appearance of authenticity, this is *not* an accurate representation of the *Scofield Reference Bible*. All of Scofield's biblical cross-references and explanatory headings have been removed. More than this, Scofield's first note on verse 1 has been removed, Scofield's second note on verse 1 (the first note here) has been drastically shortened, and Scofield's note on verse 2 has been removed (biblical references and all) and replaced with the explanatory heading from the original text, "Earth made waste and empty by judgment."

Perhaps the kindest explanation of these editorial alterations is that the curators are seeking to simplify the message for their visitors. Whatever the motivation, the removal of Scofield's biblical cross-references certainly hides the fact that those arguing for a "gap theory" of creation had a biblical argument to make. But however one understands the faux photograph, this mural signals a remarkable and dramatic fall from grace for C. I. Scofield. His *Scofield Reference Bible* reigned supreme among conservative Protestants throughout the twentieth century. Fundamentalists and evangelicals understood Scofield's notes on the Bible to be a clear exposition of "the plain sense of the text." In Scofield's Bible we had the True Interpretation. And now, at the Creation Museum, the *Scofield Reference Bible* has been explicitly and thoroughly discredited, a primary example of "Scripture Abandoned" in contemporary life, and—as is clear in the next two rooms of the museum, Graffiti Alley and Culture in Crisis—a primary contributor to the demise of both church and civilization. In short, C. I. Scofield has been—along with the architects of biblical inerrancy, Charles Hodge and B. B. Warfield— thrown under the young Earth creationist bus as part of a thorough-going purge of all conservative Protestants whose interpretation of the Bible "accommodates" the geologic ages.

Just below the Scofield Bible on the mural is the Scopes trial, described as having "caused international mockery of the Bible's history, partly because of the church's day-age view of creation." Even in the context of this peculiar mural, this is a very peculiar statement. Presumably, this reference is to the "fundamentalist church of old-earth creationists." Is it really true that the "international mockery" would have been less if Bryan had made the case that the Bible teaches that the universe was created in six twenty-four-hour days (as opposed to asserting that the Genesis "day" did not refer to "twenty-four-hour days")? Why the overpowering sense of failure, given that most fundamentalists in 1925 proclaimed the Scopes trial to be a victory, a sentiment made manifest by the burst of antievolutionist agitation over the next three years?[50]

As Susan Harding makes clear in her fascinating study, *The Book of Jerry Falwell: Fundamentalist Language and Politics*, this odd caption at the Creation Museum is very much in keeping with how many fundamentalists, including Falwell, came to understand the Scopes trial. For both fundamentalists and their secular enemies the trial's climactic moment came when Bryan agreed to take the witness stand. Darrow was determined to make Bryan's ideas look ridiculous, spending a fair amount of time quizzing Bryan about how Jonah lived inside a whale for three days, how Joshua made the sun stand still, and where Adam and Eve's son Cain got his wife. More than this, Darrow was determined to reveal that Bryan did not live up to his own claim that he simply took the Bible literally. Darrow made this point most clearly when he pressed Bryan on the day-age theory of creation:

Q – Mr. Bryan, could you tell me how old the earth is?

A – No, sir, I couldn't.

Q – Could you come anywhere near it?

A – I wouldn't attempt to.

. . . .

Q – Would you say that the earth was only 4,000[51] years old?

A – Oh no; I think it is much older than that.

Q – How much?

A – I couldn't say.

. . . .

Q – Do you think the earth was made in six days?

A – Not six days of twenty-four hours.

. . . .

Q – You think those were not literal days?

A – I do not think they were twenty-four-hour days.

. . . .

Q – You do not think that?

A – No. But I think it would be just as easy for the kind of God we believe in to make the earth in six days as in six years or in 6,000,000 years or in 600,000,000 years. I do not think it important whether we believe one or the other.

Q – Do you think those were literal days?

A – My impression is they were periods, but I would not attempt to argue as [*sic*] against anybody who wanted to believe in literal days.[52]

Not only were Bryan's day-age views of creation very much in the contemporary fundamentalist mainstream, they were understood by fundamentalists to be in keeping with biblical literalism. But as Harding points out, "the problem for Bible believers was that Darrow's interrogation converged with a rhetorical ploy that Bible-believing preachers used to police each other, a ploy which awarded the higher ground to a preacher who could establish his biblical interpretation as 'more literal' than another's." By following "the rules of fundamentalist rhetorical combat," Darrow was able to "impugn Bryan's status as a Bible believer." More than this, the fact that the verbal duel "conformed to their own rules" made it very difficult for fundamentalists, especially after the antievolution fervor died out in the late 1920s, to challenge the notion that the Scopes trial was a defeat.[53]

This is precisely how the story is portrayed at the Creation Museum: William Jennings Bryan lost at the Scopes trial because his biblical interpretation was not sufficiently literal. But thirty-six years after Scopes came *The Genesis Flood* to the rescue. As Harding observes, "Whitcomb and Morris's insistence on young-earth and strict, day-day positions enabled them to reclaim the authority of scriptural literalism with regard to the Genesis account, authority which had been lost in the Scopes trial finale when Darrow decried the inconsistency of Bryan's day-age position."[54] This is the central and pervasive argument of the Creation Museum: by understanding Genesis 1–11 as a literally accurate historical narrative, by eliminating the accommodations to modern geology and biology, the True Interpretation of God's Word, that is, the plain sense of scripture, is now locked in place, and the Bible is restored to its rightful place of authority.

Despite the Creation Museum's strenuous efforts, however, history does not stop. *Sola scriptura* continues to be a dangerous idea. The Bible's meaning remains unfixed. Readers continue to read and interpret. Protestantism still lacks a magisterium to determine The True Interpretation. Young earth creationism is not the final word.

And if/when the young Earth creationist understanding of the Bible is dethroned in conservative Protestant circles it is quite likely—given the rules of fundamentalist discourse—that it will be for the same reason that the *Scofield Reference Bible* was dethroned: it is deemed to be insufficiently "literal." Surprisingly enough, one clue as to how this might happen can be found on the Creation Museum's "Scripture Questioned" mural, near the top of the timeline, just below Martin Luther. There we find Galileo, who is there because he "argued that you can interpret the Bible based on science." It would be more accurate to say Galileo claimed that we can "interpret" nature based on reason, even if this is at odds with a surface reading of the biblical text. As Galileo argued in his 1615 letter to the grand duchess Christina of Tuscany:

> The holy Bible and the phenomena of nature proceed alike from the divine Word . . . It is necessary for the Bible, in order to be accommodated to the understanding of every man, to speak many things which appear to differ from the absolute truth so far as the bare meaning of the

words is concerned. But Nature, on the other hand, is inexorable and immutable . . . For that reason it appears that nothing physical which sense-experience sets before our eyes, or which necessary demonstrations prove to us, ought to be called in question (much less condemned) upon the testimony of biblical passages which may have some different meaning beneath their words . . . I do not feel obliged to believe that that same God who has endowed us with senses, reason, and intellect has intended [us] to forgo their use.[55]

In this letter, Galileo was responding to attacks he was receiving from conservative clerics for telescopic evidence he had gathered that supported Copernicus's argument that the Earth was in motion and the Sun is stationary, attacks that were, as Galileo noted, based on the "many places in the Bible [where] one may read that the sun moves and the earth stands still." A few months after writing the grand duchess, in February 1616, Galileo was warned by the Church that he was not to believe or argue that the Earth moved, demands to which he agreed. But in 1632, he published a dialogue in which three characters "engaged in a critical discussion of the cosmological, astronomical, physical, and philosophical aspects of Copernicanism," in which it was clear that "the arguments favoring the earth's motion were stronger than those favoring the geostatic view." The result was that in April 1633 Galileo was ordered to Rome for trial; in June the Inquisition pronounced its verdict against him, declaring him to be "vehemently suspected of heresy [for] having held and believed a doctrine which is false and contrary to the divine and Holy Scripture: that the sun is the center of the world and does not move from east to west, and the earth moves and is not the center of the world." Given the opportunity to retract his beliefs, Galileo did so, and thus the sentence ended up being house arrest for the rest of his life.[56]

Galileo was condemned for holding to ideas that were "false and contrary to the divine and Holy Scripture." AiG emphasizes obeying God's inerrant Word over human reason, the Creation Museum indicts Galileo for having allowed "science to interpret the Bible," and AiG commentators blast Galileo for making popular in Western civilization the notion "that the Bible might use figurative language and so biblical interpretation should 'accommodate' new scientific discoveries." Martin Luther himself attacked Copernicus and firmly held to a geocentric view of the universe.[57] Given all of this, and given the willingness of young Earth creationists to hold to the notion of a six-thousand-year-old universe in the face of an overwhelming scientific consensus to the contrary, one could easily assume that the Creation Museum would also proclaim and defend the idea that the sun revolves around a stationary Earth.

But such an assumption would be wrong. While there seems to be no reference to an earth-centered universe in the museum, on the AiG website one can find various critiques of geocentricism, including a negative review of the 2014 film *The Principle,* which was produced by Robert Sungenis, a Catholic apologist whose website is aptly titled galileowaswrong.blogspot.com.[58] Also found on the AiG website is a chapter from a 2008 book by AiG stalwarts Tim Chaffey and Jason Lisle entitled *Old Earth Creationism on Trial: The Verdict Is In.* Responding to old Earth creationists who have argued that the biblical arguments for young Earth creationism are akin to the biblical arguments for geocentricism, Chaffey and Lisle say that this is just another example of the poor reasoning employed by old Earth creationists. According to the authors, while "many in the medieval Church held to a geocentric view," this was because they had accommodated the biblical text to "scientific opinions of the time," which included the notion of an Earth-centered universe. The simple fact is that "the Bible does not teach geocentricism." Many of the passages used to argue otherwise are "poetic verses" that are clearly best understood figuratively.

The more "historical" passages that suggest the Bible teaches a geocentric universe are no more problematic for Chaffey and Lisle. Take, for example, Joshua 10:12–13 (NRSV):

> On the day when the LORD gave the Amorites over to the Israelites, Joshua spoke to the LORD; and he said in the sight of Israel, "Sun, stand still at Gibeon, and Moon, in the valley of Aijalon." And the sun stood still, and the moon stopped, until the nation took vengeance on their enemies. Is this not written in the Book of Jashar? The sun stopped in mid-heaven, and did not hurry to set for about a whole day.

Chaffey and Lisle dismiss out-of-hand a "hyper-literal reading" of these verses. Instead, they assert that "it is quite obvious that Joshua was simply using observational language," similar to "trained meteorologists today when they speak of the sun 'rising' and 'setting.' Surely, no one would accuse these men of believing that the earth is stationary and the sun revolves around it!"

In short, the authors of *Old Earth Creationism* assert that while "the Bible does teach a young earth," it "does not teach geocentricism."[59] While Chaffey and Lisle are primarily concerned with critiquing old Earth creationists, Danny Faulkner's review of Gerardus Bouw's *Geocentricity*[60]—originally published in 2001 in *TJ* (now the *Journal of Creation*) and now posted on the AiG website—is an attack on those young Earth creationists who hold that we live in an Earth-centered universe, an attack animated by Faulkner's sense that "their presence among recent creationists produces an easy object of ridicule by our critics." As Faulkner notes, Bouw (professor emeritus at Baldwin-Wallace College, director of the Association for Biblical Astronomy, editor of *Biblical Astronomer*) is perhaps "the chief champion of geocentricity."[61]

Thus Bouw's response to Faulkner, "Geocentricity: A Fable for Educated Man?," is quite telling. Bouw begins by scoring Faulkner and his fellow "non-geocentric creationists" for ignoring "their more radical brethren in Christ, the geocentrists." After spending a good deal of time aggressively defending "geocentric science" in response to Faulkner's attacks, Bouw turns to the biblical arguments. He critiques Faulkner for suggesting that the Bible's use of "sunset" and "sunrise" is not to be understood literally and attacks Faulkner's general approach to the Bible's poetic passages: "He feels that when God inspires poetry, that he is less bound [to] write the truth, leastwise not absolute truth." Then Bouw turns to the Bible's "strongest geocentric passages," passages ignored by Faulkner because he had "no way to refute their geocentric impact." Most prominent here is the aforementioned Joshua 10:13. Bouw points out that "God could have said 'And the earth stopped turning so that the sun appeared to stand still,' but he didn't." If it is not true that God made the sun stop in its revolution around the Earth, if it is true that in this instance God told a lie, then how can it be said that God is "the God of Truth and the Spirit of Truth?"

Bouw concludes his discussion of the Bible and geocentricity with the following shot aimed directly at young Earth creationists:

> So, if Genesis 1:1, "In the beginning God created the heaven and the earth" is a clear statement that God created, then Ecclesiastes 1:5, "The sun also ariseth, and the sun goeth down, and hasteth to his place where he arose," is just as clear a statement of geocentricity. And with that, we come to the *real* issue: Is the Scripture to be the final authority on all matters on which it touches, or are scholars, to be the ultimate authority? The central

issue is not the motion of the earth, nor is it the creation of the earth. The issue is final authority, is it to be the words of God, or the words of men.

This language should seem very familiar. All of this—the emphasis on the Bible's authority and God's truthfulness, the perspicuity of God's Word, the choice between following God or bowing to contemporary scholarship—is precisely the argument made by young Earth creationists against Christians who are old Earth creationists and Christians who are theistic evolutionists. As is always the case in fundamentalist discourse, the geocentrists' trump card is that they are the ones who are truly hewing to a literal reading of the Bible. Bouw makes this very clear at the end of his article, in terms strikingly akin to how young Earth creationists understand what happened at the Scopes trial:

> Evolutionists, atheists, and agnostics in the know can easily shame creationists on the issue of geocentricity by simply pointing out the hypocrisy of their insistence that the days of Genesis 1 are literal while the rising and the setting of the sun is not . . . We conclude that the creationist's desire to reject it can only be for the sole purpose of appearing intellectual and acceptable to the world.[62]

So it is that some day Gerardus Bouw may very well (à la George Mc-Cready Price before him) get his own Morris and Whitcomb to write the *Joshua Cosmology* in order to promote the ideas of geocentricity within the conservative Protestant mainstream. And if the *Joshua Cosmology*, like the *Genesis Flood* before it, spreads like wildfire through fundamentalism and evangelicalism, building on the 18 percent of Americans who already believe that the sun revolves around the Earth,[63] there could be an Answers in Joshua (AiJ) organization, with its own Geocentricity Museum replete with planetarium, dioramas, and placards with biblical texts. In this museum there could very well be a Scripture Questioned and Abandoned mural, with essentially the same cast of religious and intellectual ne'er-do-wells—Galileo, Descartes, Darwin, Lyell, Hodge, Scofield, and others—who have brought the Church and Western civilization to the depths of cultural depravity. But on this mural one could also find, beneath the Scopes trial on the timeline, Henry Morris and John Whitcomb and *The Genesis Flood*. Even further below, near the very bottom of the descending arc, there might very well be Ken Ham, described in an accompanying note as having "compromised God's Word by accommodating to the scientific opinions of the day."

4

POLITICS

A t the beginning of the Bible Walkthrough Experience, just after
the Dig Site with its display of an evolutionist and a creationist
uncovering dinosaur fossils, Creation Museum visitors enter
the Starting Points room. They are immediately confronted with a question stenciled upon a wall: "Same Facts, but Different Views . . . Why?"
The answer is found on a placard just around the corner. Located under
the heading "Different Views because of Different Starting Points," this
placard is divided in two by a thick black line. On the left side is a pile
of books, including Francis Bacon's *The New Organon*, Stephen Hawking's *A Brief History of Time*, Pierre-Simon Laplace's *Celestial Mechanics*, Charles Lyell's *Principles of Geology*, James Watson's *DNA*, and—at
the bottom—Charles Darwin's *On the Origin of Species*. On the right
side is a single scroll. Above the books is "The philosopher Rene Descartes said, 'I think, therefore I am'"; above the scroll is "God said, 'I AM
THAT I AM.'" Straddling the two sections of the placard is an explanatory note, in which visitors read that "broadly speaking, 'human reason'
refers to 'autonomous reasoning'—the idea that the human mind can
determine truth independently from God's revealed truth, the Bible."
While "reasoning is God's gift to humankind," God "has instructed us
to use the Bible as our ultimate starting point (Proverbs 1:7)[1] and also
to reject speculations that contradict God's knowledge (2 Corinthians
10:5)." The final three sentences drive the argument home:

> Philosophies and world religions that use human guesses rather than
> God's Word as a starting point are prone to misinterpret the facts around
> them because their starting point is arbitrary. Every person must make

a choice. Individuals must choose God's Word as the starting point for all their reasoning, or start with their own arbitrary philosophy as the starting point for evaluating everything around them, including how they view the Bible.

God's Word or human reason: there are only two possible starting points. Start with the inerrant young Earth creationist Bible, and you will have a true understanding of the origins of the world and of humans, a well-ordered life in line with God's law and eternal salvation. Start with human reason—in keeping with standard fundamentalist discourse, much more time is spent on this side of the equation—and you will have a compromised Word of God with "millions of years" and evolution, a thoroughly corrupted church, the collapse of Western civilization, and eternal damnation.

At the heart of the Creation Museum is a radical binary in which the visitor is confronted with two sets of tightly linked terms that are unequivocally opposed to each other, that is, Bible-young Earth-Eden-truth-heaven versus human reason-evolution and old Earth-sin-corruption-hell. Borrowing from social theorists Ernesto Laclau and Chantal Mouffe, these two sets of linked terms can be understood as rhetorical "chains of equivalence," a discourse of antagonism in which identifying with one set of linked terms is necessarily to reject the opposing set of terms. Intensifying the antagonism even more is that it is the very Word of God—the inerrant young Earth Word of God—that serves as the anchor (what Laclau and Mouffe call the "nodal point") of the museum's chain of equivalence.[2]

The binary is cosmic. The stakes could not be higher. The result—at the museum, on Ken Ham's blog, in *Answers* magazine, and on the AiG website—is a highly politicized and totalizing rhetoric. It presumes to speak on any and all topics of the day: the status of the United States as a Christian nation, gay marriage, the role of women, racism, climate change, public education, and alien life on Mars. Moreover, in keeping with the cosmic binary, there is no space for dissent, not even from fellow Christians. This is a culture war, yes, but it is a culture war with eternal implications. All dissenters are viewed as standing with the opposing "chain of equivalence," or, in Laclau and Mouffe's lexicon, "beyond the social." They are the opponents of Truth. They are the Enemy.

Public Schools as Sites of Evolutionary Brainwashing

It is illuminating to apply to the Creation Museum its own oft-repeated dictum that "everything depends on one's starting point," given that the museum's own starting point initiates visitors into the proper cultural and political mindset for experiencing all that is to follow. After purchasing admission tickets, and after having the opportunity to have one's picture taken with a digital *Tyrannosaurus rex*, visitors are encouraged to get in line for a twenty-three-minute film entitled *Men in White*. In keeping with the museum itself, the film has high production values and features a number of actors with Hollywood credits.[3] It plays off the film *Men in Black*, but instead of two big-gun-toting human agents (one black, one white) charged with the task of saving humanity by tracking down a globe-threatening extraterrestrial, the museum's film features two young, attractive, wing-bearing white guys, Gabe and Mike (the angels Gabriel and Michael), in white overalls and white long-sleeved shirts. They have come to Earth to save Wendy, a young woman who wants to believe there is a God who created the universe, who wants to believe that her life has meaning, but who does not want to look stupid for rejecting the evolutionary science that teaches her that "life is probably just a big accident, a predictable result of an infinite number of matter/anti-matter asymmetric collisions." Gabe and Mike try to reassure Wendy by telling her that there is a God who "invented . . . everything, the whole enchilada," and that this "God loves science" and "wants us to know how He built everything, everything from neutrinos to eagles to the Milky Way." They even provide her with an animated "film within a film" that previews what visitors will encounter once they leave the theater and begin to tour the museum, including the six days of creation, Adam and Eve's "rebellion," which resulted in a world "full of corruption, violence, and pride," and the Flood (during which time the theater seats shake from the "thunder" and visitors are spritzed with "rain").

But despite all this evidence of a God who created the universe, Wendy can't, in Gabe's words, "stop listening to the voices of her culture." The cultural voices that have put Wendy in such despair—the alien creatures, as it were, who must be combated—are located in the educational establishment. In *Men in White*, higher education is represented by biology professor J. Plumsure and Dr. Ed U. Kaded (a scientist whose disciplinary specialty is undefined). Wearing a turtleneck sweater and sportcoat and smoking a pipe, Plumsure sneers that "natural processes were simply in place," and "there was no God who had anything to do with anything." Dr. Ed U. Kaded has a more prominent role in the film. Interviewed by the blond, breathless, and brainless reporter "Suzie Teevee," who mindlessly agrees with everything the scientist tells her, Dr. Ed U. Kaded reports that "the age of zircon grains found in Australian sandstone [is] somewhere between 4.1 and 4.2 billion years old" (to which Gabe confidently responds that "new studies are suggesting radically younger ages for those rocks") and that "Dean Kenyon . . . has proven how under just the right circumstances the primordial soup turns to life" (to which Gabe responds by laughing and saying that later in life Kenyon "completely changed his mind, said it could have never happened that way.") Dr. Ed U. Kaded goes on to claim that "virtually every thinking person on the planet agrees wholeheartedly with virtually everything [Charles Darwin] ever said" and concludes with the pronouncement that "knowledge itself is attacked when those creationists try to get into our schools and question Darwin's conclusions."

The film then returns to Wendy, who sadly muses that "if you question evolution then you must be an idiot." But Gabe snorts that "what's idiotic is not to question." He then rehashes the "clues that don't support" the accuracy of radio isotope dating, which leads Mike (who at this point is playing with a yo-yo, apparently because the issue is so easy to comprehend it does not require his full attention) to ask, "Could it possibly be that, due to the evolutionists' assumptions about the past, that all those dating methods supporting the millions and billions of years might be, shall we say, wrong?" Gabe agrees and goes on to assert that "there is one place where you must never, ever bring up any of these observations, any of this science, any of these questions." Mike says, "And where's that, Gabe?" Gabe answers, dramatically, "School."

Standing in for American public education is "Enlightenment High School." Two Enlightenment High science teachers are featured, including Miss E. Certainty, a large, loud woman who excitedly (if illogically) shrieks that if you don't believe in evolution "you are in violation of the Constitution of the United States and the separation of Church and State!" But the primary villain is science teacher P. Snodgrass, a slight man who wears a white lab coat, speaks in a nasal whine, and bores his students by droning on and on about the age of the Earth while showing slides of Yellowstone. The centerpiece of *Men in White* is a scene involving the two hip Men in White—and there is no doubt that, in contrast

with the emasculated teacher, Gabe and Mike are the real men here—
sitting in the back of Snodgrass's classroom, in front of a blackboard
that contains the words "EVOLUTION" and "natural selection," as well as
a picture of Charles Darwin. Naughty and defiant, they replace Snodgrass's slide carousel with one of their own, the first slide entitled "Gabe
& Mike Productions Present: An Argument for Creationism." As they
flip through their slides, they pose a series of questions to Snodgrass designed to expose as fraudulent the notion of an ancient Earth: How do
you explain that "if the Earth were as old as you say we could practically
walk across the Atlantic because after billions of years of adding salt
it would by now be solid salt?"; "Did you know that a number of PhD
geologists are reconsidering how old the earth might be [as] there's a
lot of observational evidence out there that has convinced them that
the earth might in fact be quite young?"; "Did you know that the earth's
magnetic field is decaying so rapidly that life would have been impossible just a million years ago?"; "Did you know that there are no widely
expanded supernova in the galaxy, [which is] exactly what we would expect to find if the universe had only existed, say, a few thousand years?"
Pathetically incapable of answering the bad boys' questions, Snodgrass
repeatedly attempts to reclaim his authority and his manhood, asserting that "Charles Darwin was a great man" and that "we all know how
we evolved from those ape-like creatures—it's a very famous picture!"
Unable to respond to Gabe's objection that the picture is just "a drawing," Snodgrass is finally reduced to wailing that "there is no God who
intervenes in the world!"

It is an understatement to say that *Men in White* does not live up to
its self-description (included on the DVD case) as "an excellent summary of the major supposed 'evidences' which are used by secularists
to indoctrinate students in evolution and millions of years." Nor does it
succeed in providing "Christian/creationist answers that clearly show
the Bible's account of history in Genesis is confirmed by science." At its
most basic level, *Men in White* serves as a quintessentially fundamentalist fantasy of plain folks who—armed with only common sense and
God's Word—unmask pretentious and godless academics and educa-
tors. Every time we saw the film at the museum, it was clear that much
of the audience enjoyed the "unmasking," as many folks laughed loudly
when the bad boys exposed the inept Snodgrass as a fraud, and many
applauded at the film's conclusion.

It seems likely that teachers would fail to be placated by the explanation (as given when the film is introduced at the museum and on the DVD cover) that *Men in White* should be understood as satire. But while the film is relentless in its ridicule of public schools, its tone is moderate when compared to the "nonsatirical" commentary on public schools on the AiG website, in *Answers* magazine, and on Ken Ham's blog. Ham and other AiG commentators repeatedly assert that public schools are sites of evolutionary brainwashing, with textbooks "geared to teach evolution/millions of years as fact . . . in the government-run, education systems," and with secularists determined that evolution "be drilled into the minds of students as young as five." More than this, according to AiG, secularists are determined to keep students from knowing the problems with evolution, thus suggesting to AiG contributor Terry Mortenson that they have something to hide: "The fact that the evolutionists don't welcome the challenges but do everything they can to avoid or prevent them is strong evidence that the Neo-Darwinian 'microbe-to-microbiologist' evolution is a massive deception masquerading as proven scientific fact." Realizing that "molecules-to-man evolution is so weak in scientific terms that the only way to defend it is by suppressing any opposition," evolutionists frantically oppose "academic freedom" laws that would allow teaching the alleged scientific controversy regarding evolution: "The NCSE [National Center for Science Education] wishes to muzzle budding scientists before they are out of the proverbial cradle" as they are "afraid to let students learn how to discern" scientific truth on their own.[4]

More than this, and as exemplified by the despairing Wendy in *Men in White*, to brainwash children in evolution is to brainwash them into believing there is no God. Ken Ham argues this is precisely the point: "Secular evolutionists/atheists know that if you can indoctrinate at a young age, it paves the way to brainwashing children with a naturalistic (atheistic) worldview." In the 1960s the American Civil Liberties Union and other secularists followed this script, removing "the Bible, God, creation, [and] prayer . . . from the public schools" and replacing Christianity with "the religion of atheism" which "explain[s] the universe and life without God." As Ham dramatically blogged in January 2012, "Christians need to wake up to the fact that, by and large, public schools are churches of secular humanism and that sadly most of the teachers (even though there are some Christian missionaries in the system) are the high priests of this religion imposing an anti-God worldview on generations of students."[5]

It may be difficult to imagine actual public schoolteachers as secular humanist "high priests." But the folks at AiG go further, repeatedly stressing the point that the teachers who are indoctrinating students in atheistic evolution are, in effect, indoctrinating them into a thoroughly amoral worldview. Responding to a California law outlawing discrimination in schools on the basis of gender and sexual orientation, AiG's Georgia Purdom angrily declared that "if the public school system teaches our children that they are nothing more than highly evolved animals, then no wonder children think they can decide their gender, practice homosexuality, and be promiscuous." Ken Ham was no less emphatic on his blog: "Generations of kids (including 90% of kids from church homes) are being taught atheistic evolution and millions of years as fact," and yet "we wonder why we see increasing hopelessness, purposelessness, violence, sexual perversion, drug use, abortion, [and] homosexual behavior."[6]

For Ken Ham and his colleagues, the most dramatic evidence of the evils wrought by indoctrinating children in atheistic evolution is school shootings. They give particular attention to the April 1999 Columbine High School (CO) massacre, in which Eric Harris and Dylan Klebold killed fourteen students (including themselves) and one teacher, as well as the 2007 Jokela High School (Finland) shootings, in which Pekka-Eric Auvinen murdered the school principal, school nurse, and seven students (including himself). It is not hard to understand the focus on Columbine and Jokela. Regarding the latter, a number of AiG contributors made reference to the fact that two weeks before the killings Auvinen posted on YouTube a statement in which he described himself as a "social Darwinist" who was determined "to put NATURAL SELECTION & SURVIVAL OF THE FITTEST back on tracks!" Regarding the Columbine killings (which Auvinen apparently wanted to replicate), Ham and others have made much of the fact that on his rampage Eric Harris was wearing a T-shirt with "NATURAL SELECTION" on it. More than this, AiG contributors plug into the iconic status of Columbine for American fundamentalists and evangelicals. As Ralph Larkin points out in *Comprehending Columbine*, "from the outset, the evangelical community scripted the shootings as a struggle between good and evil, with the evangelical community assuming the role of goodness and Klebold and Harris, along with secularism and political liberalism, in the role of evil." Despite a great deal of evidence to the contrary, many conservative Prot-

estants remain convinced that Harris and Klebold targeted Christians
in their shootings and that Columbine student Cassie Bernall was killed
because she told the killers she believed in God.[7]

AiG contributors sometimes include the qualifier that it is human
sinfulness and not the belief in evolution that leads young murderers
to pull the trigger. For example, in the 2008 *Answers* article about the
Jokela shootings, "Unnaturally Selected," the unnamed author opined:
"Evolution did not make the teen turn mass murderer. He gave into his
sin nature." But such statements have a wink-and-a-nod feel to them,
given that they are inevitably and immediately followed by a clear statement of the connection between teaching evolution and school shootings: "But when society teaches that young people are just animals in a
struggle for survival and that life came about through natural, violent
processes, should we be surprised at such massacres?" AiG speaker and
author Bodie Hodge was particularly forthright in his November 2007
article, "Finland School Shootings: The Sad Evolution Connection."
After establishing that Adolf Hitler, "Karl Marx, Pol Pot, Leon Trotsky,
and Joseph Stalin . . . avidly held to Darwinian evolution to justify their
actions," he moved on to school "murderers" in Finland and the United
States:

> [They] were living a life consistent with the real teachings of atheistic
> evolutionism. Children catch on quickly and eventually put two and two
> together. When they are taught that there is no God, no right and wrong,
> people are animals, etc., then they may reason: "Why not kill or steal?"
> and so on. It is a logical connection. So long as evolutionism is forced
> onto children (no God, people are animals, no right and wrong, etc.) and
> so long as they believe it and reject accountability to their Creator, then
> we can expect more of these types of gross and inappropriate actions.[8]

Of course, not all school shootings conform to the Columbine-Jokela
narrative. For example, there was the October 2006 massacre near
Nickel Mines, Pennsylvania, in which an armed Charles Roberts entered
an Amish schoolhouse and shot ten girls (five of whom died) before he
turned the gun on himself. That an Amish school would be the scene
of carnage was stunning enough. But over the next few days the story
became even more startling as the Amish community responded to the
massacre by forgiving the killer—over half of the mourners at Roberts's
funeral were Amish—and by comforting his widow and family with vis-

its and gifts. As noted by the authors of *Amish Grace: How Forgiveness Transcended Tragedy*: "The grace extended by the Amish surprised the world almost as much as the killing itself. Indeed, in many respects, the story of Amish forgiveness became *the* story—the story that trumped the narrative of senseless death."[9]

On October 7, 2006, AiG provided its own take on the Nickel Mines massacre. The piece begins with a brief paragraph of facts on the killings, followed by a longer paragraph in which the anonymous author points out that "the root of senseless violence and, ultimately, death and suffering was the Fall of our ancestor Adam." As a result, even "the Amish, who shun modern technology, live simply and are generally very religious, cannot extricate themselves from this fallen world." The article then goes on to make brief reference to Amish grace, noting that the Amish—"who emphasize forgiveness for the killer"—have "requested that a fund be set up for the killer's widow and children." But then, in the penultimate paragraph, the article suddenly and jarringly shifts into (for AiG readers) very familiar language: while "murder can be traced back to shortly after the Garden of Eden," in the modern world "the erosion of faith in the Word of God—partially due to acceptance of Darwinism—has contributed to a growing sense of purposelessness in individuals and weakens morals to be merely subjective opinions." It is thus "no surprise to learn that the murderer of these Amish girls was reportedly mad at God." Then comes the conclusion, in which the Nickel Mines massacre completely disappears from view. Instead, the author quotes from a "father of a Colorado student murdered in the Columbine massacre" who "apparently shares our [AiG's] perspective":

> This country is in a moral free-fall. For over two generations, the public school system has taught in a moral vacuum, expelling God from the school and from the government, replacing him with evolution, where the strong kill the weak, without moral consequences and life has no inherent value.[10]

In the wake of the horrific killings at Nickel Mines, the acts of grace by the Amish prompted a remarkable national conversation on what it means, to quote the authors of *Amish Grace*, "to live a truly Christian life."[11] But the folks at AiG had no interest in participating in this conversation about what it means to "live Christian," what it means to forgive even those who commit acts of violence. Instead, for Ken Ham and his colleagues the killing of the Amish schoolgirls simply provided yet another occasion to underscore the tight connection between the teaching of evolution and school shootings.

The Decline and Fall of Christian America

At the Creation Museum the godless amorality promoted by the public schools is but one indicator of the cultural stench that is the contemporary West. A more expansive treatment of cultural corruption is to be found in Graffiti Alley,[12] which visitors enter just after having gone through the Biblical Relevance room, with its "Scripture Questioned" mural that portrays the ever-accelerating decline of church and civilization. As described earlier, Graffiti Alley consists of a narrow, dark hallway lined with what appear to be three exterior brick walls. In this dimly-lit space, visitors hear a variety of urban sounds and see a flashing neon "XXX." Plastered onto the graffiti-sprayed brick walls are fragments of articles and photos torn from newspapers and newsmagazines, most notably covers from *Time* and *Newsweek*. These fragments have been pasted onto the walls in apparently random fashion and on top of each other, with some scraps appearing multiple times. Above the walls is a placard: "Scripture Abandoned in the Culture Leads to . . . Relative Morality, Hopelessness, and Meaninglessness" (ellipses in original). Under the placard, in graffiti-style script, is the phrase, "Modern World abandons the Bible."

Graffiti Alley is profoundly and intentionally disorienting. The dim lighting, graffiti, flashing neon sign, and loud noises all underscore the museum's claim that a culture that has forsaken the Bible ends up in hopeless confusion and despair. The magazine and newspaper fragments provide the substantive evidence that the culture is awash in "relative morality, hopelessness, and meaninglessness." As these scraps of texts and photos are arrayed on the walls of Graffiti Alley in seemingly willy-nilly fashion, visitors are encouraged not to examine the textual fragments too closely. Rather, these snippets simply serve as markers for and reminders of the apparently countless instances of violence and cultural decay occurring in our chaotic culture of despair. These snippets include numerous photos of violence and tragedy, as well as large-font magazine headlines suggesting global cataclysm, including "Is Any Place Safe?" "Be Worried. Be Very Worried,"[13] and "Apocalypse Now. Tsunamis. Earthquakes. Nuclear Meltdowns. Revolutions. Economies on the Brink. What the #@%! Is Next?"

But a close examination reveals that Graffiti Alley is anything but chaotic. The Creation Museum has a very particular understanding of what constitutes evidence that American culture has abandoned the Bible, and what does not constitute evidence. There are, for example, no magazine covers or newspaper articles in Graffiti Alley pertaining to corporate malfeasance, the machinations of the financial sector, income inequality, or the persistence of widespread poverty. Such omissions are particularly striking given that, as Richard Hughes argues in *Christian America and the Kingdom of God*, care for the poor was central to "the vision of the kingdom of God" in both the Hebrew Bible and the New Testament. The Hebrew prophets proclaimed that "a kingdom ruled by God" required that Israel follow "the path of economic justice, especially for the poor who so often had been objects of abuse and exploitation." Similarly, Hughes argues, "in almost every instance where the phrase 'kingdom of God' appears in the New Testament, it is closely linked to concern for the poor, the dispossessed, those in prison, the maimed, the lame, the blind, and all those who suffer at the hands of the world's elites."[14]

But in Graffiti Alley, privileging the wealthy and neglecting or mistreating the impoverished does not constitute evidence that America has forsaken scripture. So, what does? Once again, the godless public schools. Posted on the walls of Graffiti Alley are fragments from 1960s newspaper stories reporting that the Supreme Court had removed mandatory Bible readings and the Lord's Prayer from the schools, as well as more recent articles and magazine covers on efforts to end prayer before high school football games and to remove "under God" from the Pledge of Allegiance. Much more noticeable—and in keeping with the aforementioned articles produced by AiG contributors—are headlines and photos from school shootings. As always, the Columbine massacre is very much present, including the following: fragments of front page stories from the *New York Times* and from an unidentified newspaper, the latter bearing the headline "Rampage's Toll: 15 Dead, Innocence Lost"; a photograph of a Columbine student survivor grasping her head and crying to the heavens; and fragments of a *Time* article on "The Columbine Effect" and a *Time* cover, "The Columbine Tapes." Also here is the headline from a *Time* article, "The Devil in Red Lake," on Jeffrey Weise's shooting spree at a Ojibwe reservation school in Minnesota, which left ten dead, including Weise. Graffiti Alley makes clear that shooting rampages afflict public universities as well. There is much here on the April

2005 massacre at Virginia Tech, in which Seung-Hui Cho killed thirty-
two people before committing suicide: there are numerous photos, including two photos of Cho pointing his gun at the viewer; there is a
Newsweek cover, "The Mind of a Killer"; there is also a headline from a
Time article, "Darkness Falls. One Troubled Student Rains Down Death
on a Quiet Campus," as well as, from the same issue, a quote from a
survivor: "After every shot I thought 'O.K., the next one is me.' The room
was silent except for the haunting . . . moans.'"

In Graffiti Alley school shootings are but one example of the violence
endemic to this Bible-rejecting culture. There is also legalized violence at
the beginning and at the end of life. Even before birth, there is the violence of abortion. Attached to the walls are a few newspaper articles,
including one with the headline "New York's decision to train docs in
abortion stirs debate," as well as seven copies of a photo of a woman
with her eyes shut and her mouth taped with orange duct tape upon
which is written the word "Life."[15] There are also a number of covers
from *Time*: "Abortion: The Battle of 'Life' vs. 'Choice'"; "The Abortion
Pill"; and "The Abortion Campaign You Never Hear About." The latter
cover features a picture of tiny model fetuses as well as the following
under the headline: "Crisis pregnancy centers are working to win over
one woman at a time. But are they playing fair?" In Graffiti Alley, the
words "playing fair" have been scratched out.

Not only does this Bible-rejecting culture end life before it has a
chance to start, but it also seeks to hasten the end of life, by way of
assisted suicide and euthanasia. In Graffiti Alley, there are article fragments regarding the assisted suicide law in Oregon as well as two newsmagazine covers (each of which is posted twice): *Time's* straightforward
"How to Die" and *Newsweek's* shockingly titled "The Case for Killing
Granny." Even more noteworthy is the material on Terri Schiavo, the
Florida woman who in 1990 collapsed into what became a fifteen-year
coma, during which her parents (Robert and Mary Schindler) fought
her estranged husband's efforts to have her feeding tube removed. The
Christian Right gave vocal and very public support to the Schindlers's
efforts, including ongoing protests outside Schiavo's hospice, in the process arguing that Schiavo was not really in a vegetative state and that
this was a matter of defending the right to life. Adding his support to
the effort, Representative Tom DeLay claimed that "Terri Schiavo is not
brain-dead. She talks and laughs, and she expresses happiness and discomfort." In March 2005, the federal courts finally allowed her to be

taken off life support, which, as Michelle Goldberg describes in *Kingdom Coming*, sent the Christian Right into the rhetorical stratosphere: "Calls for the mass impeachment of judges were followed by demands that the courts themselves be abolished, then prayers to deliver judges to Satan, and finally coy hints about murdering them." While the subsequent autopsy revealed that Terri Schiavo's brain had severely atrophied, in keeping with what would be expected of a person in a persistent vegetative state,[16] such counterevidence—as with the Columbine shootings—has not diminished the iconic status of the Schiavo case for the Christian Right. In Graffiti Alley one finds a *Time* cover, "The End of Life—Who Decides?," as well as an article from that issue, "Lessons of the Schiavo Battle," with quotes from Pat Robertson referring to Schiavo's death as "murder" and Tom DeLay describing it as "act of medical terrorism." Accompanied by a magazine headline, "Who Has the Right to Die?" there are also four photos of Terri Schiavo, three of which were taken after her brain injury, and two of which give the impression she was mentally alert.

Graffiti Alley presents contemporary America as a culture of death, the twisted creation of liberals and secular humanists who deny the Bible's authority, want evolution taught in the public schools, support women's right to an abortion, and promote assisted suicide and euthanasia. In this culture, and as vividly depicted by magazine covers and photographs, children are taught that the universe originated in a "Big Bang?," that "The Origin of Life" is simply a matter of chemical reactions that could be replicated (as suggested by an accompanying photograph) in a set of test tubes, that "Man Evolved" "Up From the Apes," and that "Humanity's Beginnings" are to be found in "Two-million-year-old fossils." Not only do atheistic evolutionists proclaim that it is "God vs. Science," not only has it become reasonable to ask if "Life [was] Inevitable, or just a fluke?," not only is NASA actually engaged "In Search of Aliens,"[17] but central Christian doctrines are being questioned and abandoned, for example, "Is God Dead?, and "What If There's No Hell?" Moreover, there is a "War On Christians" (complete with a picture of a bloodied Jesus) that includes attacks from outside America's borders. In keeping with the Christian Right's demonization of Muslims as terrorists as well as contemporary fundamentalist understandings of the role of Islam in the "End Times,"[18] Graffiti Alley makes clear that Muslims have targeted America and the West for destruction. There are numerous photographs of the September 11, 2001, attacks on the World Trade

Center towers and the Pentagon; there are four copies of a *Newsweek* cover with the headline "The Next Jihadists" above a photo of a dark-skinned boy holding a gun; there are even photographs of Iranian president Mahmoud Ahmadinejad accompanied by the headline "Denying the Holocaust, Desiring Another One."

The story told on these darkened walls is the story of "The Decline and Fall of Christian America." This April 13, 2009, cover of *Newsweek*—which features black type on red in the shape of a cross—figures prominently in Graffiti Alley. What does not figure prominently in Graffiti Alley is the article to which this cover refers, that is, Jon Meacham's "The End of Christian America." In this piece Meacham sought to make sense of the 2008 American Religious Identification Survey, which found that the number of self-identified Christians in the United States had dropped from 86 percent to 76 percent since 1990. With Barack Obama's 2008 presidential victory as backdrop, Meacham asserted that "while we remain a nation decisively shaped by religious faith, our politics and our culture are, in the main, less influenced by movements and arguments of an explicitly Christian character than they were even five years ago." Meacham was not at all distressed by this development. He said that "the decline and fall of the modern religious right's notion of a Christian America creates a calmer political environment."[19]

But Ken Ham and his AiG colleagues do not view "Christian America" as a "notion." That is, they (and the Christian Right in general) see their understanding of American history not as a particular and ideologically freighted interpretation, but simply as incontrovertible truth. Thus, they argue in *Answers* and elsewhere: "Most of America's founding fathers" believed "the Bible was true from beginning to end," and "the only source upon which to establish the new, independent nation"; our "culture developed as the result of regular Christians walking faithfully before God, submitting to His Word on a daily basis"; America's "general worldview (including moral issues) was built on the foundation of the authority of the Word of God"; throughout much of its history America was "a great Christian witness" to the world.[20] But since the 1960s, anti-Christian forces have succeeded in steadily eroding America's biblical foundation. While Jon Meacham lauded the benefits of the "separation of church and state," Ham sees this phrase as a "club to 'beat down' and eliminate Christianity from public places." The result is that today "the Bible, prayer, creation, and the Creator God are almost totally eliminated from the government education systems." But the takeover of

the public schools has not satisfied the increasing and increasingly aggressive horde of secularist attackers who want the entire public square
scrubbed clean of God, Christ, and Christianity "in favor of a religion
of humanism/atheism." So the cultural campaign continues unabated,
with aggressive efforts to remove nativity scenes from city property, with
the "Ten Commandments being ripped out of public places," and with
efforts to deny "the right of our military personnel to say 'God Bless you'
at military funerals."[21]

The decline and fall of Christian America, indeed. Of course, one
might respond that it makes sense to end Christianity's privilege in the
public square, given that, as the 2008 American Religious Identification
Survey made clear, twenty-first-century America is a religiously diverse
place, filled with people of very different faiths and of no particular faith.
In fact, this point has been made in various ways by Barack Obama. In
AiG articles and on his blog, Ken Ham has taken note, repeatedly quoting from Barack Obama's *Audacity of Hope: Thoughts on Reclaiming
the American Dream*: "Whatever we once were, we are no longer just a
Christian nation; we are also a Jewish nation, a Muslim nation, a Bud-
dhist nation, a Hindu nation, and a nation of nonbelievers." While Obama's comments would seem to be benignly descriptive of a religiously
pluralistic United States of America, the AiG CEO laments that the
president's claims are prima facie evidence that "we are in big trouble—
big trouble spiritually." As Ham noted in his videotaped lecture, "Revealing the Unknown God," "The president is saying we no longer build
our thinking on the Bible. We have a different starting point. But what
he's really saying is that it's a change from one God to many gods."[22]

In his 2010 *Answers* article, "One Nation under . . . ?," Ham goes
even further, denying the very premise of Obama's remarks by rejecting
the notion of many religions: "Ultimately there are only two religions—
one starts with God's Word and the other starts with man's word." Such
comments further underscore Ham's intense discomfort with America's increasing religious pluralism, a discomfort that is reflected in
much of American evangelicalism.[23] More than this, and very much in
keeping with the rhetoric of the Christian Right, Ken Ham and AiG
in general understand the demographic realities of pluralism and the
legal structures that protect such pluralism as having turned the nation upside-down. In what was and should be Christian America, true
Christians are in the minority, true Christians are the downtrodden and
the persecuted, true Christians are portrayed as the enemy, and true

Christians are seen as "fair game" for "brazen" attacks that "are vicious, slanderous, and full of lies and hatred." And it is only going to get worse. As Ken Ham put it in *The Lie: Evolution/Millions of Years*, the day is fast approaching when true Christianity will be illegal:

> Christian absolutes—those truths and standards of Scripture that cannot be altered—are becoming less and less tolerated in society because increasingly God's Word is no longer the foundation for building a person's worldview. Eventually this must result in the outlawing of Christianity—a possibility that seems more and more real with legislation that not only restricts Christian activities even in America but that also lays a foundation for Christians to be viewed as criminals because of the ways hate crimes legislation and other laws can ultimately be used.[24]

As Graffiti Alley makes clear, what we have now in America is a war on Christians. Moreover, Ken Ham asserts, "the increasing anti-Christian attacks in America should be a warning . . . that the chasm is widening between what is Christian and what is not in this nation." Those who have tried to "straddle both sides" will "have to finally decide which side they are on!" This is not simply a culture war: "The real war being waged is a great spiritual war—God's Word versus autonomous human reasoning, Christian absolutes (built on the Bible) versus moral relativism (man determining his own rules)." In this cosmic battle, anti-Christian enemies use false ideas such as tolerance—which "really means an *intolerance of the absolutes of Christianity*" (emphases in original)—to deceive and muzzle Christians. In this cosmic battle there are and can be no neutral positions, no neutral parties, no neutral public square, and no neutral educational system. As Ken Ham repeatedly proclaims, "There is no neutrality when it comes to Jesus Christ—either you're for Him or you're against Him . . . Eliminating Christianity [from the culture] does not produce a neutral result, but an anti-God one!"[25]

Here we have the cosmic binary, the chain of equivalences in which truth, morality, and the inerrant young Earth creationist Word are arrayed together on one side of the battle against the great "lie" (evolution/millions of years), amoral relativism, and atheism. Of course, when one thinks of contemporary America in this fashion, when one understands oneself as with God in war against the anti-God forces, it makes great sense to conclude, as Ken Ham does, that Christians must reject the lie that "they have no right to impose their views on society." Instead, American culture must "return to the authoritative Word of God as the foundation for all thinking."[26] It is either Christian America—where true Christians again run the show and where people of all other faiths or of no faith know their place—or it is Godless America. There is no middle ground. The choice is clear. The stakes could not be higher.

Eden as Heterosexual Paradise

In her book, *Pray the Gay Away: The Extraordinary Lives of Bible Belt Gays*, Morehead State University (KY) sociologist Bernadette Barton tells the story of taking upper-division students on a field trip to the Creation Museum. Including Barton, there were fifteen in the group: ten women (five of whom were self-identified lesbians, including two couples) and five men. Most of the students were from Barton's class on religion and inequality, in which they "had spent much of the semester analyzing manifestations of Christian fundamentalist and conservative Christian thought." While the students had been upbeat in the van ride to the museum, most found the museum—particularly Graffiti Alley and the Culture in Crisis room—to be an extremely unpleasant place. As Barton observes, "student reactions to the museum included feeling paranoid, angry, glared at, unwelcomed, and generally tense." One of the students—a woman wearing a long shirt and leggings—reported that as she was approaching the Noah's Ark section of the Walkthrough a security guard with a dog circled her twice, "making it a point for me to know that I looked out of place"; when the guard left, a man with two children approached her and said, "The reason he did that is because of the way you're dressed. We know you're not religious, you just don't fit in." The two lesbian couples felt the greatest discomfort, "experiencing a strong sensation of homophobic surveillance." While they worked hard not to show any signs of affection, this did not—as another student observed— alleviate the anxiety: "I know I was just being paranoid, but I couldn't shake the feeling that people could look at us and somehow automatically see that we were more than just two girls who were friends."[27]

Barton presented a version of this chapter at the 2010 American Sociological Association meeting, which was picked up and featured in a LiveScience.com article entitled "Creation Museum Creates Discomfort for Some Visitors." AiG editorial staff members responded almost immediately with a detailed rebuttal, in which they observed that "if someone is offended by the museum, this says far more about that person's standing before God than anything about the museum." As noted in the rebuttal, the museum "displays the true history of the universe as recorded in God's Word," and "the truth sometimes makes people uncomfortable, particularly those people who have been living in constant rebellion against God"; more than this, the tense and unsettled Morehead State students now had a small taste of what Christians experience on a regular basis in public universities and other secular institutions. As regards the aforementioned story of the security guard and his dog, the editors could not understand the fuss, remarking that the "dog goes throughout the museum sniffing all sorts of things . . . [and] is capable of finding children who become separated from their parents." Moreover, given that "this group was *looking* for something to criticize," there were real questions "about the authenticity of this story." In fact, the AiG editors concluded, Barton and her students were so determined to attack the museum that they invented "anti-gay messages that do not exist," as "there is no signage that discusses gay marriage or homosexuality."[28]

Interestingly, their choice of language—"museum signage"—conveniently omits Graffiti Alley, an omission defended by the AiG editors on the basis that Graffiti Alley simply contains "décor [that] comes from *secular* sources."[29] Notwithstanding the disingenuousness of this claim— as if the museum curators did not select which clippings to display or not to display—it is not at all surprising that Bernadette Barton's students were particularly discomfited by Graffiti Alley. Plastered to the darkened walls are newsmagazine headlines proclaiming "The Battle over Gay Teens," "The War over Gay Marriage," "The Religious Case for Gay Marriage," and "The Conservative Case for Gay Marriage." There is also a photograph of a lesbian couple with children, a headline from the Syracuse (NY) *Post-Standard* reporting that "Mass. Opens Doors to Gay Marriages," another newspaper headline declaring that "Ford makes donation to gay advocacy group promoting same-sex marriage," as well as a few snippets regarding "Homosexual Bishop" Gene Robinson.

The museum's Garden of Eden reinforces these antigay messages. Early along the path visitors encounter the life-size, naked, and beautiful (by Western standards) Adam and Eve. They are holding hands and looking longingly into each other's eyes. In front of this scene one finds a signboard entitled "God Forms Eve from Adam's Side" and containing the following text: "The Lord God said, 'It is not good that the man should be alone; I will make a helper for him.' . . . And the rib, which the Lord God had taken from man, he made into a woman and brought her to the man.—Genesis 2:18, 22." On either side of this text are interpretive placards. On the left placard, under the heading "Male and Female," one reads: "Have you not read, that he who made them at the beginning made them male and female?—Jesus Christ." On the right placard, under the heading "One Flesh," one reads: "For this cause a man shall leave father and mother, and shall cleave to his wife: and they two shall be one flesh. Therefore, what God has joined together let not man separate.—Jesus Christ."[30] Below the verse on the right placard is the heading, "Doctrine of Marriage," and the following commentary: "The special creation of Adam and Eve is the foundation for marriage: one man and one woman. The fact that they were one flesh is the basis for the oneness of marriage."

While it is true that these signs do not explicitly mention homosexuality or gay marriage, the subtext is readily available, close enough that AiG editors coyly acknowledge that "seeing the basis for marriage as one man and one woman in the [museum's] Garden of Eden might make someone uncomfortable who is rebelling against God's created order and laws regarding marriage." Working hard to get around "the inconvenient fact," noted by Randall Balmer, that "according to the New Testament, Jesus himself said nothing about . . . homosexuality," AiG spokespersons repeatedly use these verses to make the case that the Son of God was strongly and clearly against gay marriage. In fact, in reading AiG materials one could easily conclude that opposing gay marriage was Jesus's top priority while he was on Earth. See, for example, *The Seven C's of History*, a booklet by Ken Ham and Stacia McKeever. In the section on the fifth C, "Christ," the authors—ignoring the Sermon on the Mount as well as the dozens of Jesus's parables in the Gospels—mention but one specific "teaching" of Jesus: "He even used the book of Genesis to explain that marriage is between one man and one woman (Matthew 19:3–6, quoting Genesis 1:27 and 2:24)."[31]

Once again, the Creation Museum (and AiG) insists upon an absolute, irreconcilable binary, in this case the gender binary. Writing in *Answers*, an attorney connected with the Alliance Defending Freedom (a Christian Right legal organization originally named Alliance Defense Fund) asserts that the public square no longer attends to the Genesis doctrine that human beings were "created in two distinct complementary, yet equally dignified genders, male and female." Responding to a YouTube video, "When Did You Choose to Be Straight?," Roger Patterson puts it more clearly: "I did not choose to be straight; God made me straight. All people are born within the male and female order of creation. If I choose any other possibility . . . I am choosing a perversion of God's good design." According to Tim Chaffey, that "a homosexual can choose to become heterosexual" is suggested by the fact that "there are many thousands of former homosexuals and numerous ministries . . . devoted to helping them."[32] That said, whether or not someone is predisposed to homosexuality is, in the end, irrelevant, given that homosexual behavior is a sin. AiG's Elizabeth Mitchell underscores the point: "Dressing up this sin as natural, normal, and unavoidable does not change [the] biblical and historical reality" that homosexuality is "a sin that has been particularly associated with the slippery slope of sinful attitudes and lifestyles leading to ever greater rebellion, social degradation, disease, and spiritual blindness throughout history."[33]

Of course, there are those brave individuals who have resisted America's moral slide downward, including Chick-fil-A president Dan Cathy, whose comments in opposition to gay marriage and whose financial contributions to organizations opposing LGBTQ rights exploded into a national controversy. Not only did Ken Ham laud Cathy and Chick-fil-A for "boldly support[ing] biblical values as an organization," but AiG staffers headed to the local Chick-fil-A for lunch on the Christian Right–designated "Chick-fil-A Appreciation Day." Notwithstanding courageous resisters like Dan Cathy, by 2013 the downward slope leading to new depths of social degradation and spiritual blindness had become much more slippery. As reported by Ham on his blog, first it was the February ruling by the Massachusetts Department of Elementary Secondary Education allowing transgender students to use restrooms in keeping with their gender identity: "Sadly, these school authorities don't recognize the sinful heart of man . . . [What] about the potential for high school boys to pretend to 'identify' as a female just so they can have access to the girls' restroom and, maybe, to their locker room—winking to their friends as they do it?" Then, in the same month, eHarmony's Neil Clark Warren, a self-described "passionate follower of Jesus," buckled in response to "persecution from the world" and allowed gays access to the online dating service. In May, the Boy Scouts of America (BSA) lifted its ban on gay youth membership. In response, Ham suggested to Christian parents that they should consider pulling their youth out of the BSA: "We can't expect our children . . . to always know how to respond in situations . . . where they're mixing with other boys who may be expressing homosexual tendencies."[34]

Then, on June 26, 2013, the Supreme Court struck down the Defense of Marriage Act and ruled that married same-sex couples had
rights to federal benefits while at the same time it refused to overturn
a trial court's ruling against California's Proposition 8 (a ballot initi-
ative that banned gay marriage in the state). The next day Elizabeth
Mitchell posted the AiG response. In an article entitled "Defining
Marriage—*Supreme Court v. Sovereign God*," Mitchell proclaimed that
the Supreme Court, "by deigning to declare God's definition of marriage
inadequate and unfair, legitimized a counterfeit form of marriage for
our country." The eventual result will be the "dissolution of the family
unit" in America, which really is "the goal of those seeking the right to
'gay marriage.'" While there was still time for Americans to repent and
demand that this decision be reversed, there was no question that the
Court's decision "only adds legal legitimacy to our individual and national rebellion against God," a rebellion that "destroys lives, families,
and—historically—nations."[35]

Mitchell's strident, even apocalyptic response to the Court's ruling
was moderate compared with what Ken Ham had to say, particularly in
his hour-long opening address at the July 2013 Answers Mega Conference. Ham began his talk, entitled "The Great Delusion: The Spiritual
State of the Nation,"[36] by noting that while America had been a great
Christian nation, we now see "catastrophic spiritual decline" and a thor-
oughgoing wickedness akin to the "days of Noah." After a discussion of
Obama's endorsement of America's move from God's Word to man's
word as its foundation, Ham launched into a thirty-minute diatribe
that had as its focus the claim that the acceptance of homosexuality and
gay marriage is the clearest evidence that America has abandoned the
Bible. Obama remained the target of Ham's diatribe. According to Ham,
from the beginning of his administration, the president and his con-
federates had followed a "deliberate agenda to push the gay marriage
issue across this nation" and to "condone homosexual behavior." Their
success in promoting the gay/lesbian agenda in the courts and in the
culture would eventually mean the legalization of polygamy and more.
In light of such wickedness, Ham again asserted that the sad condition
of contemporary America mirrors the condition of the world in "the
days of Noah." According to Ham, the evidence is unassailable: the "lie"
that "God's Word is not true" permeates our culture; the legal system
requires that our children be taught that "they are animals that evolved

over millions of years"; and "moral relativism" rather than "the absolutes of Christianity" pervades the country.

Ham then asked the audience a rhetorical question: "What does
God think of a nation like this?" Here is Ham's angry reply:

> People, a sign that God is giving a culture over to judgment, a sign that
> God is withdrawing the restraining influence of the Holy Spirit, is that
> sign of homosexual behavior, gay marriage, that is permeating the na-
> tion. Therefore it is my assertion that AMERICA IS UNDER JUDGMENT
> [accompanying slide]. It is under judgment by an almighty God who
> looks upon this culture that has thrown God out of the culture.[37]

Ham then raised the stakes by asking a second, even more pointed question:

> If . . . America is under judgment, then how should we view the presi-
> dent of the United States, who has promoted gay marriage, pushed the
> gay marriage/homosexual agenda in a big way, has condoned the killing
> of 55 million children that makes what Hitler did at the Holocaust pale
> in comparison?

After a few quotes from the Bible—including Daniel 2:21, "'He removes
kings and raises up kings'"—Ham provided the dramatic answer:

> President Obama has been appointed by God to be where he is. Scrip-
> ture makes it clear. IF AMERICA IS UNDER JUDGMENT, THEN . . . THE
> LEADER IS THERE FOR AMERICA'S JUDGMENT [accompanying slide].
> And that's a sobering fact, if it's true, and I believe it is. If the leader is
> there for America's judgment. Wow. We need to fall on our knees before
> a holy God. . . . This nation needs to fall on its knees before a holy God
> and repent.

According to Ham, God is withdrawing from America, turning it
over to judgment and using Barack Obama, who can be legitimately
compared to Adolf Hitler, as his tool to facilitate America's judgment.
It was a frightening, even menacing, pronouncement. Ham returned to
it at the end. After an extended pitch on behalf of the Ark Encounter
project, in which he asserted that the primary purpose of rebuilding the
Ark was to make clear to visitors "the door of salvation," Ham then reminded conference attendees that only eight people walked through the
door of Noah's Ark to safety and—with an animation of Noah and the

Ark on the screen behind him—"nobody else came in." The animated door shut, accompanied by the sound of a slamming door: "And God shut the door." Slam. "Let me do that again." Slam. "And then the people outside," presumably while their lungs were filling with water, "realized something was wrong." Then Ham again made explicit the connection with contemporary America, pointing out that while, thanks to Jesus Christ, the door of salvation is still open, this moment would pass: "One day, like Noah's Ark, that door is going to shut. And then the Judgment will come."

Ham devoted the final seconds of his talk to calling on attendees to use this conference to "get equipped with answers" so as not "to be intimidated by the secularists," so as to "stand tall and not compromise." True Christians must be "preachers of righteousness," bringing others into the ark of salvation. But they must hurry. America is under judgment, as evinced by the cultural and legal acceptance of homosexuality and homosexual marriage. Judgment Day is nigh, Judgment Day for those who have abandoned God's Word for man's word, Judgment Day for gays and lesbians who refuse to repent of their vile affections, Judgment Day for Barack Obama[38] and his minions who have helped carry out the homosexual agenda. As it was in the time of Noah, so it is now. Judgment Day is almost here, and Ken Ham, safely inside the ark of salvation, is ready to watch God pour down his terrible and violent wrath on this wicked president and this wicked culture.

Eden as Patriarchal Paradise

At the Creation Museum, heterosexuality and heterosexual marriage are the norm, established by God in the Garden of Eden. In more subtle fashion, the museum makes clear that God also ordained a particular form of heterosexual marriage. Let us return to the scene in which the beautiful Adam and Eve look upon each other for the first time. There, as noted above, the visitor encounters a placard titled "Male and Female," accompanied by a quote from Jesus: "Have you not read, that he who made them at the beginning made them male and female?" Below that is the museum's take on this all-important gender binary: "Eve (like Adam) was specially fashioned by God and did not come from an animal. Eve was not made from dust but from the side of Adam. God made male and female fit for different roles from the beginning."

"Fit for different roles from the beginning." While this phrase certainly suggests that the division of humans into male and female carries with it significant social implications, nowhere does the museum elaborate upon what these implications are. But one certainly gets a clue from the fact that women are largely absent or relegated to secondary roles in the museum. The ubiquitous voiceovers are male. Most of the talking heads on the various video screens are male, with Georgia Purdom the most prominent exception. *Men in White* features, well, men in white sent to rescue Wendy, the damsel in distress. The Dig Site features two male archeologists actively bringing the dinosaur skeleton to the surface, followed soon thereafter by the model of a small female ape, "Lucy," trapped in a glass case. The Biblical Authority room is filled with male authority figures from the Bible, followed in the Biblical Relevance room by the figure of Martin Luther. In the Ark Construction room we see Noah and his male craftsmen performing an almost inconceivable creative act as they construct the Ark, while nearby the only two women in the scene sit quietly by (neither is animatronic) weaving baskets. In the Voyage of the Ark we see women quietly engaged in domestic tasks on behalf of Noah's Ark project, listening to Noah speak authoritatively, or drowning in the Flood.

There is one apparent exception to the rule that men drive the museum narrative: Eve. She is half of the perfect couple, loving Adam in paradise. She is the one who is tempted: near a model of the glowering serpent visitors find a placard entitled "God's Word Questioned," with a quote from Genesis: "The serpent said to the woman, 'Has God really said you shall not eat of every tree of the garden?'"[39] And as seen in the next diorama, in which Eve is giving the fruit to Adam and which is accompanied by a placard entitled "God's Word Is Abandoned," she is the one who succumbed to temptation and who successfully enticed her man to join her in sin.

It turns out, however, that, in the museum's telling, Eve played at most a supporting role in the drama of the Fall. It was Adam's sin that mattered, not Eve's. This crucial point is made clear by the seven Cs placard that appears along the walkway just before the serpent. It is entitled "Rejection of God's Word Led to Corruption," and it explains that "the first man, Adam, disobeyed the Creator, bringing death and corruption into the creation." This observation is reinforced by a quote from Romans 5:12: "By one man sin entered into the world, and death by sin." Then it is on to the serpent, the diorama of Eve giving the fruit to Adam, a door upon which has been scratched, "The World's Not Safe

Anymore," and a gallery of photos making vivid what Adam's sin has wrought, including a malnourished child, a nuclear explosion, a collection of skulls, and a woman in the agony of childbirth. Then the visitor comes upon the large "Curse" placard where the message of the seven Cs placard that introduced the Corruption epoch is reinforced:

> The moment Adam disobeyed, he died spiritually, separated from the God who gives life to all things. Sin separates all descendants of Adam from God. Because of man's rebellion against His word, God cursed the creation.

The Creation Museum does not explain why Adam's sin and not Eve's brought down the divine Curse. But AiG commentators are not reticent on this point, explaining that Adam's responsibility for the Fall was connected to his God-assigned position in the First Family. As Georgia Purdom asserts in her article, "Genesis, Wifely Submission, and Modern Wives," the "plain reading of God's Word makes it clear that Adam's original created role was to be a leader in the family." God made Adam first, God gave Adam responsibility to name the animals, and, most striking, God gave Adam power "to name his wife," which "in Old Testament times . . . was a sign of authority for the person doing the naming." God underscored Adam's authority over Eve by creating her to be her husband's "helper." But while "Adam was responsible for his wife Eve," he "shirked that responsibility by following her leading in disobedience to God." As AiG contributor Steve Golden observed, God called Adam to account "for the moral actions of his family" because he had failed to live up to the responsibility entrusted to him in the "God-given system of male headship."[40]

While AiG is quite clear that "from the beginning" God established that husbands are to rule over wives and children, one is hard pressed to find this idea anywhere in the Creation Museum. To the contrary, it would be easy for visitors to come away with a quite different understanding of the origins of the patriarchal family. On the aforementioned "Curse" placard, below the assertion that it is because of Adam's sin that God cursed his creation, there are specifics as to how this curse affected the "Serpent" ("you shall go upon your belly, and you shall eat dust") and "Adam" ("cursed is the ground for your sake," and "in the sweat of your face you shall eat bread"). As regards "Eve," visitors are provided the text from Genesis 3:16: "I will greatly multiply your sorrow and your con-

ception; in sorrow you shall bring forth children; and your desire shall
be to your husband, and he shall rule over you."

A museumgoer reading this placard could reasonably conclude that
female submission to male authority, along with pain in childbirth, is
not part of God's divine plan, but is, instead, one of the unfortunate consequences of sin entering the world. As Margaret Lamberts Bendroth
observes in her book, *Fundamentalism and Gender*, this has certainly
been the standard reading in dispensational premillennialism. For example, in C. I. Scofield's *Reference Bible—the* Bible for dispensational
premillennialists—the editor commented that, in response to the first
couple's sin, God established the "Adamic Covenant [that] conditions
the life of fallen man" and that includes burdensome toil, physical
death, and "the changed state of the woman . . . in three particulars:
(a) multiplied conception; (b) motherhood linked with sorrow; (c) the
headship of the man . . . [as] the entrance of sin, which is disorder,
makes necessary a headship, and it is vested in man." Bendroth points
out that for dispensationalists, "Adam's—and especially Eve's—original
sin" rendered these curses "irreversible"; as Scofield notes, all of these
"conditions" occasioned by the Fall, including men's rule over women,
"must remain till . . . the kingdom age."[41]

Note the implications of this traditional dispensational premillennialist reading of Genesis 3. God intended that there be male-female
equality; in Eden there had been such equality; in the millennial age
men and women will return to a gloriously egalitarian state. Even if
one acknowledges patriarchy as an unfortunate byproduct of human
sin, it does not seem an enormous stretch to argue that Christians—
seeking to live as God wants human beings to live, seeking to follow the
guidance of the Holy Spirit—are called to practice gender equality. As
Paul proclaimed, "there is no longer Jew or Greek, there is no longer
slave or free, there is no longer male and female; for all of you are one in
Christ Jesus" (Galatians 3:28).

Directly borrowing from the late-nineteenth-century holiness
movement, evangelical feminists have been making this argument since
the 1970s. In response to such arguments, and drawing upon Calvinist
theology more than traditional dispensational premillennialism, much
of conservative Protestantism (including AiG) has adopted the Calvinist position that roots "sexual inequality in the original created order."
Of course, if male headship was established "from the beginning" and

was not the result of human sin, then how are we supposed to understand the aforementioned Genesis 3:16, which on the face of it seems a pretty clear statement that one result of sin is that husbands shall "rule over" wives? In response to an *Answers* reader who makes precisely this point, W. Gary Phillips explains that while Genesis 3:16 "is often misinterpreted as the institution of headship," the reality is that "male headship was intended by God when He created woman." What God's Curse has wrought is not submission, but, instead, the inability of wives to rest easy with their God-ordained submission. In his article, "Is Male Headship a Curse?," Steve Golden puts it this way: "Because of the Curse and the new sin nature," the "wife will attempt to usurp" her husband's "God-given authority." These rebellions by sinful wives seeking to dethrone their rightful rulers are made worse by the fact that their husbands are tempted to react to these female rebellions by exercising too much or too little authority, i.e., "to respond with ungodly domination or to remain passive and allow" their wives "to rule."[42]

As Georgia Purdom observes, if married couples would follow the plan established by God in the Creation, if husbands would accept their role as leaders, and wives would accept their role as submissive helpers, then marriages would flourish. According to Purdom, one reason women resist submission is that they equate it with inferiority. But the very nature of God provides a very instructive analogy: "Jesus and God the Father have different roles within the Trinity, but in their personhood they are equally God. Likewise, a wife and husband have different roles in marriage, but they are equally loved by God and equally bear His image." Gary Phillips makes even clearer the connection between the nature of God and proper roles in marriage:

> Eve was equal to Adam before God but created to be Adam's helper, implying his functional headship. This personal equality but functional headship is a reflection of the nature of the trinity—the Son is equal in being but subordinate in role to the Father, and the Spirit is equal and submissive to both.[43]

In these claims Purdom and Phillips are echoing a cohort of conservative Calvinist theologians who have connected the argument for male headship within marriage, and within the church, to the claim that the Trinity is hierarchical and that this hierarchy serves as a model for men and women in the family and in the church. See, for example,

Wayne Grudem's book *Evangelical Feminism and Biblical Truth*, which has attained a certain notoriety for its astonishingly detailed list of roles that women may and may not fill in the church—for example, women may chair a church committee or sing a solo in Sunday morning worship, but they may not serve as "permanent leader of a [home] fellowship group" or teach the Bible "to a college-age Sunday school class." In this book Grudem, the co-founder (along with John Piper) of the Council on Biblical Manhood and Womanhood, asserts that "just as God the Son is eternally subject to the authority of God the Father, so God has planned that wives be subject to the authority of their husbands." Bruce A. Ware, T. Rupert and Lucille Coleman Professor of Christian Theology at the Southern Baptist Theological Seminary, goes to even greater lengths to connect authority and subordination in the Trinity with authority and subordination in the family and the church. In his book, *Father, Son, and Holy Spirit: Relationships, Roles, and Relevance*, Ware calls on husbands to emulate God the Father, who "possesses the place of supreme authority" in the Trinity, in leading their families, work that includes "the privilege and duty, before the Lord, to lead in the spiritual direction, training, and growth of their wives and children." Ware calls on wives to "render to [their] husband[s] . . . a joyful, heartfelt, willing, glad-hearted submission," in doing so "realizing that the submission required of them as wives is itself reflective of the very submission eternally given by the Son to his Father, and by the Spirit to the Father and the Son."[44]

Such notions raise weighty theological questions regarding the equality of the Father, Son, and Holy Spirit, which was established as orthodox Christian theology at the Councils of Nicaea (325 CE) and Chalcedon (451 CE). Ware and his colleagues have sparked controversy within conservative evangelicalism, the argument turning on whether their notion of eternal subordination equates to "ontological" subordination. If this were the case, which the proponents forcefully deny, it would open them to charges of Arianism (Jesus is not God in the same way the Father is God), a theological position declared to be heresy at Nicaea.[45]

This is not the place to explore this theological question. For our purposes, what is striking is the determination to connect the supposed authority-subordination structure within the Trinity to "proper" gender hierarchies in the family and the church. In fact, in reading Grudem,

Ware, and AiG contributors it is easy to conclude that the whole point
of this theological exercise is to cement once and for all that women are
to be subordinate to men. Indeed, if patriarchy in the home and patriarchy in the church can be established as the reflection of patriarchy
within the Trinity, then any claims for gender equality would be forever
rendered out of order.

After Eden: The Propriety of Incest

Before we leave the topic of gender and power, we need to make one
more stop in the Bible Walkthrough Experience. After the "Curse" signboard there is a series of placards detailing the difference between life
before sin and life after sin, a difference underscored by a diorama with
the now-slithering serpent and an animatronic *Tyrannosaurus rex*
threatening to chomp on the heads of visitors. Visitors then come upon
another diorama featuring the First Couple, in which the handsome and
muscular Adam is tilling the soil, two young boys (presumably Cain and
Abel) are providing assistance, and the beautiful and pregnant Eve is
contentedly watching. While the display of Adam's labor and Eve's labor-
to-come reminds visitors of the curse wrought by sin, the scene is almost
idyllic. But the idyll is short-lived, as the adjacent diorama portrays the
aftermath of Cain's murder of Abel, with Abel face down in the dirt and
Cain standing over him with upraised arms, looking into the heavens.

   Immediately across the path, we find a large placard entitled "Where
Did Cain Get His Wife?" While Cain's marital status is not necessarily
the obvious question raised by these dioramas, the Creation Museum
clearly finds this a pressing matter. And the placard forthrightly provides the answer: given that "all humans are descendants" of Adam and
Eve, and given that "Genesis 5:4 teaches that Adam and Eve had sons
and daughters," then "brothers had to marry sisters." In short, Cain married his sister. But as some visitors might find this answer disconcerting,
the remaining five-sixths of the placard is devoted to explaining why
we should not be bothered by this incestuous relationship. For start-
ers, "all humans are related" and thus "whenever someone gets married,
they marry their relative." Moreover, the historical account in the Bible
makes clear that it was not until many generations after Abraham, who
himself "was married to his half sister," that God finally "instructed the
Israelites not to marry close relatives." Biologically, God's delay in banning incest makes great sense, given that "at the time of Adam and Eve's

children, there would have been very few mutations in the human genome." While "sexual activity outside the bounds of marriage, whether
between close relatives or not, has been wrong from the beginning,"
marriage "between close relatives was not a problem in early biblical
history," as long as "it was one man for one woman (the biblical doctrine of marriage)." The placard concludes by noting that "since God
is the One who defined marriage in the first place," then "God's Word
is the only standard for defining proper marriage"; those "who do not
accept the Bible as their absolute authority have no basis for condemn-
ing someone like Cain marrying his sister."

Even in the context of the Creation Museum, this is one strange
placard. Here we have an argument on behalf of incest that includes
questionable claims regarding human genetics, an attack on those who
criticize incest on any grounds other than the Bible, and the suggestion
that incest is not as bad as it seems, given that all humans engaging in
sexual activity are having sex with relatives. Here we have an argument
that renders a near-universal moral taboo wrong primarily on instru-
mentalist grounds, an argument explained in more detail in Ken Ham,
Carl Wieland, and Don Batten's *One Blood:*

> By the time of Moses (about 2,500 years [after Creation]), degenerative
> mistakes would have accumulated to such an extent in the human race
> that it would have been necessary for God to bring in the laws forbidding
> brother-sister (and close relative) marriage . . . Also, there were plenty
> of people on the earth by now, so close relations did not have to marry.[4]

How are we to understand this elaborate effort by the Creation Museum to defend the notion that—according to their six-thousand-year
historical timeline—for the first third of human history incest was fine,
as long as it was in the context of heterosexual, monogamous marriage?
The most benign explanation is that the Creation Museum, AiG, and
young Earth creationists in general have much at stake in extinguishing
all possible challenges to a literal reading of the Bible. So, for example,
Ken Ham and his colleagues have produced two volumes of *Demolishing Supposed Bible Contradictions*, books which promise to disprove
the notion that there are "contradictions or errors in the Bible" so as to
"reassure people that the Bible is inerrant."[47] In this same vein, then, we
can understand this placard as an effort to answer once and for all one
of the standard "village atheist" questions, one which Clarence Darrow
posed to William Jennings Bryan at the Scopes trial:

In a scene that appears between the Fall and the Flood along the Bible Walk-through Experience, Adam tills the soil with assistance from sons Cain and Abel as Eve descends the stairs pregnant with, presumably, the female child who provides the answer to the question that the museum's literal reading poses—namely, where did Cain get his wife? Photograph taken by Susan Trollinger. For more images of and information about the museum, go to creationmuseum.org.

> Q—Did you ever discover where Cain got his wife?
> A—No, sir; I leave the agnostics to hunt for her.
>
> . . . .
>
> Q—The Bible says he got one, doesn't it? Were there other people on the earth at that time?
> A—I cannot say.
> Q—You cannot say. Did that ever enter your consideration?
> A—Never bothered me.
> Q—There were no others recorded, but Cain got a wife.
> A—That is what the Bible says.
> Q—Where she came from you do not know. All right.[48]

While Bryan had no idea "where Cain got his wife," the Creation Museum has a definite and airtight answer: Cain's wife was also his sister.

Once again, the Creation Museum is excising the ghost of the Scopes trial (and William Jennings Bryan's failure to stand firmly on the literal Word of God). In the process, the museum and AiG offer up an extended defense of incest as part of God's divine plan. Such an argument is disturbing for all sorts of reasons, including that the museum and AiG place such great emphasis on men ruling over their families, and given that much incest involves fathers (and uncles and brothers) taking advantage of and exerting power over their daughters (and nieces and sisters). The placard becomes even more disturbing when one looks back at the diorama of the First Family, at Adam, Cain, Abel, and . . . the pregnant Eve. The very logic of the Walkthrough—in which dioramas provide visual "peep holes" into the past and visual depictions of material on accompanying placards—strongly suggests that the curators have created a diorama with a mother (Eve) who will soon give birth to her son's sister and future sexual partner. As museumgoers learn from the placard, not only was this just fine, but to question the propriety of the impending incest (and the millennia of incest to come) is to place oneself in opposition to Almighty God.

The Question of Race

The final stop on the Bible Walkthrough Experience is devoted to the Tower of Babel. On the wall across from the Babel video, visitors find a large signboard, two-thirds of which is devoted to making the case that "According to God's Word . . . We're All One Race—'One Blood.'" A phrase from Acts 17: 26, "God has made of one blood all nations," is quoted repeatedly. Other Bible verses are offered as evidence that "We're All Created by God," "We're All in God's Image," "We're All One Family," and "We're All Loved by God." References to genetics are combined with verse snippets to establish that "Biological Differences Are Superficial," "We Came from One Woman," "We Came from One Man," and "We Are Fully Human from Conception." There are photos of smiling people of various ages and ethnicities looking directly at the camera—including one photo of a couple with their twin daughters, one white and one brown—as well as four photos of fetuses. In the bottom right segment of the signboard there is a phrase in bold, "God's Word Condemns the Abuse of Others," followed by text that concludes with the following: "God's Word condemns a long list of abuses: the abuse of the unborn, the abuse of the young, the abuse of the old, the sick, and the poor. Principles derived from God's Word also condemn discrimination based on language, culture, gender, or skin tone."

It is easy to imagine museumgoers—having viewed *Men in White* and walked through Graffiti Alley—bewildered, even flummoxed, by such apparently progressive sentiments on the matter of race. But Ken Ham and AiG make much of this exhibit. As Ham proudly noted on his blog, "We have a large section of our museum that is devoted to combating racism." Sometimes their enthusiasm carries them into the hyperbolic stratosphere, as when Mark Looy claimed that this exhibit not only provides "the only true remedy for racism (the Bible's teachings)," but it is also the "best exhibit in our region on anthropology."[49]

But when visitors turn their attention to the left third of the signboard—away from the array of smiling faces and Bible verses confirming that we are all "One Race—'One Blood'"—they encounter material much more in keeping with the rest of the museum. There are two large-font headings: "According to Human Reason . . . Everyone Decides What Is Right in His Own Eyes," and "Recent Excuses to Reject God's Word . . . Evolution over Millions of Years" (ellipses in original). In smaller font there are a few sentences devoted to making the point that when "people abandon the authority of God's Word" they turn to reason, which has provided "a multitude of excuses to justify abuse." More text is devoted to the point that in the modern world evolution is the primary excuse for humans to do evil. Bolstering the case is a Stephen Jay Gould quote (highlighted in bold font) that is oft-repeated in AiG materials: "Biological arguments for racism may have been common before 1859, but they increased by orders of magnitude following the acceptance of evolutionary theory."[50] All of this is underscored by the visual evidence: two iconic photos, unidentified by the museum, of a lynching and of a slave bearing on his back the scars of a horrific whipping, with the heading "Racism"; two familiar, and also unidentified by the museum, photos from the Holocaust, with the heading "Genocide"; and, under the heading "Abortion," an unidentified photo of an anti-abortion banner, behind which there is a field and rows of crosses.

Below the signboard is a shallow glass case containing a collection of items constituting, according to a small placard inside the case, a veritable "Hall of Shame" from a "History Marked by Abuse and Racism." According to the placard, "evolution . . . is not the cause of racism—sin is the reason for such horrible behavior," but, in a move reminiscent of AiG articles on school shootings, the very next sentence states that "Dar-

winian evolution has certainly fueled racism." Most items in the case
are designed to underscore the horrific effects of Darwinism. There is
a pair of emu-feather shoes worn by Australian aborigines during surprise attacks on their enemies. Unrelated to the shoes, the placard notes
that, thanks to "Darwinian evolution," aborigines were "considered to
be primitives who were close to their ape-like ancestors," an idea which
"fueled a particular type of racism against them"; importantly, this placard elides the colonial "genocide" that took place in Tasmania and Australia in the decades before the publication of *Origin of Species*.[51] There
is also a picture of Ota Benga, an African pygmy who was housed in the
anthropology wing of the 1904 World's Fair in St. Louis, after which
he "was displayed in the primate house of the Bronx Zoo." Most prominent are three books that make explicit the connection between Darwinian evolution and Nazism: an 1874 edition of Charles Darwin's *The
Descent of Man*, with commentary stating that "Darwin lamented the
foolishness of society for caring for its 'weak members' . . . [and] also
wrote that the Caucasian 'race' was further from the apes than other
supposed lower 'races.'"; German biologist Ernst Haeckel's 1876 *History of Creation*, with the accompanying quote from Haeckel that "A
sharp boundary line . . . has to be drawn between the most highly developed and civilized man on the one hand, and the rudest savages on the
other, and the latter have to be classed with the animals"; and finally, a
1939 edition of Adolf Hitler's *Mein Kampf,* along with the assertion that
"perhaps the most infamous application of evolutionary theory to justify
racism was Adolf Hitler's Nazi regime." Next to *Mein Kampf* is an array
of Nazi paraphernalia, including a swastika and a "Cross of Honor"
awarded to "genetically fit" mothers who had six or seven children.

Much of the material in this exhibit is found in the chapters Ken
Ham wrote for *Darwin's Plantation: Evolution's Racist Roots*, a 2007
book for which A. Charles Ware also contributed chapters.[52] The "we're
all one blood" Bible verses and accompanying phrases are in the book;
so is the paragraph (quoted above) proclaiming that "God's Word condemns a long list of abuses"; and so is most of the text on the left side of
the signboard, including the aforementioned Stephen Jay Gould quote,
which appears a number of times in *Darwin's Plantation*. The Australian aborigines and their "death shoes" appear in the book, again without
reference to the killing of Australian and Tasmanian aborigines in the
first half of the nineteenth century. Ota Benga is also here, although the
book makes more explicit that "racist attitudes fueled by evolutionary

thinking were largely responsible for the inhumane abuse" endured by the African pygmy. Finally, much of the text on the placard accompanying the Nazi paraphernalia is found (with slight revisions) in *Darwin's Plantation*, including: "Perhaps the most infamous application of evolution to justify racism was Adolf Hitler's Nazi regime, which promoted a master race and sought to exterminate so-called inferior races."[53]

According to Ham, there are "vast amounts of material that show the connection between evolutionary thinking and Hitler's genocidal slaughter of innocent human beings." Ham's Darwin-to-Hitler narrative is standard fare within creationism: as one scholar has observed, "in seeking to link Darwinism to the Holocaust, there is a sense of expectation that by doing so one might discredit Darwinian theory as a whole." Making such linkages was a staple in the writings of Henry Morris, co-author of *The Genesis Flood* and founder of the Institute for Creation Research. As Morris proclaimed in *The Long War against God*, "Hitler himself became the supreme evolutionist, and Nazism the ultimate fruit of the evolutionary tree."[54] Then there's Northwest State Community College's Jerry Bergman, armed with a PhD in measurement and evaluation from Wayne State University as well as a PhD in human biology from Columbia Pacific University, the latter (as Ronald L. Numbers has established) "an unaccredited correspondence school that recruited students with the lure of a degree 'in less than a year.'" Bergman, who is featured on the AiG website, has been a prolific contributor to the Darwin-Hitler genre, including books such as *The Dark Side of Charles Darwin* (2011) and *Hitler and the Nazi Darwinian Worldview* (2012). From the intelligent design camp, and much more in the academic mainstream, California State University historian and Discovery Institute Center for Science and Culture Fellow Richard Weikart has produced two books making the Darwinism-to-Nazism argument: *From Darwin to Hitler: Evolutionary Ethics, Eugenics, and Racism in Germany* (2004), and *Hitler's Ethic: The Nazi Pursuit of Evolutionary Progress* (2009).[55]

Many scholars would agree with historian Paul Weindling that "the assumed linkage between a reclusive Victorian naturalist and twentieth-century demagogues and Nazi perpetrators of genocide is curious, strained, and—when historically scrutinized—tenuous." As Weindling points out, part of the problem with neatly linking Darwin and Hitler has to do with the fact that "Darwin's and Hitler's actual views on evolution and race were poles apart," as "Darwin had variable views on race . . . and did not focus exclusively on the annihilatory struggle in nature." More than this, there was a great variety of "Darwinisms" that emerged after *Origin of Species*, including the "culturally and scientifically distinctive" German Darwinism, best exemplified by the biologist Ernst Haeckel. But perhaps the most important critique of the Darwin-to-Hitler trope is that Nazi ideology drew upon a great variety of sources, including anti-Semitism, fascism, nationalism, racism, and the *völkisch* movement. According to historian of science Robert Richards, Weikart and other historians who have made the Darwin-to-Hitler argument "have not . . . properly weighed the significance of the many other causal lines that led to Hitler's behavior" and thus "have produced a mono-causal analysis which quite distorts the historical picture."[56]

Karl Giberson has wryly observed that "Aryan Germans were not happily playing soccer and eating bagels with Jewish Germans until Darwin convinced them that this was a bad idea"; put more directly, the "Holocaust would have happened with or without Darwin," but social Darwinism sometimes provided the Nazis with "*rhetoric and rationalization*" (emphasis in original) for their murderous crusade against the Jews. In similar fashion the Anti-Defamation League has vigorously critiqued the Darwin-to-Hitler trope, pointing out that such an argument, usually "offered by those who wish to score political points in the debate over the teaching of intelligent design," neatly erases the multiple factors that led to the Holocaust, including a Christian anti-Semitism that long preceded Charles Darwin. Focus on Darwin-to-Hitler, and the slaughter of German Jews by eleventh-century Crusaders, the Spanish Inquisition and its persecution of Jewish converts, and the history of Church teachings versus the Jews conveniently disappear. By focusing on the role of evolution in leading to the horrors of Nazi Germany, one does not have to consider the historical import of Martin Luther's venomous words in "On the Jews and Their Lies":

> "First . . . set fire to their synagogues or schools, and . . . bury and cover with dirt whatever will not burn, so that no man will ever again see a stone or cinder of them . . . Second, I advise that their houses also be razed and destroyed. Third, I advise that all their prayer books and Talmudic writings, in which such idolatry, lies, cursing, blasphemy are to be taught, be taken from them . . . Fourth, I advise that their rabbis be forbidden to teach henceforth on pain of loss of life and limb . . . Fifth, I advise that safe-conduct on the highways be abolished completely for

the Jews . . . They are a heavy burden, a plague, a pestilence, a sheer misfortune for our country."[57]

While the Darwin-to-Hitler trope dominates the "One Race—'One Blood'" exhibit, there are also two references to American slavery. As noted above, on the left side of the signboard there is a large photo of a slave. Its proximity to the aforementioned Stephen Jay Gould quote—"Biological arguments for racism may have been common before 1859, but they increased by orders of magnitude following the acceptance of evolutionary theory"—could easily lead the historically unaware visitor to conclude that Darwinism had something to do with the 250 years of slavery in North America, even though *Origin of Species* was published just six years before slavery was abolished. Readers of *Darwin's Plantation* may also reach a similarly confused conclusion not only because of the book's title but also because the cover has a photo of (presumably) a Nazi concentration camp which blurs into a photo of (again presumably) African American slaves working the cotton fields.

In the shallow glass case mentioned above there is a pair of antebellum slave shackles, accompanied by a small placard entitled "'Blacks' Labeled as Cursed." From reading this placard visitors learn that "in the mid nineteenth century, various distortions of Bible passages were used to try to justify slavery." Some people simply "denied the biblical truth that all humans are descended from Adam and Eve, claiming that 'blacks' were not human." Others "distorted what the Bible teaches to argue falsely that dark skin color was a curse upon Noah's son Ham," when the curse was actually "pronounced upon Ham's son Canaan and it had nothing whatsoever to do with skin shade." Here the museum is addressing the infamous "Curse of Ham" drawn from Genesis 9: 25—Noah said, "Cursed be Canaan; lowest of slaves shall he be to his brothers"—which was popularly used to justify the enslavement of Africans and, after the end of slavery, the segregation of African Americans.[58]

The placard alludes to the exhibit's "one blood" theme but provides no antislavery Bible passages. Responding to Paul Taylor's December 2006 *Answers* article about British abolitionist William Wilberforce, in which Taylor asserted that "Wilberforce and the other abolitionists were driven by their belief in the inerrancy of Scripture, acknowledging that God had made all nations of one blood (Acts 17: 26)," a reader angrily wrote AiG: "Just because human beings are from 'one blood' doesn't mean that the bible is anti-slavery. The bible supports and regulates slave ownership and doesn't say that owning a slave is wrong . . . Find me one verse in the bible that condemns owning a slave. I dare you." Taylor and Bodie Hodge responded with an article entitled, "The Bible and Slavery." After establishing that the curse on Canaan had nothing "to do with skin color," they asserted that "neither slavery in New Testament times nor slavery under the Mosaic covenant" was anything like the "harsh slavery" found in the Americas and elsewhere. In fact, biblical slaves are better understood as "bondservants" who "were being paid something, and were therefore in a state more akin to a lifetime employment contract." In the case of Hebrew bondservants who were in servitude because they had "lost themselves to debt," they "made a wage, had their debt covered, had a home to stay in, on-the-job training, and did it for only six years," which (the authors jocularly noted) "almost sounds better than college." Not only did the Bible give "regulations for good treatment by both masters and servants," but the apostle Paul instructed Christian masters to treat bondservants "with respect and as equals." Hodge and Taylor conclude by asserting that it was "Biblical Christians [who] led the fight against slavery."[59]

Hodge and Taylor ignore the fact that Jesus did not denounce slavery; they elide the biblical evidence that, whatever legal protections Hebrews slaves may have had, these protections did not apply to heathen slaves. More than this, and despite the letter writer's dare, they do not provide what the correspondent requested, that is, an antislavery Bible verse. Eugene Genovese would not have been surprised: the distinguished historian of American slavery reportedly offered an A to any of his students who could come up with such a verse (an achievement he considered impossible). While the museum placard is right to note that there is nothing in the Bible connecting Ham's son Canaan with blackness, the museum is silent on the point that Noah's curse clearly established enslavement as legitimate punishment. Abraham had slaves, and slavery was firmly embedded in the Hebrew legal code. As regards the New Testament, not only was Jesus, who frequently referred to slaves, silent about the institution of slavery, but regardless how Paul instructed Christians to treat slaves, he did not condemn the institution, and he ordered the slave Onesimus to return to his master.[60]

The historian Molly Oshatz has observed, "Using the Bible to argue against slavery in itself is, objectively speaking, a difficult task, and if one believes the entire biblical text is inspired, it is quite possibly an impossible task." In response to this challenge, Hodge and Taylor are forced to claim that "slavery" in the Bible has a very different meaning from how we understand the word "slavery" today. In making this argument the authors abandon the "plain sense" reading of the Bible that AiG so passionately advocates when it comes to, say, Genesis 1:1, or the question of homosexuality. In contrast with the AiG writers, in antebellum America millions of white Christians (in both the North and the South) held tight to a "plain-sense" reading of the Bible, one which, as Mark Noll has pointed out, emphasized "the natural, commonsensical, ordinary meaning of the words" in order to construct a powerful argument justifying the enslavement of African Americans. These white Christians stood on their literal reading of the Word of God to issue forth a raft of proslavery polemics and to deliver an almost-infinite number of proslavery sermons; in the South, Elizabeth Fox-Genovese and Eugene Genovese observed that "evangelicals, having cited chapter and verse, successfully enlisted the Bible to unify the overwhelming majority of slaveholders and nonslaveholders in defense of slavery as ordained of God." These white Christians argued that opponents of slavery, who struggled mightily to combat the straightforward biblical arguments of the proslavery advocates, were undermining the authority of the Bible with their unbiblical antislavery arguments that depended more on Christian experience, humanitarianism, and morality than on the "literal" meaning of the text.[61]

Contra Hodge and Taylor, it is much more accurate to say that prior to the Civil War, "Biblical Christians," those holding to plenary verbal inspiration and a commonsense reading of the Bible, led the fight *for* slavery. Not surprisingly, almost a century after the Civil War—when the civil rights movement challenged the Jim Crow system of white supremacy in the South—supporters of segregation used biblical literalism to bolster their campaign against integration and racial equality. In a remarkable example of the interpretive pluralism inherent in inerrancy, historian Mark Newman has documented that "biblical segregationists" in the Southern Baptist Convention—in keeping with the "One Race—'One Blood'" exhibit at the Creation Museum—"cited Acts 17: 26 . . . more frequently than any other biblical text"; but unlike the Creation Museum (which only quotes the first few words) Southern Baptist segregationists quoted the entire verse: "**God** hath made of one blood all nations of all men for to dwell on all the face of the earth, and **hath determined** the times before appointed, and **the bounds of their hab-**

itation" (the text in bold understood as "confirm[ing] God's segregation plan.") Standing on the literal Word of God, these Southern Baptist
segregationists resisted efforts at integration, as did the fundamentalist
Bible Baptist Fellowship (BBF), which declared in 1957 that we "pledge
ourselves to contend against integration with every legal and Scriptural
means at our command."[62]

In her book, *Mississippi Praying: Southern White Evangelicals and
Civil Rights Movement,* historian Carolyn Renee Dupont puts it bluntly:
Mississippi's white "evangelicals fought mightily against black equality,
proclaiming that God himself ordained segregation, blessing the forces
of resistance, silencing the advocates of racial equality within their own
faith traditions, and protecting segregation in their churches." While
many white mainline Protestants and white Jews joined the movement
for civil rights, a host of white evangelical and fundamentalist ministers and leaders vehemently attacked them for having "dangerously
perverted both the Bible and the divine plan." So in 1958 BBF minister Jerry Falwell thundered from his Lynchburg, Virginia, pulpit: "The
Hamites . . . were cursed to be servants of the Jews and Gentiles . . . If
we persist in tearing down God's barriers" between the races "God must
punish us for it." When it became clear that the United States government meant to enforce integration of the public schools, white fundamentalists, including Falwell, started segregation academies throughout the South to ensure that their children would not attend school with
black children. In fact, Randall Balmer has argued that the origins of
the Christian Right are not to be found in *Roe versus Wade,* but in the
anger over the Internal Revenue Service's efforts to remove tax-exempt
status from Christian schools that discriminated on the basis of race.[63]

Jerry Falwell and other evangelicals and fundamentalists eventually
(and quite belatedly) recanted their opposition to civil rights for African
Americans. By the twenty-first century, as Michelle Goldberg has noted,
"racial prejudice" has become "increasingly taboo among the religious
right, and in many places evangelical culture is remarkably integrated."
While Goldberg points out that in 2004 "48.3 percent of white conservative Christians said they would disapprove if their child wanted
to marry a black person," even this taboo seems to be fading. For instance, in *Darwin's Plantation* Ken Ham devotes a chapter to critiquing
Christians who assert that the Bible forbids interracial marriage (while
simultaneously attacking Darwinism): "When Christians legalistically
impose non-biblical ideas such as no 'interracial' marriage, they are

helping to perpetuate prejudices that have often arisen from evolution-
ary influences." Ham's co-author on *Darwin's Plantation* was A. Charles
Ware, African American author of *Prejudice and the People of God: How
Revelation and Redemption Lead to Reconciliation* and president of
Crossroads Bible College in Indianapolis, which has as its mission "to
train Christian leaders to reach a multiethnic urban world for Christ."[64]

The social climate has changed. Segregation and blatant racism
have become increasingly unacceptable. A new understanding of what
the Bible "literally" means has appeared. As Susan Harding has noted,
"the strict Bible inerrancy polemic cover[s] up ... the speed with which
interpretations, including official ones, can be revised—or forgotten
altogether."[65] They are forgotten indeed at the Creation Museum, and
for good reason. If the Creation Museum were to attend to history it
would bring forth a slew of unacceptable questions. If the Bible is clear
and should be read in a commonsensical fashion, as the museum and
AiG assert again and again, then how might we explain the millions of
biblical literalists who were convinced that they were standing on the
Word of God, who were convinced that they were upholding biblical
authority, and yet who turned out to be so wrong when it came to slavery and segregation? More dangerous, if millions of biblical literalists
in the not-so-distant past were so wrong about what the Bible had to
say about slavery and segregation, is it not possible that in twenty-five,
fifty, or a hundred years we will have a great host of biblical literalists
fervently arguing that egalitarian marriages and gay rights are the obviously biblical position? Perhaps most threatening, if millions of biblical
literalists were on the wrong side of history when it came to slavery and
segregation, and if "less literalist" Christians were on the right side of
that same history, what happens to the absolute, good versus evil, God's
Word versus Human Reason binary that undergirds the Creation Museum, AiG, and young Earth creationism?

Paying attention to history is out of order at the Creation Museum,
and not simply because it would bring to light the ways in which interpretations of the inerrant Word are everchanging. The almost complete
silence on the history of racism in the United States, notwithstanding
Ken Ham's claim that "we have a large section of our museum that is
devoted to combating racism," is part and parcel of the Right's emphasis
on a "color-blind" society in which civil rights for racial minorities have
already been achieved such that there is no need any more to attend to
questions of race and institutional racism. As Daniel Rodgers has noted

in *Age of Fracture,* "in the 'color-blind' society project, amnesia [is] a conscious strategy, undertaken in conviction that the present's dues to the past had already been fully paid."[66]

Ken Ham and AiG are fully committed to the "color-blind" project. Thus, in their publications they repeatedly put "race" in quotation marks, refer to "skin color" instead of "race," and argue that we should replace "race" with "people group."[67] More than this, there is a palpable lack of concern with institutional racism in contemporary America, a lack of concern matched by the obsession with gays and gay rights. Three examples will suffice. First, *Darwin's Plantation* devotes only one and a half pages to the civil rights movement, and virtually all of this is devoted to a sketchy description of the 1965 march in Selma (just to mention one omission, there is no reference to Martin Luther King Jr.) In contrast, eighteen pages at the end of the book are devoted to a section entitled "Hijacking the Civil Rights Bus—Homosexuality and the Scriptures," in which it is asserted that "the homosexual agenda is extending its tentacles throughout the United States' culture" as homosexual activists use "similar-sounding words" and appeal to "similar heartfelt emotions" in order "to draw parallels" between the civil rights and gay rights movements.[68]

Second, while Ken Ham and AiG immediately launched a series of attacks on the June 26, 2013, Supreme Court decision that struck down the Defense of Marriage Act, they had nothing to say regarding the Supreme Court decision that came down one day before, a decision which, to quote the *New York Times,* "effectively struck down the heart of the Voting Rights Act." This striking contrast was weirdly repeated in the summer of 2015: while Ken Ham and other AiG contributors published a raft of angry articles in response to the June 26, 2015, Supreme Court decision legalizing same-sex marriage, they were (as far as we can tell) silent as regards the Confederate flag controversy that erupted in the wake of the June 17, 2015, massacre of nine African American churchgoers in Charleston, South Carolina.[69]

Then there is the matter of "Ebenezer the Allosaur," the dinosaur fossil that the Creation Museum received as a gift from the Peroutka Foundation in 2013 and put on display in May 2014. According to an AiG press release, at the dedication of the Allosaur exhibit Ham thanked "the Peroutka Foundation, which wants to use this great fossil in a God-honoring way"; Michael Peroutka pronounced that "Ebenezer" is "a testimony to the creative power of God in designing dinosaurs" and "lends evidence to the truth of a worldwide catastrophic flooding of the earth in Noah's time." The press release did not mention that Peroutka—the 2004 Constitution Party candidate for president—is a Christian Right activist who has repeatedly called for the impeachment of Barack Obama; suggested that the Maryland General Assembly is "no longer a valid legislative body" because it has "tried to redefine 'marriage,' tried to restrict the right of the people to keep and bear arms . . . [and] declared that little girls must share bathrooms with older men who are 'gender confused'"; and asserted that to "enslave a Christian people . . . you must take [children] away from their parents and every day, day by day, indoctrinate them to reject and forget the Christian ideas and habits of their fathers and mothers, their grandfathers and their grandmothers. **This is precisely what government schools were designed to do. This is what they have done and continue to do**" (emphasis in original).[70]

All of this is standard fare in AiG publications. So is Peroutka's assertion in a 2014 article published on Martin Luther King Day that "all elevation or denigration of individuals or groups based on skin color is immoral and shameful." But just a few months earlier Peroutka had joined the Board of Directors of the League of the South. According to its president, Michael Hill, the League of the South stands for the "survival, well-being, and independence of the Southern people. And when we say 'the Southern people' we mean white Southerners."[71]

It is hard to imagine better confirmation of the point that the Right's "color-blind project" involves "amnesia as a conscious strategy." In a ten-minute televised presentation on Peroutka's gift of Ebenezer to the Creation Museum (and the Ark Encounter) MSNBC's Rachel Maddow (with clips of Peroutka speaking in front of Confederate flags) described Peroutka as "a neo-Confederate" who was now "running for his local county council in Maryland as a Republican . . . on a quasi-secessionist platform." In Ham's response, entitled "Rachel's Rant," he blasts Maddow's emphasis on the tax incentives used to help finance the Ark Encounter; as regards Peroutka as a neo-Confederate (not to mention Peroutka's connection with the League of the South) Ham had nothing to say, except to charge that Maddow, who is clearly "angry at God," used "a sleazy tabloid approach in her attempts to bring disrepute to creationists."[72]

There is nothing surprising about Ham's and AiG's silence about their donor's neo-Confederate commitments, just as there is nothing surprising about their silence about the 2013 Supreme Court decision gutting the Voting Rights Act, their silence about the Confederate flag controversy, and their silence about the persistence of institutional racism in the United States. Whatever Ham and company might say about "One Race—'One Blood,'" at the end of the day the AiG enterprise (including the Creation Museum) is best described as a Christian Right arsenal in the culture war. The museum's pervasive amnesia is invaluable in helping to create and sustain that culture war, invaluable in establishing the Christian Right's moral superiority in that war. Once history—be it geological history, the history of biblical interpretation, or the history of American racism—is erased, then lines of cultural battle are easily drawn. These lines can be fixed anywhere one likes, established as absolute, and made permanent. Indeed, once the messiness of history has been erased, once changes in "literal" interpretations have been forgotten, once the various ways in which "biblical Christians" have used the Word to line up with forces of oppression have been elided, the binary locks into place. One can know who the true Americans are, and who they are not. One can know who the true Christians are, and who they are not. One can know who is holding to the True Word of God, and who is not. One can know who is saved, and who is damned.

This culture war is a war for keeps, and the outcome is not in doubt. Ken Ham and AiG make clear again and again that they are on the right side of history. They are already safe and snug on the Ark. And the door is shutting. Slam. "Let me do that again." Slam.

5

JUDGMENT

When visitors to the Creation Museum complete the Bible Walkthrough Experience, they enter a small room with walls that look like the stone and brick walls of an ancient street. Upon the walls are hanging what the museum designates as the final three Cs of history: "Christ: The Promise of God's Word," "Cross: The Answer of God's Word," and "Consummation: The Fulfillment of God's Word." This room is the waiting area for admission into the Last Adam Theater.[1] Every twenty minutes, museum employees open the doors leading into a small auditorium with three large screens up front. Once everyone is seated, a staff member goes to the front and briefly introduces the film, sometimes pointing out that the movie includes the final three Cs. Then the lights are darkened, and it is show time.

The Last Adam,[2] which was shot at the Holy Land Experience theme park in Orlando, and (as with *Men in White*) has high production values, opens with a crackling campfire. A paleontologist sits on a nearby rock, holding a fossil and a Bible. Speaking gently, he observes that the Bible—"written by someone who was actually there"—not only describes the creation, but also explains why humans must endure "pain, suffering, and death." As the paleontologist narrates the story of original sin, viewers see drawings of Adam and Eve and Cain and Abel, as well as snippets of biblical text. Actors playing Abraham and Moses appear next, along with more biblical text; then there are shots of a small white lamb, a priest, a knife, and lots of blood on the lamb and the priest's gown. Interspersed are scenes from the campfire, with the paleontologist explaining that these sacrifices symbolized the Messiah, "the ultimate and perfect sacrifice" who "was born of a young virgin named

Mary who already knew something about sacrifice." The film shifts to the adult Mary sitting by a fountain and then—after a fade to black—an ancient village where a young Mary watches a man (presumably her father) give a small white lamb to the priest. The adult Mary narrates how her entire family watched the annual killings—"My parents insisted that all of us were there. They wanted to make sure we each understood how terrible sin is and just how much it costs to cover it"—while there are images of the young Mary petting the lamb's head and close-up images of bloody hands and bloody knife. Then the adult Mary narrates the story of the angel who, even though she was a virgin, told her she was going to give birth to the Son of God, who "would be called something else as well—a lamb." The scene cuts to an image of a bloody lamb, whose eyes are open and looking into the camera.

Back at the campfire, the paleontologist talks about how "Jesus' life was everything the prophets foretold." Viewers watch John welcoming Jesus to be baptized, which is followed by very brief scenes of Jesus healing and teaching nicely dressed people. Fade to black, the paleontologist abruptly shifts the narrative, and the audience now sees brightly uniformed Roman soldiers taunting and torturing Jesus. Fade to flames (in metal bowls outside a Roman structure) and then a centurion (a tough guy, with a scarred face) walks onto the scene to tell his story. On "assignment in Jerusalem," he was given the routine task of assisting in the crucifixion—a method of execution "not meant to be pretty"—of "two thieves and one religious rebel," the latter referred to as "The King of the Jews." As he speaks, there are shots interspersed of Jesus being flogged, soldiers forcing him onto the cross, and soldiers hammering nails into his hands. There is much blood and many slow-motion shots. The centurion continues his narration, reporting that while Jesus was hanging on the cross he said "Father, forgive them"—leading the centurion to look "around for his dad"—and then yelled out, "Teleo!" explained by the centurion as "it is finished, like a, like a receipt stamp, you know, paid in full." Jesus dies. There are more images of Jesus hanging on the cross, a close-up of Jesus's left hand nailed to the cross, and people who appear to be crying, as well as sounds of thunder rumbling and lightning cracking. The centurion ends his story: "I am a soldier of Rome. I'm not afraid of anything. But . . . I've never seen anything like it. This man was—truly he was—the Son of God." The film cuts to a black-and-white image of the centurion's face, looking up at the cross with tears in his eyes.

Fade to black and then a Bible verse from Revelation appears in white on a parchment background: "And whosoever was not found written in the book of life was cast into the lake of fire." Then it is back at the campfire with the paleontologist, who informs the audience that we are all sinners. Fade to black and then this verse from Romans appears: "The wages of sin is death; but the gift of God is eternal life through Jesus Christ our Lord." The film goes back to the paleontologist at the campfire, who—bolstered by a series of additional Bible verses that appear on the screen—concludes the film by informing viewers that while "the first Adam polluted" God's creation, the "last Adam"—who rose from the dead—gives us the opportunity to start over, spotless, forgiven, loved. If we believe in the Lord Jesus Christ, we shall be saved.

Thus *The Last Adam* comes to an end. The theater lights come back on. A museum employee goes to the front and expresses his or her desire that this salvation message is what visitors take away from the film while also inviting audience members to pick up one of the "Gospel booklets" that are available just outside the theater. While it is impossible to assess how effective this fifteen-minute film is as a conversion tool, in his blog, Ken Ham occasionally mentions success stories (usually involving children from Christian families). There was, for example, Esther, who "asked to give her life to Christ after seeing *The Last Adam*," understanding—as her mother said—"about the lamb depicted in the film and that Jesus became a sacrifice for us." There was also ten-year-old Kyle, who—according to a letter sent by his father to Ken Ham—"had prayed to receive Christ when he was four years old," but who realized after watching *The Last Adam* that he was not really saved. As his father reported, "we all had a seat on a bench in the museum, and in a simple child-like way Kyle asked Jesus to be Lord of his life."[3]

Michael Matthews, the Creation Museum's content manager, has stated that the goal was to place *The Last Adam* "in the context of everything we said in the museum, about science, about the purpose of our life, about history as given in the Bible."[4] While this might seem an odd claim—the content of the film seems removed from much of the content of the museum (to give one obvious example, the film makes no reference to Noah or to a global flood)—upon closer inspection it is clear that *The Last Adam* and the Creation Museum are of a piece. For one thing, the museum's relentless emphasis on binaries is visually replicated in the film's dramatically contrasting scenes. On the one hand, there are scenes that are constructed in very bright, positive ways: the peaceful campfire scenes, framed in a warm glow with the narrator's soothing voice; the scene with Mary, with very clean clothes, pretty face, and beautiful flowers; the scenes about Jesus's ministry, with their bright sunshine, clean clothes and hands, and very happy responses to healing. On the other hand, there are scenes that are very dark and violent, particularly the torture and crucifixion of Jesus: slow-motion shots of a soldier driving the nail into Jesus's hand, a blood-orange sky, a black cross, knives and hands dripping with blood, blood all over a white garment, slow-motion shots of Jesus in pain on the cross (all of which is very much in keeping with Mel Gibson's *Passion of the Christ*).

While there is more about Jesus in *The Last Adam* than in the rest of the museum, it is very much in keeping with the museum that—in a segment that comes immediately after Jesus's baptism—only thirty-two seconds are devoted to Jesus's ministry and teachings. (In contrast, three minutes and forty-five seconds are given to his flogging and execution.) What precisely do audience members learn in these thirty-two seconds? They learn that "during his ministry, Jesus healed the sick, restored sight to the blind, preached good news to the poor, and told the people that the Kingdom of God was at hand." There is no elaboration as to what "good news" was given to those in poverty; perhaps viewers are to infer that Jesus informed the poor (or perhaps friends of the poor, as in the film the people to whom Jesus gives the news seem quite prosperous) that while they will suffer on Earth, they could eventually end up in Heaven. There is also no explanation as to what Jesus meant by the "Kingdom of God"; again, perhaps viewers are to infer that he is referring to the afterlife. In short, in their brief glimpse of his ministry, viewers of *The Last Adam* learn that Jesus performed miracles but apparently had nothing to teach us about how we should live our lives.

It is also noteworthy that the film devotes significant time—one minute, thirty-five seconds—to an extrabiblical story about the youthful Mary and her family viewing the annual sacrifice of a lamb. Given the commitment to the inerrant Word of God, it might seem strange to forego all the available material on the life of Jesus that is contained in Matthew, Mark, Luke, and John for a story that does not actually appear in the Bible. But the producers have an argument to drive home. Through the eyes of Mary, viewers are instructed in the sacrificial system. While the slaughter of the little lambs "always broke [her] heart," Mary came to accept the fact—from her repeated viewings of these killings—that God required perfect, innocent, and bloody sacrifices as the payment for sin. Properly instructed, Mary was then prepared to see her own son brutally and bloodily tortured and murdered as the necessary, "ultimate, and perfect sacrifice for the sin of the world."[5]

If we believe in Jesus, the perfect sacrifice, our sins are atoned, and we are saved. But for those who do not get on board, judgment awaits. In his discussion of *The Last Adam*, content manager Michael Matthews noted that one of the challenges for the filmmakers was to "bring up hell in a tasteful way so nobody even mentions it." As Matthews explained, this is why the verse from Revelation—"And whosoever was not found written in the book of life was cast into the lake of fire"—appears on the screen immediately after the crucifixion, with no narration.[6] What Matthews did not mention is that while viewers read but do not hear these words of damnation, they do hear the crackling flames (before the film shifts back to the campfire). In fact, viewers see flames and hear crackling fire throughout the film, including when the paleontologist talks about how some people see the Bible as an old relic, when the centurion makes his appearance, and when the narrator reads Romans 5:12 ("By one man sin entered into the world, and death by sin, and so death passed upon all men, for that all have sinned"). The metaphor of fire is impossible to miss in *The Last Adam*. The "tasteful" threat of hell undergirds the entire film. The flames are crackling. Judgment is coming.

Arsenal for the Culture War

After museumgoers have stopped in at the Museum of the Bible exhibit, after they have strolled through Buddy Davis's Dino Den and examined the dead bugs in Dr. Crawley's Insectorium, they pass through a hallway, climb a set of stairs, and enter the Dragon Hall Bookstore. While readers uninitiated in young Earth creationist apologetics might be baffled by the bookstore's name, museumgoers should have no trouble making the connection. When visitors first enter the museum they immediately encounter the Dragon Legends exhibit, a series of brightly illustrated displays devoted to dragons, including one signboard devoted to such legendary creatures as "the Lernaean Hydra" of Greek mythology, the "Red Dragon of Wales," and "the feathered serpent Quetzalcoatl" worshiped by the Aztecs. More striking is the signboard devoted to dragons described by John of Damascus ("goldish eyes and horny protuberances on the back of the head" and "a beard [protruding] out of the throat") and Marco Polo ("great serpents" who "have two forelegs near the head, but for foot nothing but a claw," and whose "mouth is large enough to swallow a man whole, and is garnished with great [pointed] teeth"). According to the placard, "The animals described here are believable, and these men intended to relay information about what they had discovered. Did these men actually see or hear about real dragons—creatures that are today called dinosaurs?" While these particular reports come from the eighth and thirteenth centuries, respectively, the museum's dragon legends exhibit also includes a display highlighting an article from the April 26, 1890, *Tombstone* [AZ] *Epitaph*, in which two ranchers reported shooting and wounding a "winged monster resembling a huge alligator with an extremely elongated tail and an immense pair of wings." Again, the question is posed: "Is it possible that these cowboys encountered the legendary Thunderbird or Piasa of Native American lore, a creature whose descriptions remind one of a pterosaur?"

This exhibit is devoted to making the remarkably creative suggestion that dragon legends provide evidence that human beings—perhaps even as recently as the late nineteenth century—have encountered living dinosaurs. Of course, it is important to note the careful use of the interrogative in these placards—"did these men actually see . . . ?" "is it possible that these cowboys encountered . . . ?" The same tack is taken on the "Were Dinosaurs Dragons?" signboard. Here museumgoers learn that while some dragon legends are somewhat or completely mythical, in many other instances "the dragons were viewed as real animals," thus suggesting the real possibility that "some of these dragon legends actually speak of dinosaurs." But as with the flood geology "models," the point of these rhetorical questions is simply to advance what visitors are to understand as a plausible hypothesis. Anything akin to substantive evidence is unnecessary, given that this "plausible" hypothesis— dragon legends are evidence that human beings have encountered living dinosaurs—is confirmed by the Bible. As explained on the "Were Dinosaurs Dragons?" signboard, "In the Bible, God told Job about two creatures that could be considered dragons: the Behemoth and the Leviathan."[7] While "some Christians suggest these are mythical monsters," this argument does not make sense, given that the reference to these two creatures "follows the descriptions of about a dozen real animals." More important: "Why would God tell Job to consider two beasts that did not even exist." In short, a plain reading of the Bible clearly establishes that humans and dinosaurs inhabited the Earth simultaneously.

The Dragon Legends exhibit devotes a large display to St. George.

This is not surprising given that, according to the placard, this "devout Christian and Roman military officer" saved a princess and a city from a dragon, only asking in return that the city dwellers "believe in the God who strengthened my hand to gain this victory." There is also a large display devoted to *Beowulf*, the epic poem written in Old English that, as noted on the signboard, "is named for its hero and tells of his mighty deeds in sixth-century Scandinavia." *Beowulf* does much work for young Earth creationists: as literary scholar Eve Siebert has observed, they have transformed this poem (written sometime between the eighth and tenth centuries) into a historical work because of its references to Noah and the Flood, and because the "monsters" that Beowulf fights can be construed as dinosaurs. Siebert notes that the most complete creationist interpretation of *Beowulf* is to be found in Bill Cooper's *After the Flood: The Early Post-Flood History of Europe Traced Back to Noah*. Cooper claims *Beowulf* was written prior to the Anglo-Saxon arrival in Britain from Scandinavia in the fifth century; while Siebert gives a number of reasons that this argument is ludicrous (including the fact that *Beowulf* was written in Old English, a language that was not developed until "after various tribes had migrated [to England] from their continental homelands"), she explains that "Cooper needs to push the date of *Beowulf* as far as he can into the past," in part "so he can argue that it is a poetic but basically historical account of real people, real events, and real animals."[8] While the Dragon Legends exhibit is silent as to when *Beowulf* was written, the "Dragon Legends around the World" map places *Beowulf* in Scandinavia, and the *Beowulf* display states that the poem "contains accurate historical information" regarding sixth-century Scandinavia. In classic Creation Museum fashion, the placard concludes with the interrogative: "Did these men or their ancestors actually fight dinosaurs and pterosaurs? This idea would be consistent with the Bible. But those who believe dinosaurs lived millions of years before man cannot adequately explain why cultures around the world have dragon legends whose creatures often match descriptions of dinosaurs."

Both St. George and Beowulf are featured prominently in the medieval-themed Dragon Hall Bookstore, St. George with a midrelief sculpture in which he is in the midst of killing the dragon, Beowulf brandishing a sword in the form of a freestanding sculpture over the store entrance. These warrior heroes reinforce the bookstore's function as a fully stocked Christian Right arsenal for the culture war. Below is

Visitors consider the offerings in the medieval-castle-themed bookstore located just off the Main Hall as a statue of Beowolf, whom the Creation Museum suggests battled dinosaurs, stands above them on the back wall. Photograph taken by Susan Trollinger. For more images of and information about the museum, go to creationmuseum.org.

a tiny but representative sample of what was for sale (besides the assortment of Creation Museum T-shirts, bumper stickers, toy dinosaurs, and the like) on a July 16, 2013, visit: Bibles and Bible study tools, including *The King James Version Bible, The* [John] *MacArthur Study Bible,* and (of course) *The Henry Morris Study Bible;* books written and edited by Ken Ham and various collaborators, including *Demolishing Supposed Bible Contradictions; How Do We Know the Bible is True?; The Lie: Evolution/Millions of Years; One Race One Blood; Raising Godly Children in an Ungodly World; The True Account of Adam and Eve; War of the Worldviews; Why Won't They Listen: The Power of Creation Evangelism;* books by Henry Morris, including *The Biblical Basis for Modern Science, Biblical Creationism, The Genesis Flood, The Long War against God: The History and Impact of the Creation/ Evolution Conflict;* books by Christian Right authors, including Jerry Bergman, *Slaughter of the Dissidents: The Shocking Truth about Killing the Careers of Darwin Doubters;* Tim LaHaye, *Faith of Our Founding Fathers: A Comprehensive Study of America's Christian Foundations;* John MacArthur, *Right Thinking in a World Gone Wrong: A Biblical Response to Today's Most Controversial Issues;* R. Albert Mohler, *The Disappearance of God: Dangerous Beliefs in the New Spiritual Openness;* John Piper and Wayne Grudem, editors, *Recovering Biblical Manhood and Womanhood: A Response to Evangelical Feminism;* Christian Right DVDs, including *Erosion of Christian America: State of the Nation 2 with Ken Ham; Global Warming: A Scientific and Biblical Expose' of Climate Change; Godless in America: State of the Nation with Ken Ham; IndoctriNation: Public Schools and the Decline of Christianity in America.*

Also for sale is a large selection of curricular material for homeschooling families, Christian schools, churches, small groups, and the like. For example, there is *America from the Beginning* ("This new 'hands-on' American history curriculum . . . starts at creation" and makes obvious "God's grace toward the United States") as well as *History Revealed* ("Unlike the 'politically correct,' evolution influenced public school curriculum, this biblically-based curricula shows students the real history of the world!") There is *Chemistry 101/Biology 101*, which are designed both for homeschooling and as supplements to public school curriculum (Biology 101 is "excellent for teens taking or about to enter a science class where the Bible's clear explanation of how life began, is ignored"); there is also *God's Design for Life* (these "textbooks help you teach science from a biblical, creationist perspective . . . [and] help strengthen your student's faith by showing how science consistently supports the Bible's written record"). Then, of course, there is *Ken Ham's Foundations Curriculum,* a "top-quality DVD-based curriculum" designed for Sunday School classes and Bible studies, which offers "inspiring (and faith strengthening!) teaching on Genesis and biblical authority" and discloses "the impact of important world events," such as President Obama's public declaration "that America is no longer a Christian nation.[9]

Also for sale in the Dragon Hall Bookstore is Sebastian Adams's *SynChronological Chart or Map of History,* a chronological chart that stretches twenty-five feet and chronicles "Ancient, Modern, and Biblical History" from Adam and Eve's creation in 4004 BC to 1870 (the year before the chart's publication.) Reissued in 2007 by Master Books, it is sold at the museum not as a lavishly illustrated nineteenth-century oddity, but—according to the AiG website—as a "monumental work of history . . . based on James Ussher's *The Annals of the World*" that "is perfect for homeschool settings" and "Sunday School walls." The AiG website even includes a video trailer for the *SynChronological Chart,* where viewers are informed that this book has been "reproduced in [its] Original, Evolution-Free Large Format," and where a "Mr. Bob L. Warner" is quoted as describing it as "TRUTH IN HISTORY! THE SPIN STOPS HERE" (emphases in original).[10]

What can be purchased in the Dragon Hall Bookstore is but a fraction of what can be purchased online at the AiG store. There one finds a plethora of young Earth creationist and Christian Right material for sale. One can even purchase enrollment for Answers Education online courses, including Creation Apologetics and Biology, Creation Apologetics and Geology, Creation Apologetics and Astronomy, and Foundations of Creation Apologetics; students who take the latter course can, as of 2014, receive three college credits through God's Bible School and College (Cincinnati).[11]

Given the enthusiasm for producing and selling young Earth creationist materials, it might be tempting to conclude that AiG is first and foremost a money-making enterprise. It makes sense, however, to heed Ken Ham when he stresses that the organization has a much larger purpose. As Ham sees it, the situation in America is dire: "The spiritual war . . . is raging around us more than ever," and "the battle lines are in our homes, churches . . . schools (whether public or private), workplaces, courts—in fact, *everywhere* in the culture" [emphasis in original]. AiG headquarters is responding by producing some of the most "advanced 'weaponry' designed to counter the enemy's attacks in this era (e.g., books, DVDs, curricula, websites, the Creation Museum, the coming Ark Encounter, radio programs, and so on)." Fortunately for beleaguered Bible-believers, these "Christian 'patriot missiles'" are having an effect. As Ham noted in June 2014, AiG is playing a major role in ensuring that

> there is a considerable creationist influence in this nation . . . Tens of millions of people this year will access the AiG website; well over 2 million people (including hundreds of thousands of kids) have visited the Creation Museum; millions of pieces of biblical creationist literature have been distributed by AiG throughout the culture; thousands of presentations by AiG speakers have been given in churches, colleges, conferences, etc.[12]

These numbers, inexact as they are, should be sobering news for those who fantasize that young Earth creationism will soon be following the dinosaurs into extinction. The reference to AiG presentations is particularly significant. As has been the case with fundamentalism from its very beginning, AiG has devoted much attention to spreading its message at the grassroots level. Randall Stephens and Karl Giberson have observed that Ken Ham's dream is to "convince millions of ordinary people to reject evolution in the hopes that this grassroots groundswell would change society and work its way up, or at least marginalize the apostate eggheads at the top."[13] As of August 2014, the AiG Outreach Department had twenty-five speakers (plus seven in the United Kingdom) available to make presentations. And they are busy. According to the AiG website, in one four-month stretch (August–November 2014) Answers Outreach speakers presented at forty-eight separate events. Five of these involved many speakers; four of these were at the museum, including Creation College, Children's Ministry Conference, Answers for Pastors, and College Expo, and one was held in West Bromwich, England (UK Creation Mega Conference). Thirty-nine of the remaining forty-three conferences were held at local churches, the majority of which were led by one person from the AiG Outreach Department (but some included two or more AiG speakers). Baptist churches hosted almost half of these conferences, with nondenominational churches hosting most of the rest (although there were two meetings at Assemblies of God churches and one apiece at a Church of Christ, an Evangelical Covenant church, and a Free Methodist church). These church conferences took place from coast to coast, from Washington and Idaho in the West to Maine, New York, and New Jersey in the East. Not surprisingly, the vast majority were in the Midwest and the South, with nine in Ohio.

To get a feel for these local outreach efforts we attended the Saturday evening sessions of the July 26–27, 2014, AiG conference at Cornerstone Community Church, located in the rural countryside outside Millersburg, Ohio.[14] Approximately seventy people were in attendance. The sessions were held in the church sanctuary, a rectangular carpeted room with a cathedral ceiling and windows facing the parking lot along one side. In this multipurpose room (with a basketball hoop on the back wall) there was a platform in the front with a podium, a piano, an American and a Christian flag, and a few artificial potted plants. To the left there was a large projection screen. The attendees sat in blue upholstered chairs in neat rows facing the platform. Tables loaded with

AiG materials were lined up along the front exterior wall. Cornerstone's pastor, Pat Weaver, served as the event's emcee, and his wife, Sue, played the piano during intermission; both were quite friendly, and they each made a point of greeting us.

After welcoming the attendees and telling them that he had "long been interested in the cosmic battle between Truth and Error," Pastor Weaver turned the evening over to Bob Gillespie, a trim, bespectacled, middle-aged man, appropriately wearing a black oxford shirt with an embroidered Creation Museum logo. A Cedarville University (OH) and Baptist Bible Seminary (PA) graduate, he and his wife, Lois, previously taught in Christian schools in Ohio and in schools for missionary children in Africa. According to his AiG Outreach biography: "Bob's goal today is to teach the teachers (i.e., parents, Sunday school teachers, youth directors, and pastors) the same things he has taught his students in schools for many years: to uphold the authority of God's Word from verse one, to proclaim the gospel of Jesus Christ, and to confront the forces of secular humanism that are taking over our culture."[15]

At the podium, Gillespie immediately launched into his first presentation, Dinosaurs and the Bible, by asking how many in the audience had been to the museum; despite the fact that Millersburg is approximately three and a half hours driving time from Petersburg, Kentucky, about two-thirds of the audience raised their hands. Gillespie did not thank the audience for their support, but instead admonished them not to rush past the Starting Points room when they visit, for if they do they will "miss the whole point of the Museum," which is that there are only two perspectives, and they both can't be right. Explaining why some people do not believe in God, Gillespie argued that "once there is a God, who makes the rules? God. That's why they don't want to believe."

Then it was on to dinosaurs.[16] While they are being used by evolutionists to indoctrinate very young children, Gillespie asserted that "if we start with the Word of God and the seven Cs of history, you then get the true history." While people make fun of the idea that Adam and Eve could have ridden dinosaurs, it really could have happened, given that, before the Fall, all animals were vegetarians. Moreover, Gillespie explained, while people say that two of every species could not have fit on the Ark, it really was only two of every "kind," which means there were but two thousand plus animals on the Ark. Clearly, that was not a problem for the Ark. While people say that Noah could not have fit dinosaurs onto the Ark, Gillespie pointed out that there were really only fifty "kinds," the average dinosaur was the size of a sheep, and—regarding the big dinosaurs—young adult dinosaurs could have been put on just before their growth spurt began. It all makes sense, assured Gillespie, once "we put our biblical glasses on."

As regards humans walking the Earth with dinosaurs, Gillespie referred to evidence from the Bible (references to "Behemoth" and to "dragons"), dragon stories (St. George and the Dragon, Marco Polo's stories), and pictographs of dinosaur-like creatures drawn by Native Americans. As Gillespie raced toward his conclusion, he asked, "What happened to the dinosaurs?" He commented that the evolutionists have no clue: maybe the dinosaurs had indigestion, maybe they ate too much, maybe the dinosaurs ate too little, maybe a meteor hit the Earth. But Gillespie knew what happened: "They died. Animals go extinct all the time." After another appeal for the audience to use their "biblical glasses," Gillespie pronounced that "we are taking the dinosaurs back— they are our 'missionary lizards'" that help make clear that Genesis and the whole Bible is true.

After a brief intermission, during which Lois Gillespie staffed the AiG table, Bob Gillespie began the second session, Science Confirms the Bible, by noting that in school students are taught that religion is belief ("we just believe Jesus on blind faith"), and science is fact. However, "God is big on evidence, and the Word of God is an eyewitness account." He then went on to distinguish between observational and historical science, the latter just a guess about the past unless you have a reliable source or eyewitness account. Again he addressed the question of why some people refuse to believe in God: "It's not an evidence problem; it's a heart problem." Gillespie pointed out that as the Bible is the true history book of the universe, the evidence ought to conform to it, and it does. Observational science will always catch up with the Bible: it happened with the fossilized tooth used at the Scopes trial (which turned out to be from a pig, not an ape-creature), it happened with so-called junk DNA (which has turned out not to be junk at all), and it will happen with the "distant starlight" problem (which at the moment suggests a very old universe).

Then it was on to what he referred to as the various types of evolution, the goal being to assess which ones were "just belief" and which ones were facts established by observational science. Into the former category Gillespie placed "cosmic evolution" (big bang), "chemical evolution" (origins of chemical elements), "stellar planetary evolution" (gas particles congealed to become stars), and "organic evolution" (life was

formed "in the primordial soup"); regarding the latter, Gillespie pronounced that this could not have happened by chance: "I think we could
stop right now and have a praise and worship time for our great Creator!" Then it was onto "macroevolution" ("cells to people"), which is impossible because this would require the addition of "new information."
Here Gillespie showed a clip from an *Ancient Aliens* episode, which
suggests humans got their intelligence from aliens,[17] after which he exclaimed, "How about we go with the infinite God instead?" Finally, there
is "microevolution," which is observational science since we can observe
changes within animal "kinds" in the present. In short, biblical creation
and evolution are both beliefs, but "observations conform with the Bible
and contradict evolution." Gillespie then showed a video that, according
to AiG, "refute[s] evolution in less than three minutes."[18] Gillespie concluded by noting that "it is no trouble to believe the Book if you believe
the author."

Bob Gillespie was remarkably good at staying "on message," so good
that it was hard to keep in mind that we were not actually at the Creation Museum. The importance of starting points ("put on your biblical
glasses"), observational versus historical science, the seven Cs of history,
and the Bible as eyewitness account are all favorite AiG tropes. Ditto
for vegetarianism before the Fall, the emphasis on "kinds," and the use
of dragon legends as evidence that humans walked the Earth with dinosaurs. More than this, Gillespie's often mocking tone and his breezy
confidence that he was presenting Truth that quickly and completely
demolishes evolution were eerily reminiscent of *Men in White*, wherein
smart-alecky Gabe and Mike at the back of the classroom intellectually
annihilate poor Mr. Snodgrass. In both cases, the posture taken is that
young Earth creationism is so obvious that the only ones who do not
see the Truth are evolutionists, and that is because they do not want to
follow God's rules. Finally, Gillespie's rapid-fire delivery, the inclusion
of peculiar video clips (e.g., the segment from *Ancient Aliens*, a show
described by one reviewer as "some of the most noxious sludge in television's bottomless chum bucket"),[19] and the overabundance of slides,
few of which were explained in any detail, were quite reminiscent of the
aural and visual overload one experiences at the Creation Museum.

The latter point is highlighted by a slide Gillespie showed in the
"Science Confirms the Bible" session after making the point that observational science will someday provide the answer to the "distant starlight" problem. Entitled "*Hundreds* of Physical Processes Set Limits on

the Age of the World" (emphasis in original), the slide contained three columns listing not hundreds of processes, but fifty-six. The individual processes were in small type, and the slide was only on the screen for ten to fifteen seconds. So we were only able to write down the slide's title and note how the slide was formatted, which included a jagged line underneath the columns, as if the page had been ripped and there was a missing fragment somewhere that contains one hundred and fifty or more additional age-limiting processes. But we located an identical slide, complete with the "rip," on a presentation entitled "Evidence for a Young World," created by Heinz Lycklama and found on the website of the NW Creation Network (which is devoted to "defending Biblical history.") On the slide, Lycklama—who "lectures regularly with PowerPoint presentations that will help the believer defend the Bible's teaching on Creation"—credits "Russell Humphreys, Ph.D." as his source. Humphreys is a physicist who worked for the Institute for Creation Research and then Creation Ministries International; among his publications is the article, "Evidence for a Young World," in which he identifies fourteen specific evidences of a young Earth while also noting that "much more young-world evidence exists." Humphreys is also an advocate for "galactocentricity," the notion that our galaxy is at the center of the universe. Interestingly, while AiG representatives attack the notion of geocentricity, the organization apparently has a much more favorable view of galactocentricity. Not only is Humphrey's article "Our Galaxy Is the Centre of the Universe" on the AiG website, but the argument for galactocentricity is featured in the AiG publication *Evolution Exposed: Earth Science*.[20]

On the *"Hundreds* of Physical Processes Set Limits on the Age of the World" slide we find the following: "16. Tight bends in rocks," "23. Human civilizations," "32. Interplanetary dust removal," "41. Peat bog growth," "50. Niagra [*sic*] Falls erosion," "51. Stone age burials," and "54. Squashed radiohalos." Even if attendees had adequate eyesight to read this chart, and even if Gillespie had given them time to read the chart, it is inconceivable they could have made any sense of this "evidence." More than this, it is impossible to imagine what Gillespie would have said if anyone had pressed him to explain one or more of these fifty-six allegedly age-limiting processes. Of course, this was not going to happen. There was no time allotted for questions in either session, and Gillespie gave absolutely no indication that questions were welcome. In this regard too his presentation was similar to what one experiences at the Creation Museum. It is, after all, Answers in Genesis, not Questions about Genesis.

In fact, we were struck by how little effort Gillespie made to establish rapport with the audience. There was no comment at the beginning about how happy he was to be in Millersburg, no statement about how beautiful the setting was (and it is indeed beautiful), even no statement of appreciation for Pat and Sue Weaver, his church hosts who apparently had primary responsibility for publicizing the event locally. But in Gillespie's defense, that he did not engage in such rhetorical niceties or invite questions or explain his slides in any detail seemed quite irrelevant. The attendees were remarkably focused on Gillespie, with many nods of agreement and occasional verbal affirmations as he was speaking. Some members of the audience worked very hard to convey that they were with him. For example, in the row in front of us a woman we took to be the mother of the children sitting with her strenuously prompted them to shout out the right answer when Gillespie asked a question, such as, "What day of creation were dinosaurs made?" By the end of the second session we were struck by the overwhelming sense that the folks in the audience very much wanted to signify to the AiG spokesperson that they too are good Christians, that they too hold to the Word of God, that they too understand evolutionists to be the enemy.

Purifying the Church

While Bob Gillespie emphasized the battle between evolutionists and Christians, there were a couple of moments at Millersburg when he let it slip that many Christians, perhaps 50 percent, actually hold "to millions of years and evolution." Such division in the Christian ranks not only renders problematic the oft-repeated cultural binary, but it also poses challenges for AiG in its war against atheistic evolutionism. Not surprisingly, Ken Ham understands very well the gravity of the situation: "Yes, there is a culture war happening in our nation, but there is a much greater battle that needs to be fought—a battle to call the church back to an uncompromising position on the authority of the Word of God." In his 2013 book devoted to this topic, *Six Days: The Age of the Earth and the Decline of the Church*, Ham laments the fact that so many Christians "want to be unified around man's beliefs in evolution and/or millions of years—instead of being unified on what God's Word teaches so clearly in Genesis." These Christians think that it is not necessary or even important to teach Genesis 1–11 as literal history, not "comprehend[ing] that every major doctrine, like marriage, sin, why we wear clothes,[21] and so on, is founded ultimately in the Book of Genesis." While "salvation is not conditional upon what you believe about the age of the earth," the moment the Church began allowing the possibility of "millions of years" a "door was unlocked . . . that said we do not have to take God's Word as written from the beginning." The result is "a mass exodus of young people from our churches," as youth "begin to doubt the Bible in Genesis, based on the compromised teachings of church leaders . . . , reject it as God's Word, and then leave the Church."[22]

Instead of a church that compromises with the culture, what is needed is a church that stands "unashamedly on the authority of the Word of God." But one big reason that the church does not do so is because many Christians "are not equipped to answer the skeptical questions of our age," and thus "they shrink back or just do not get involved and let the world continue on its destructive Christianity attack course." But Ham reminds readers that this is why "the Lord has raised up ministries like Answers in Genesis . . . to equip believers with answers" in order to equip Christians with "resources so that [they] will be able to respond" to the incessant attacks.[23]

As the Creation Museum makes clear, Ham is not interested in getting people simply to read and trust the Bible. The AiG resources he touts are designed to inculcate Christians with the young Earth creationist interpretation of God's Word, to persuade them that such an interpretation is the One True Understanding of God's Word, and to provide them with the answers that will enable them to rebuff any and all challenges to this One True Understanding of God's Word. That is to say, the AiG apologetics ministry is designed to give people the Truth and make sure they are so fully indoctrinated in the Truth that they do not budge from the Truth. To use the warlike imagery so beloved by Ken Ham, AiG provides not only the "Christian 'patriot missiles'" with which to attack the enemy, but also (and perhaps more importantly) a suit of armor that will ensure complete protection from the enemy, including an extraordinarily thick helmet that will render the mind impervious to dangerous ideas.

Thanks to AiG, parents can take their children to the Creation Museum and watch Christian Right DVDs with them at home; they can homeschool their children with AiG curricula or send them to Christian schools that use the same materials; they can take their children to fundamentalist churches where they hear young Earth creationism preached from the pulpit and where they are taught from AiG curricula in Sunday School. At some point, however, it will be time for the children to leave home and go to college, where it might turn out that their well-polished young Earth creationist armor is inadequate to withstand the myriad of enemies lurking in the halls of higher education.

Here again, however, AiG seeks to be of assistance. Seeking "to help parents and students find a college or university," AiG's jack-of-all-trades, Bodie Hodge, created a list of fifteen questions they could ask an administrator at "a Christian college to see if the institution really believes what the Scriptures say, [to see] whether they adhere to biblical authority or not." Some of the questions include the following: "4. Do you believe the days in Genesis are literal approximate 24-hour days?" (Correct answer: "Yes!"); "5. Is it proper to interpret Genesis to conform to what one already believes about the past?" (Correct answer: "No . . . Question 5 is important because you can determine how the institution thinks—whether biblically or 'humanistically.'") "7. Do you believe there are intelligent alien life forms?" (Correct answer: "No."); "11. Why don't people today believe the Gospel when we boldly preach it?" (Correct answer: "People today have the wrong foundational beliefs (an 'evolutionary/millions of years' foundation) that teaches [sic] the Bible is wrong."); "15. Are there any legitimate contradictions in the Bible?" (Correct answer: "No.")[24]

It is hard to imagine what, say, a twenty-three-year-old admissions representative, even one at a conservative Christian college, would do if handed this list of questions—"intelligent alien life forms?" and all—as a test for ascertaining whether or not his or her institution is adequately biblical and doctrinally safe. But when it comes to assisting prospective college students and their parents, AiG does not limit itself to the Hodge questionnaire. On its website it also provides a list of "Creation Colleges," defined as schools "whose presidents have affirmed in writing their personal agreement with AiG's statement of faith." While AiG includes a disclaimer that such affirmation by a school's president is "no guarantee that all professors/textbooks/courses etc., take the same stand on God's Word including Genesis as they should," the list is still "a start for parents wanting a short list to research." In addition to the list, the Creation Museum hosts an annual "College Expo" to give parents and students the opportunity to meet with representatives from various "Creation Colleges," as well as to attend sessions led by AiG luminaries on topics such as "Creation vs. Evolution: Is Genesis Relevant?," "Dragons and Dinosaurs," and "Is Evolution Compatible with Biblical Christianity?"[25]

As of June 2015 the list of "Creation Colleges," that is, schools where "high regard for the inerrancy of God's Word is dominant over evolutionism and other millions-of-years beliefs," included ten Bible schools, forty-two colleges, and nine seminaries. Most of these schools are unknown to anyone outside the world of fundamentalist higher education. The list includes Arrowhead Bible College (MT), Front Range Bible Institute (CO), Landmark Baptist Theological Seminary (TX), New Tribes Bible Institute (MI and WI), Piedmont International University (NC), and Puritan Reformed Theological Seminary (MI). Perhaps the best known schools are two institutions historically identified with politicized fundamentalism: Liberty University (Lynchburg, VA), a school of more than twelve thousand residential students established and presided over, until his death in 2007, by Moral Majority founder Jerry Falwell and succeeded by Jerry Falwell Jr.; and Bob Jones University (Greenville, SC), which has more than five thousand students, and which actively resisted the Internal Revenue Service's efforts to remove its tax-exempt status because of its policy (abandoned in 2000) against interracial dating, a battle that helped spark the emergence of the Christian Right.[26]

But of all these "Creation Colleges," AiG lavishes the most love and attention on Cedarville University (Cedarville, OH), so much so that one could legitimately think of Cedarville as Answers in Genesis U. An independent Baptist school of more than three thousand students, Cedarville is located just eighty-five miles from the Creation Museum. When the museum opened in May 2007, the school's radio station was there to provide a live broadcast. In the years since, the museum has often printed advertisements for Cedarville on the back of admission tickets, and Cedarville has offered Creation Museum tickets for those who scheduled visits to the campus. Cedarville frequently sponsors events at the museum, including the 2014 Creation 4 Conference, an event that, according to Ken Ham, would give attendees the opportunity to "learn in-depth information from leading creationists on how to refute pseudo-science claims in our culture."[27] Joy White, the wife of Cedarville president Thomas White, spoke at the museum's Answers for Women 2014: Women Living after Eve conference organized by AiG stalwart Georgia Purdom, the 2015 Cedarville Alumna of the Year. Cedarville has hosted various creationist biology and geology conferences, including the first "Conference on Creation Geology" meeting in 2007, which included AiG representatives such as Andrew Snelling.[28]

More significant, in the fall of 2009 Cedarville rolled out its undergraduate major in geology; as an unnamed contributor enthusiastically gushed in *Answers*, "plans to refine the creation model just took a giant leap forward" with the establishment of Cedarville's geology program, the first to be offered by a college that "teaches a young-earth and worldwide flood cataclysm." Leading the program is John Whitmore. Armed with a M.S. in geology from the Institute for Creation Research and a PhD from Loma Linda University in California (a Seventh-day Adventist school), Whitmore has been very tightly connected with AiG, writing articles for *Answers*, contributing to *The New Answers Book 2*, and participating in an April 2014 AiG-sponsored Google chat about catastrophic plate tectonics. He also leads AiG-sponsored Grand Canyon raft trips, "highly recommended" by Ken Ham for "the solid teaching . . . in geology and the Bible" that makes clear how the Grand Canyon confirms "what the Bible teaches about the Flood and its after-effects." More than this, AiG invited Whitmore "to investigate the collection site in Colorado" where the museum's Allosaurus specimen was found. Whitmore took seven of his Cedarville geology students with him. According to the Cedarville press release, "the work in Colorado, which allowed the students to learn more about the rocks where the dinosaur was found, provided a better interpretation of how Ebenezer was buried."[29]

Then there is AiG's impresario, Ken Ham, who has been a significant presence on the Cedarville campus for years. His *Answers . . . with Ken* video series was filmed there in the late 1990s. Until Cedarville's radio network was sold in 2011, it broadcast *Answers with Ken* every weekday afternoon; and his February 2014 debate with Bill Nye was "streamed live" in the school's Dixon Ministry Center. In the classroom, Ham's DVD presentation, *Genesis: The Key to Reclaiming the Culture*, has been required viewing in John Whitmore's "Principles of Earth Science" course.[30]

Most notably, and despite his busy travel schedule, Ham has made a number of in-person appearances on campus, be it for a "Worldview Weekend" conference designed to "teach participants how to understand the world from a biblical context," at a Pastor's Appreciation Day luncheon, or as part of a special two-day AiG conference featuring both Ham and Georgia Purdom. Ham has clearly enjoyed something akin to celebrity status at Cedarville. After his August 31, 2010, chapel presentation to three thousand students and faculty, Ham reported that not only did students buy AiG's entire "truckload of books," which was sent to Cedarville as part of Ham's visit, but "many students lined up to talk with me for quite a long time . . . overwhelm[ing me] with their positive comments." Cedarville's press release regarding Ham's visit included quotes from a few of these students: "It was an honor to have Ken Ham at Cedarville! . . . It was exciting to hear him speak so definitively about the truth of Scripture"; "Hearing him speak in person was a privilege . . . His message needs to be heard around the globe." The latter student also reported that "Ham's work with Answers in Genesis inspired [him] . . . to pursue a degree in geology."[31]

Not all Cedarville faculty welcomed the arrival of AiG's CEO on campus; as one former professor commented, "Whenever Ham showed up and spoke in chapel there were many of us in the Biblical and Theological Studies Department who cringed, shook our heads in embarrassment, and actively pushed back in our classes against his horrible hermeneutics, flawed logic, and culture wars mentality."[32] Of course, active faculty resistance to Ken Ham and his young Earth creationist fundamentalism did not fit with Cedarville as the model "Creation College." Not to put too fine a point on it, to be the sort of school that could legitimately claim the label "Answers in Genesis U" requires constant vigilance to ensure that faculty and administration do not stray from the Truth.

Cedarville's crackdown on perceived doctrinal deviance began in January 2012, when the school's board of trustees adopted "white papers . . . designed to clarify and elaborate" on the school's strongly fundamentalist doctrinal statement, which faculty and staff sign annually as a condition of employment; the white paper on "God and Creation" explicitly asserted "all positions which endorse Darwinian evolution, or which deny the historicity of Adam and Eve are incompatible with the Cedarville University Doctrinal Statement." In August 2012, theology professor Michael Pahl was removed. Questions had been raised about his book on Genesis and Revelation, *The Beginning and the End*. In response, Pahl affirmed the historical Adam and Eve, but for theological reasons and not for reasons of biblical exegesis. Cedarville administrators judged this distinction to constitute a failure "to concur fully with each and every position" of Cedarville's doctrinal statement.[33]

In October 2012 Cedarville's president, William Brown, announced his resignation, to be followed in January 2013 by Carl Ruby's resignation as vice president for student life. While neither man explained his decision, one member of the board of trustees who resigned in response to these decisions noted that both Brown and Ruby "were considered problematic by the faction of trustees fearful of what they perceive as a creeping liberalism." In Ruby's case, this included having too much compassion for those "people struggling with gender identification" (i.e., LGBTQ students). Later in January, and in the wake of a furor over an anti-Romney-for-president editorial penned by two philosophy professors, the board eliminated the philosophy major, the official (but vehemently contested) reason being that it was not "financially sustainable." By June 2013 the board of trustees had been drastically revamped, with the new board now including Paige Patterson, a leader of the fundamentalist takeover of the Southern Baptist Convention in the 1980s. Finally, Thomas White from Southwestern Baptist Theological Seminary, where Patterson serves as president, was appointed as Cedarville's new president.[34]

As is always the case in fundamentalist crackdowns, the doctrinal purification of Cedarville required a purge of insufficiently conservative faculty and staff. As of the summer of 2014, there had been an "exodus" of forty-three administrators, faculty members, and staff members, some of whom were forced out (having signed nondisclosure statements) while others quit and moved on to less hostile professional and religious climes. Add to this the departure of fifteen trustees, many or most of whom departed in response to the fundamentalist crackdown, as well as the summer 2013 removal of an additional twenty-nine staff members in what was explained as a financially driven "reduction-in-force." As is almost always the case in fundamentalist crackdowns, the Cedarville purge focused on clearing out the Biblical and Theological Studies Department (fourteen of the faculty who left Cedarville were from this department and were replaced in good part by faculty from Southern Baptist fundamentalist seminaries).[35]

The Cedarville purge included the departure of a number of important women on campus, including a Bible professor. In line with the "complementarian" position that women are not to teach men in theological/biblical matters and that they are to submit to their husbands, President White and his administrative colleagues determined that biblical and theological studies classes taught by women could no longer include any male students. In a blog entry entitled, "I Will Not Suffer a Woman," alumna Sarah Jones responded:

> If this is the path Cedarville chooses to take, it won't be a college any longer, it'll be a glorified Sunday School. That's fine if you want to produce graduates who can only function in fundamentalist echo chambers, but it certainly doesn't prepare them for the real world. It doesn't even encourage them to empathize with their fellow Christians. Here's what it *does* do: train half the student body to disregard the other half and treat them as if they're incapable of holding worthwhile opinions on the religious tradition that defines their entire lives. To quote the school's infamous marketing slogan: That's so Cedarville. I just wish it weren't.[36]

There is no way to know if Ken Ham and AiG provided specific input to the Cedarville administration during its fundamentalist crackdown; from what we can tell, Ham made no public statement as events were proceeding at AiG's model "Creation College." But there is no question Ham's organization remained very much a presence at Cedarville in these years, and Cedarville remained quite visible at the Creation Museum, even sponsoring—as of August 2015—the "Cedarville University Mining Company sluice," where museumgoers can "'mine' their own fossils and gemstones and then identify them."[37] More than this, Cedarville's increasingly hardline fundamentalist policies were very much in line with what AiG advocates. Certainly, Ham and AiG would strongly affirm the "God and Creation" white paper of the board of trustees, with its insistence that to deny "the historicity of Adam and Eve . . . brings into question the foundational doctrines of the inspiration and inerrancy of Scripture, and the deity of Christ."[38] It should go without saying that AiG's focus in "ministering" to LGBTQ individuals would be to insist they abandon their homosexual and otherwise gender-bending behavior. Finally, Cedarville's understanding of women as confined to a strict "complementarian" role in the family and the church and in the Biblical and Theological Studies Department is fully in keeping with AiG publications and other materials for sale in the Dragon Hall bookstore and at the AiG online store.

Cedarville's willingness to enforce a staunch young-Earth-creationist fundamentalism stands in stark contrast to many Christian colleges in the United States. In their 2011 book, *Already Compromised: Christian Colleges Took a Test on the State of Their Faith and The Final Exam Is*

In, Ken Ham and his co-author Greg Hall, president of Warner University (FL), announce (as noted on the back cover) their "stunning revelation about the *nation's Christian colleges!*" At the heart of the book is a summary of the results of an AiG-commissioned survey of 312 university presidents, academic deans/vice-presidents, religion department chairs, and science department chairs at Christian or religiously affiliated higher education institutions. Curiously, Ham and Hall do not inform readers how many schools are represented in the survey (we counted 196 from AiG's online list of responding institutions), how many people responded from each school (it is obvious that in most cases it had to be just one or two individuals), or how many from each job category. The authors do note that 223 of the respondents (71 percent) came from schools associated with the Council for Christian Colleges and Universities, an organization of 120 schools that "hire only persons who profess faith in Jesus Christ as full-time faculty members and administrators." According to *Already Compromised,* the results revealed that (among other things) while 57.1 percent of science department chairs believe in a young Earth, only 14.8 percent of religion department chairs do; 86.5 percent of college presidents interviewed believe there was a worldwide flood, but only 42.9 percent of vice presidents do; and 38.5 percent of these Christian college faculty and administrators do *not* believe that the world was created in six literal twenty-four-hour days.[39]

While some readers may find it astounding that 59.6 percent (apparently a few did not answer the question) of these faculty members and administrators *do* accept the notion that the world was created in six twenty-four hour days, Ham and Hall are not cheered by the overall results. As they dramatically summarize the findings, "only 24 percent of the 312 people surveyed answered every question correctly," which means—in a statistical leap of faith, given the small sample size and given that administrators as well as faculty were surveyed—that "if you send your students to a Christian college . . . three out of four times" they will have "teachers who have a degraded view and interpretation of Scripture." Despite all of these faculty members with "degraded" views, administrators routinely fail to do what Cedarville did, that is, purge AiG-defined doctrinal malefactors from their schools. According to Ham, "we have seen very few instances where faculty were fired for compromise teachings," and "in those cases it usually has to be something really outlandish." When it comes to faculty fudging the "details of 'inerrancy,' 'infallibility,' or even evolution/creation/millions of years," there are no consequences. The results are in: "while professing to be institutions of truth, many [Christian] schools have become the propagators of confusion."[40]

Horrified by the damage perpetrated by these institutions of confusion, Ham has repeatedly used his blog to call them out, naming the compromising institutions and detailing the doctrinal offenses of their faculty. It probably goes without saying that he is almost silent regarding mainline Protestant schools or Catholic universities: not only have these institutions made their peace with mainstream science, but their constituencies are not his constituencies, and thus would take no notice of Ham's criticism (or may even consider a "Ham attack" a badge of honor). Instead, Ham focuses on evangelical and fundamentalist colleges, which are much more vulnerable to pressure from Ham and his supporters. Among his favorite targets are Nazarene schools, where a 2014 poll revealed that a strong majority of faculty hold to theistic evolution. In his blog Ham has blasted faculty and administrators at Eastern Nazarene College (MA), Southern Nazarene University (OK), Northwest Nazarene University (ID), Olivet Nazarene University (IL), and Point Loma Nazarene University (CA): "If your child goes to a compromising Christian college like Point Loma Nazarene, don't be surprised if they begin to doubt and ultimately reject the Word of God! God hates compromise—and so should you!" Actually, according to Ham, the entire Church of the Nazarene has been badly compromised by the fact that denominational leaders and school administrators have permitted faculty members to teach students "that they can't trust the Bible when it comes to Genesis 1–11 and that fallible man can reinterpret" the Bible, teaching that greatly assists "the devil" in his effort "to cause [youth] to doubt and subsequently disbelieve the Word of God."[41]

Ham has also had a few things to say about Bryan College (originally William Jennings Bryan University), established in the wake of the 1925 Scopes trial and located in the town where the trial took place. Bryan appeared on Ham's radar in July 2010, when *USA Today* published an article featuring an interview with Rachel Held Evans, who grew up in Dayton (TN) and was a 2003 graduate from Bryan. Evans had just published a memoir, *Evolving in Monkey Town*, about her changing understanding of what it means to be a Christian: while she had been instructed by some Bryan professors that to question creationism indicated that one was not a "true Christian," she had since "learned you don't have to choose between loving and following Jesus and believing in evolution." The article also highlighted Brian Eisenback, a Bryan biology professor whose pedagogical approach, which Evans praised as preferable to what she had endured, was to teach evolution "straight from the textbook" while also having "separate discussion[s] about other views," all the while not endorsing "any particular belief." The article concluded with Evans's plea for peace: "My generation of evangelicals is ready to call a truce on the culture wars . . . We are ready to be done with the whole evolution-creation debate. We are ready to move on."[42]

Six days later Ken Ham responded, in a blog entry entitled "'Move on From Evolution-Creationism Debate?'—No, It Is Heating Up!" Ham instructed "Rachel" that "the coming generations are not 'ready to move on.'" Instead, "they are increasingly seeking answers to the bankrupt compromise positions taught by many churches and Christian colleges" while "gearing up to be reformers like Martin Luther, so they can call church and culture back to the authoritative Word of God." Regarding her alma mater, he quoted from the article's description of Brian Eisenback's biology classroom and then issued a threat: "There are many colleges/seminaries—like Bryan College—across the nation with professors who compromise God's Word in Genesis . . . It's about time that these colleges were held accountable for such undermining of Scripture."[43]

Less than four years later, in February 2014, Bryan's president (Stephen Livesay) and its board of trustees made good on Ham's threat. Appearing to follow the Cedarville script, they issued a "clarification" to the Statement of Belief that must be signed by all faculty and staff as a condition of employment: while the original statement asserted "that the origin of man was by fiat of God in the act of creation as related in the Book of Genesis," the "clarification" stated, "We believe all humanity is descended from Adam and Eve. They are historical persons created by God in a special formative act, and not from previously existing life forms." The tiny campus erupted in controversy with the faculty voting thirty to two (with six abstentions) that they had no confidence in Livesay (who had come to Bryan from Liberty University), and with almost three hundred of the eight hundred students signing a petition asking the trustees to reconsider their action.[44]

In response to the news reports about "a massive reaction at [Bryan] because the administration has said that the college takes a stand on a literal Adam and a literal Eve," Ken Ham sniffed that it was simply additional evidence as to "how far from God's Word certain Christian academics are going—including many instructors at Bryan College." Notwithstanding the controversy, Livesay and the board held firm. By the end of spring 2014, at least nine instructors (out of forty-four full-time faculty members) were gone; the departures included Eisenback, who landed a position at another school, as well as two faculty members who refused to sign the amended statement and instead sued the board for altering the faith statement despite the fact that "the school's charter says the statement can't be altered." In keeping with the Cedarville purge, four disaffected board members resigned in the summer of 2014, and there were additional staff cuts, explained as necessary because of the school's financial woes. On the plus side, at least from the administration's perspective, Bryan College had now worked its way into the good graces of AiG. Not only did the administration sign the AiG faith statement whereby the school was included on AiG's list of "Creation Colleges," but in November 2014 Bryan College was for the first time, as far as we can tell, represented at the annual College Expo at the Creation Museum.[45]

But over the years Ken Ham's favorite target has been Calvin College, the Christian Reformed school of more than four thousand students in Grand Rapids, Michigan, that is regarded by many as the academically strongest conservative Protestant college in the United States. Notwithstanding its elevated reputation, Ham has seen Calvin as one of the worst offenders among "compromised" Christian colleges, devoting four pages to the school in *Already Compromised.*[46] As he noted in a 2009 blog entry, "Send Students to Calvin College to Be Indoctrinated in Evolution," the school's faculty had taught "evolution and liberal theology to generations of students," in the process taking "students who believe the Bible and systematically destroy[ing] that belief." In his 2010 "State of the Nation 2" address, Ham critiqued Calvin College for teaching evolution and Calvin religion professor Daniel Harlow for denying inerrancy. When a writer for the Calvin student newspaper (*Chimes*) defended his school while charging Ham with "lies, inaccuracies, and caricatures," Ham waded into battle with his nineteen-year-old adversary: "Mr. Camacho, document the lies! In fact, I publicly challenge you (as you wrote this public article) to a debate—even at Calvin College if you want—in regard to your accusations." This was followed by a sarcastic op-ed by the *Chimes* editor (which Ham reprinted and excoriated in a blog entry entitled, "You Won't Believe What Came Out of Calvin College Last

Week"), emails to Ham from a third student (which Ham reprinted, under the heading "More Calvin College Sadness"), and a pro-inerrancy letter to the *Chimes* (which Ham also reprinted) from a former Calvin College board member: "The recent attacks on Ken Ham make it clear that Calvin's own history is not only no longer adhered to but in an age of relativism is declared never to have even existed in the CRC [Christian Reformed Church] and at Calvin College."[47]

A few months after Ham's battle with Calvin students, conservatives in the Christian Reformed Church were up in arms after it was reported in the denominational magazine that Harlow and his colleague in Calvin's religion department, John Schneider, had published articles in the journal *Perspectives on Science and Christian Faith* suggesting that developments in genetic science made it very difficult to accept the notion of a historical Adam and Eve. While these unhappy constituents were barraging the Calvin College administration with complaints, Ham helped stoke the fires of controversy with blog entries such as "False Teaching Rife at Calvin College."[48] That summer, Schneider was forced out of his position at Calvin; Harlow managed to hold on to his position, while acknowledging that he felt very much in danger: "There are a lot of people in my denomination who are glad that Schneider is gone, and would like to see me gone as well." Not appeased by Schneider's removal, Ham continued unabated in his verbal assaults on Calvin College. As he noted in a July 2014 blog entry, "The Secularization of a Christian College":

> The sad point . . . is though the very vocal and more radical compromis-
> ers [such as Schneider] may sometimes end up leaving a Christian col-
> lege, that by no means suggests that this compromise disappears with
> them. No, such compromise is rife through most Christian colleges. In
> fact, BioLogos, the leading promoter of compromise with evolution and
> millions of years to the church, is now located at . . . Calvin College!"
> [ellipses in original][49]

For AiG, BioLogos is the arch-enemy. Founded in 2007 by Francis Collins, who at the time was the director of the Human Genome Project, with a $2,028,238 grant from the Templeton Foundation, the BioLogos Foundation understands its mission as inviting "the church and the world to see the harmony between science and biblical faith as we present an evolutionary understanding of God's creation." The BioLogos website was launched in 2009 and remains the organization's centerpiece, providing articles, videos, and other resources "on contemporary issues at the science/faith interface." Over time, BioLogos has expanded its offerings in the "actual" world, hosting conferences and workshops, offering grants to scholars and church leaders in an effort "to promote Christian scholarship that addresses the theological and philosophical concerns commonly voiced about evolutionary creation," and providing "professional development opportunities for Christian middle and high school science teachers."[50]

Notwithstanding its support for evolutionary science, BioLogos is a quintessentially evangelical organization, appealing to the same conservative Protestant constituency as does AiG. In keeping with other evangelical organizations, BioLogos has a faith statement, which begins, in good evangelical fashion, with a statement regarding the Bible: it "is the inspired and authoritative [albeit not inerrant] word of God." It goes on to include an affirmation of the "historical death and resurrection of Jesus Christ, by which we are saved and reconciled to God" as well as the other "miracles described in Scripture." Moreover, BioLogos' leadership has been staffed with scientists from evangelical colleges. When Collins established the organization, physicist Karl Giberson, then at Eastern Nazarene College, and geneticist Darrel Falk, from Point Loma Nazarene College, served as executive vice president and executive director, respectively. Soon after Collins's 2009 appointment as director of the National Institutes of Health, Falk became sole president, and BioLogos moved its headquarters to Point Loma. After Falk resigned his presidency in 2012, astronomer Deborah Haarsma of Calvin College became the president, biologist Jeff Schloss from Westmont College became the senior scholar, and (as noted above) BioLogos moved its headquarters to Calvin.

As part of its mission, BioLogos seeks to avoid unnecessarily alienating evangelical creationists, striving "for humility and gracious dialogue with those who hold other views."[51] But Ken Ham and AiG have no interest in "dialoguing about how the human soul could come about if evolution were true!" They are not going to dialogue with folks who are so completely wrong: "BioLogos is wrong. What BioLogos teaches was written by fallible, fallen, finite humans! There is only one way to sum up this organization: 'Let God be true but every man a liar.' "[52] Instead of conversation, much less affirmation of Christian solidarity, with his fellow evangelicals, Ham has devoted great energy to monitoring and attacking BioLogos, with at least forty-one separate blog blasts of various sorts between May 2009 (two weeks after the advent of the BioLogos website) and November 2014. Not only has Ham repeatedly castigated the colleges that have housed BioLogos (Point Loma Nazarene and Calvin), but he has vilified all of the aforementioned scholars who have led the organization. About its esteemed founder, Ham has written, "It is compromisers like [Francis] Collins who cause people to doubt and disbelieve the Bible—causing them to walk away from the church"; about its most recent leadership team, Ham has said that "what Haarsma and Schloss are promoting are not evolutionary ideas alone—what they're doing is helping to promote the blending of Christianity with an atheistic religion, and consequently are undermining the authority of the Word of God!" In fact, any scholar or pastor who has had connections with BioLogos is fair game for Ken Ham. This includes biblical scholar Peter Enns, who had been a Senior BioLogos Fellow and who is frequently on the receiving end of Ham's wrath in blog entries such as "Peter Enns—Mutilating God's Word."[53]

When BioLogos president Deborah Haarsma made an October 2014 plea that she and Ham meet for dinner in order to jump-start a less-acrimonious conversation, Ham peremptorily refused, claiming that Haarsma and BioLogos really want "me to say that they have a view of Genesis that's valid . . . she does not really want me to judge her view against Scripture, and she does not want AiG . . . to warn people about those who undermine the authority of God's Word." For the CEO of AiG, BioLogos is not a friendly evangelical competitor that simply has a different understanding of creation, but a deadly enemy seeking to poison the hearts and minds of American Christians. In fact, as Ham sees it, AiG's conflict with BioLogos is but another front in the great cosmic war between God and Satan:

> BioLogos (an organization dedicated to promoting evolution and millions of years to Christians) has made it clear that they're going after our youth—young children though [sic] high school. The Bible makes it clear that God's people are to train up their children in truth, so they won't be led astray by the evil teachings of this world. But the devil also knows this teaching, so he works hard to indoctrinate children at a young age in false teaching. Sadly, I do believe that organizations like BioLogos—even if staffed by Christians—are helping the devil in leading this and coming generations away from the truth of God's Word . . . It's interesting to note that as biblical creation and apologetics organizations like AiG have risen up in influence in the church and culture, we

also see those who are opposed to taking God's Word as written in Genesis increase in influence. I believe this is no coincidence and reflects the spiritual battle roaring around us—a battle for the hearts and minds of this and future generations.[54]

In AiG's Great Binary, there is no neutrality, no middle ground. There are but two warring camps. At the end of the day, those academics who promote evolution and/or an old Earth are serving the forces of Satan. But why do Christian scholars place themselves on the wrong side of the cosmic battle? Ham's standard explanation is that these "compromising" Christian academics are afflicted with an overweening sense of "academic pride." This is personal for AiG's chief: in *Already Compromised* he complained that he has been "belittled and cut down by professors at 'respected' Christian universities" whose "attitudes are laced with an arrogance and a condescending attitude that looks down on other Christians in a voice that says, 'Don't worry about it, you wouldn't understand anyway.'"[55]

Belying his claims that AiG does not engage in ad hominem attacks, Ham has frequently used his blog to denounce these supercilious scholars, like Wheaton College's John Walton, who possess an "academic pride, largely from academic peer pressure, because [they] ultimately 'loved the praise of men more than the praise of God' (John 12:43)." As he noted in a scathing condemnation focused on Deborah Haarsma, "I see so much pride in academia today . . . This academic pride causes many to become so puffed up with all their supposed knowledge . . . that they make themselves the authority instead of God's Word." Ham implored Haarsma and other prideful scholars to "get back to the child-like faith that we all need to have." He similarly admonished Peter Enns— whose "'special knowledge' he has to share about not taking Genesis as history" betrays "clear academic pride"—to "humble himself as a child and repent of this before a holy God." In short, "compromising Christian academics need to," as he instructed a host of BioLogos scholars in 2011, "fall on their knees before a Holy God and repent of their attack on the *Word*" (emphasis in original).[56]

Hell

What if these scholars do not heed these directives from the AiG boss? According to Ham, there will be a day of divine reckoning for the arrogant academics who undermine the authority of God's Word with their devotion to evolution and "millions of years" and for church leaders who promote this poisonous message among Christian laypeople. On that day, these compromising professors and preachers will have to answer to God for what they have taught their students and their parishioners. On that day, Southern Nazarene University's Mark Winslow and other evolution-promoting professors will have to stand "before the almighty God" and account for "having shattered generations of students' faith in God's Word by exalting the word of fallible sinful man." On that day, Martin Thielen and other prominent ministers who are "undermining the authority of the Word of God" by promoting theistic evolution will discover that it is (quoting from Hebrews 10:31) "a fearful thing to fall into the hands of the living God."[57] On that day, Nazarene church leaders and others who are "spread[ing] evolution and millions of years in the church," in the process "leading many away from the Christian faith," will discover that "God will judge—and He will have the last word!" On that day, a Frankfort (KY) pastor who claimed that Genesis is not a historical account or science textbook (in effect launching "an attack on the Word—an attack on Jesus Christ") will discover the truth of Jeremiah's warning: 'Woe to the shepherds who destroy and scatter the sheep of My pasture!'" On that day, one of Ken Ham's favorite targets of vituperation will discover just how terrifying it is to await the Almighty's judgment:

[Karl] Giberson is actually . . . editing and rewriting God's Word! Wow! He has to answer to God for that one day, and there will be a day of reckoning! Can you imagine standing before the holy, infinite God the Creator one day and saying, "God, I hope you like my rewrite of Genesis. I wanted to correct the way you had it written to fit with what we humans believe about evolution and millions of years. Your version was outdated and didn't work for academics like me in the twenty-first century." I would not want to be in his shoes![58]

According to the arcane intricacies of dispensational premillennialism, and as articulated by Henry Morris in his *Study Bible,* there will be not one but two Judgment Days: the Great White Throne Judgment, when unbelievers will be judged and then cast into the lake of fire; and, the Judgment Seat of Christ, when believers will be judged on the quality of their "works" and receive "rewards or loss of rewards" before they are ushered into heaven. Leaving aside questions that would seem obvious to those outside dispensational premillennialism (e.g., "What exactly are these 'rewards,' and what does it mean to claim that heaven contains different status levels?"), the salient point here is that Ken Ham is never explicit as to which form of divine judgment compromising academics and church leaders are facing. Suffice to say that in Ken Ham's vitriolic condemnations of Christian leaders who promote evolution and millions of years it is impossible not to hear threats of eternal damnation. This is particularly the case when it comes to his blasts on Christians who are willing to tolerate same-sex marriage, as he makes clear in this 2014 rant against Presbyterian Church USA (PCUSA) officials who agreed to permit "gay weddings" in denominational churches: "Leaders in the PCUSA have essentially declared war against the holy Creator God with . . . their 'evil deeds.' But God always has the last say! Every one of these leaders will one day die and have to face the God against whom they have declared war . . . These leaders need to be reminded of this warning: 'It is a fearful thing to fall into the hands of the living God.'"[59]

The Creation Museum very much reiterates Ken Ham's emphasis on damnation. Here the story of the past has been reduced to original sin and the Fall and the spread of pervasive wickedness, followed by the global flood, with the rescue of a tiny remnant and the drowning of millions of creatures. Of course, this past is prologue to an equally simplified present, with its rapid decline of Western civilization, the moral filth of contemporary America (as evinced in such examples as gay marriage, school shootings, and rampant secularism), and the collapse of true Christian witness. The past is also prologue to the future. While this future is not specifically foretold at the museum, the museum's controlling and repetitive narrative of disobedience and punishment, especially with its emphasis on the global flood and Noah's Ark, makes it clear that judgment for America, for the West, for all humanity is forthcoming, and with it the rescue of a faithful remnant and eternal damnation for the rest of humanity.

That is to say, for many (perhaps most) human beings the future means hell. For Ken Ham and AiG, and for fundamentalism in general, the very essence of God's nature requires hell, and an extraordinarily robust hell at that. As Tim Challies, fundamentalist pastor and occasional speaker at AiG conferences, proclaimed in a 2012 *Answers* article, if "you want a God who is good—truly good—and if you want a God who is just and holy, then you must have . . . this God who condemns people" to hell. According to Challies, the Bible makes clear that hell must be eternal, because "when you sin against an infinite God—and all sin is primarily oriented toward God—you accrue an infinite debt." Hell must also involve torment, because God is "truly, purely holy," and "His holiness is like a gag reflex that acts out in wrath against all sin." Moreover, hell must contain sinners who are fully aware of their torment, because, as "active rebels" who "sinned consciously," they "must also bear their punishment consciously." In short, it is the very nature of God, or his "transcendent holiness," that requires that sinners not only "be kept out of his presence" but that they be cast into a hell where they will endure eternal, conscious torment.[60]

Leaving aside other questions regarding the logic that undergirds such an argument, it is not clear how the requirement that sinners must be kept far, far away from God squares with the Jesus of the Gospels, who not only ate with prostitutes and tax collectors, but who seemed much more at home with sinners than he did with the righteous. But then again, and as we have seen in various ways, it is difficult to find much reference to Jesus at all either in the museum or in AiG materials, except to say that he was the sacrifice required to bear the holy wrath of God the Father in payment for the sin of human beings. One is hard-pressed to find anything from the life and teachings of Jesus, except that he allegedly issued an edict against gay marriage. One will look long and hard in the flood of apologetic text produced by AiG to find anything of substance on Jesus's Sermon on the Mount, Jesus's admonition to love our enemies, or Jesus's repeated calls for us to care for those who exist on the bottom rungs of the social ladder.

The absence of Jesus is particularly noticeable when it comes to the Last Judgment. For all of the preoccupation with this day when we will stand before God and await our fate, it is striking how little attention Ken Ham and his AiG colleagues give to the one place where Jesus spoke in depth on the topic. That can be found in Matthew 25, where Jesus delineated who will receive eternal life and—in verses 41–46— who will receive eternal punishment:

> Then he will say to those at his left hand, "You that are accursed, depart
> from me into the eternal fire prepared for the devil and his angels; for
> I was hungry and you gave me no food, I was thirsty and you gave me
> nothing to drink, I was a stranger and you did not welcome me, naked
> and you did not give me clothing, sick and in prison and you did not
> visit me." Then they also will answer, "Lord, when was it that we saw
> you hungry or thirsty or a stranger or naked or sick or in prison, and did

not take care of you?" Then he will answer them, "Truly I tell you, just as you did not do it to one of the least of these, you did not do it to me." And these will go away into eternal punishment, but the righteous into eternal life. (NRSV)

As delineated in the lengthy notes accompanying this passage in the *Henry Morris Study Bible*, Jesus's words here do not apply to us. The only people who will face the judgment described in Matthew 25 will be the very small group of people who somehow manage to live through the horrific seven years of tribulation *and* the battle of Armageddon at the end of history. According to Morris, this tiny remnant will be judged as to whether they tried to help refugees who were fleeing the forces of the Antichrist. Those "who try to befriend and care for these refugees during these seven frightful years will be recognized as 'sheep' and allowed to continue their natural physical lives into the millennial kingdom." By contrast, "those who had not done this, turning the refugees away, and perhaps even reporting them to the authorities, will be sent 'away into everlasting judgment.'"[61]

It is hard to imagine how such an esoteric explanation of Matthew 25—with so little grounding in the actual biblical text—squares with the notion of reading the Bible in a commonsensical, literal fashion. As opposed to this fantastical interpretation from the father of contemporary flood geology, a "plain sense" reading of Jesus's teachings in Matthew 25 would suggest that, at the very least, how one treats the thirsty, the hungry, the naked, the sick, and the imprisoned will be an essential component of how one's life will be judged. Theologian Michael Himes puts it thusly: "The only criterion for that final judgment, according to Matthew 25, is how you treated your brothers and sisters." In fact, there is a good deal of historical evidence that—in the first few centuries after Jesus's death—Christians understood Jesus's teaching in Matthew 25 as the standard for how they should live, with an emphasis on punishment for those Christians who failed to care for "the least of these."[62]

But the Jesus who supped with sinners and taught followers that they must take care of "the other" is nowhere to be found at the Creation Museum. On the contrary, the Creation Museum and AiG are Christian Right sites that relentlessly and aggressively promote a highly ideological and radically politicized young Earth creationism as true Christianity. What is sad—to use one of Ken Ham's favorite words when he is talking about "compromising" academics and church leaders—is that millions of Americans who are seeking to be good Bible-believing Christians have bought the message that AiG is selling. Such a message may satisfy some deep desire for the comfort of certainty, may offer a way to respond to what can feel like the unrelieved elitism of the academic and scientific powers-that-be, may provide a place to stand in what seems like an increasingly decadent culture, may reinforce the conviction that America really was and could again be God's nation. But in the end, the ideological and politicized young Earth creationism of the Creation Museum and AiG has little to do with the Jesus of the Gospels. It has little to do with the Hebrew prophets. It has little to do with Christianity's rich intellectual and social justice tradition, little to do with Augustine and Aquinas, Barth and Bonhoeffer, Day and King. It has little to do with faith and hope and love.

Sad indeed. For all of us.

EPILOGUE

O
n February 27, 2014, Ken Ham held an "online press conference," as he called it, in the Legacy Hall auditorium at the Creation Museum where the Nye–Ham debate had occurred just
twenty-three days prior. Surprisingly, no members of the press appeared
for the event. None can be seen on the AiG video of the event posted
on the AiG website, and no questions were invited or taken from any
members of the press (or any audience members) during the recorded
event. Instead, Ham asked most of the questions and, perhaps not surprisingly, regularly provided the answers.

The big news at the "press conference" was that AiG's fundraising efforts over the last few years along with a successful bond offering meant
that construction of the Ark Encounter project could begin.[1] In the course
of the "press conference," Ham, with assistance from Patrick Marsh (vice
president of attractions) and Joe Boone (vice president of advancement),
described the massive physical and financial size of the project.

At the heart of the project is an ark whose construction will match as
closely as possible, both in materials and in size, the Ark as described in
Genesis 6–8. This gigantic wood-frame structure will stretch five hundred and ten feet in length, eighty-five feet in width, and fifty-one feet
in height. When complete, it will be "the biggest timber frame building
in the United States." Amazingly, the whole structure will sit several feet
above the ground so that visitors can get a glimpse of its underside. It
will be held in place by three eighty-foot cement towers, and some four
million board feet of timber will go into its construction.[2]

The interior of the ark will consist of three floors that together will
feature one hundred and thirty-two wooden bays with an exhibit in each
bay.[3] Plans are for the first floor to focus on biblical "kinds"—that is,

the pairs of animals that, according to AiG, boarded the Ark and later through reproduction and natural selection became the millions of species that populate the Earth today. These animals will appear in live, animatronic, and static forms. The second floor is expected to feature exhibits and displays that answer practical questions about how Noah built the Ark and how he and his family fed and took care of the animals. Bays on this floor will display large and small animals in cages along with exhibits featuring, for instance, Noah's workshop, animal feeding, the Ark's door, ancient man, and the pre-Flood world. Visitors will also find on this floor a snack shop and restaurant. The third floor is expected to devote a number of bays to Noah's living quarters, who Noah was, and more on how Noah built the Ark. It will also include cages with small animals and exhibits on such topics as flood geology, the Ice Age, the Tower of Babel, fossils, and flood legends. In addition, the third floor is expected to feature the Christ-the-Door Theater. Plans for areas outside the ark include additional exhibits, a petting zoo, camel and pony rides, and more.

The design and construction of the ark represents only the first phase of a much larger project. Future phases of Ark Encounter promise the addition of such attractions as a Tower of Babel, a first-century Middle Eastern city, a journey through biblical history, a children's area, an aviary, and a large petting zoo. This expansive multistage project will be located on eight hundred acres of land about forty miles south of Cincinnati just off Interstate 75 at exit 154 near Williamstown, Kentucky, and forty-five minutes from the Creation Museum. AiG expects that the museum's proximity to the Ark Encounter will increase the number of visitors to the museum by as much as 50 percent. Not surprisingly, given such expectations, Ken Ham has already begun making plans to expand the Creation Museum in order to accommodate more visitors.[4]

According to Ham, as of July 2015 the "Lord has blessed us in providing $70 million of the approximately $90 million needed to complete the life-size Ark project," which is the first phase of the theme park. Those interested in supporting the construction of the first phase have several options. They may purchase a peg (for one hundred dollars), a plank (for one thousand dollars), or a beam (for five thousand dollars), the latter which brings with it, among other benefits, one's name on the ark's recognition wall, a replica of the ark signed by Ken Ham, and four free passes to behind-the-scenes events. Prior to the completion of the first phase, contributors may also purchase a "Charter Boarding Pass" for the whole family (a three-year pass costs fifteen hundred dollars, and a lifetime pass costs three thousand dollars), for an individual (one thousand dollars for a three-year pass or two thousand dollars for a lifetime pass), or for a grandparent (three thousand dollars for a lifetime pass allowing the grandparent to bring up to four grandchildren to the park each visit).[5]

On May 1, 2014, just a few months after the press conference, AiG supplanted the traditional groundbreaking event by hosting a "Hammer and Peg Ceremony" inside the Creation Museum wherein John C. Whitcomb, co-author of *The Genesis Flood*, along with AiG founders and its board of directors pounded pegs into beams with wooden mallets. In the months that followed, construction of Ark Encounter began as "over a million cubic yards of dirt" were moved. By late November 2014, "the first of many loads of concrete was poured."[6]

The Troyer Group, a design and construction firm located in Mishawaka, Indiana, is building the ark with help from two Indiana Amish brothers who, according to AiG, are "amassing a construction crew from various Amish families in a handful of states." According to AiG, given the enormity of the project "during our peak construction phase, nearly 100 workers will need housing." As of this writing, expectations are that their efforts will culminate in the grand opening of the first phase, which features the ark, in July 2016.[7]

What is the point of this massive project? Ken Ham and others at AiG have been explicit that the primary purpose of the Ark Encounter is evangelism, as Noah's Ark serves as a powerful call to the salvation offered by Christ Jesus. In his statements at the "online press conference" Ham made this clear, with particular emphasis on the Ark's door:

> When you think about the message of salvation, the cross of course is the greatest reminder of the message of salvation. But other than the cross, I believe the Ark of Noah is the greatest message, the greatest reminder of the message of salvation. Because you think about it: God at the time of Noah said he's going to send a global flood because of the wickedness of man, and so he had Noah build this great big ship and that representatives of the land animal kinds were to go on board this particular ship, and then Noah and his family went on board. They had to go through a doorway to be saved. Actually, Noah's Ark is a picture of salvation as Noah and his family had to go through a doorway to be saved. So, we need to go through a doorway. Jesus Christ said, "I am the door. By me if any man enter in he will be saved."[8]

For Ham, the door on Noah's Ark served as a passageway to salvation. Salvation came to those who crossed its threshold; damnation to those who did not. More than this, and based on John 10:9—wherein Jesus says that he is a door—Ham reasons that the door of the Ark is like Jesus. Both provide passage to salvation, as anyone who wishes to be saved is obliged to make that passage and cross that threshold. That being the case, Ham continues, then the door of the Ark serves as a great symbol for the Christian message.

All of this explains why—at least as of the summer of 2015—the theater in the ark at Ark Encounter will be named "Christ-the-Door Theater." In the course of his description of the ark during the "online press conference," Patrick Marsh gushed, "One of the most special things on this floor is what we call Christ-the-Door Theater," which "is the *most important exhibit* inside the ark" (emphasis added). According to Marsh, this theater will serve as "an evangelistic outreach . . . that's about bringing people to Christ." With enthusiasm, Marsh said that "this will be the full gospel message."[9]

The reader will recall that the door to the Ark figures prominently at the Creation Museum as well. In the Voyage of the Ark room, visitors are powerfully positioned in relation to that door. As they stand before one of the miniature dioramas depicting the floodwaters rising to mountaintops, they observe the final horrific moments in the lives of animals, adults, and children as they are about to perish on those mountaintops. And they see the Ark, impossible now for them to reach and sealed up tight, floating by them. From the perspective of another miniature diorama, visitors are positioned inside the Ark. They are standing behind Noah and his family and looking with them through the open door of the Ark just before God shuts the door for good. The placard below this diorama says the following: "God shut Noah's family into the Ark. This ended any opportunity for people outside the Ark to be saved."

At the Creation Museum, the door of the Ark stands open for a time. While open, it serves as the passageway to safety and to salvation. Crossing the threshold secures salvation. But then, once the Flood begins, God shuts the door. Once God shuts the door, it remains shut. Indeed, according to a blog entry on the Ark Encounter website, "From a careful reading of Scriptures, it appears that the door was never opened again." Given the importance of this point, the blog entry goes on to work out a scenario by which Noah, his family, and all the animals managed to exit the Ark without opening that door. Based on Genesis 7: 13, the blog entry speculates: "Did Noah remove the wooden planks—the 'skin' of the Ark—and build off-loading ramps?"[10] Importantly for Ken Ham and AiG, once God shuts the door to the Ark, it stays shut forever. Once shut, it bars forever all those who are on the other side of it.

Does a door like that work well as an analogy for Jesus? At least when looking at the verse in John that inspires Ham's analogy, that is not so clear. The complete verse (in the King James Version) says the following: "I am the door: by me if any man enter in, he shall be saved, *and shall go in and out, and find pasture*" (John 10:9, KJV, emphasis added). The door Jesus describes seems to be the sort that swings back and forth; one can walk through it in either direction repeatedly, and find pasture on either side. By contrast, Ken Ham's door stands open for a while and then shuts forever, an eternal barrier to all who find themselves on the wrong side. Far from finding pasture on either side, one finds the happy righteous (patriarchal) family on one side and global destruction, suffering, and death on the other.

Ham calls the Ark and his idea of the door "a picture of salvation." On closer examination, it appears that the Ark Encounter and its door are very much like the Creation Museum, that is, much more about judgment than salvation. In another online video Ham—mixing Jesus-as-ark and Jesus-as-door metaphors—says: "You see, there's a coming judgment by fire (not by water) next time, and God's providing an Ark of salvation for us—the Lord Jesus Christ. Jesus said 'I am the door. By me if any man enter in he will be saved.' Jesus is our Ark of salvation. And just as surely as there was a Flood, and we see the evidence all over the earth in the fossil record, just as surely there's going to be coming judgment by fire."[11] According to Ham, last time around, the righteous handful found their way through that Ark door and were saved while the rest of humanity perished. Next time, the righteous who claim Jesus—who make it through before the door shuts—will find salvation just like Noah's family did. As for the rest, they are sure to find themselves on the wrong side of that shut door, burning in Hell for eternity.

Reporting on her visit with Mark Zovath (a co-founder of AiG), and Patrick Marsh, *Atlantic Monthly* contributor Amanda Petrusich noted:

> When I visited this fall, I was shown a few scale models of their ark, which, compared with the delightful wooden boat pictured in many a children's book, is a terrifying-looking thing: it has no portholes or open decks, and except for a single door that God is supposed to have

slammed behind Noah ("And Jehovah shut him in") and some very narrow openings for light and ventilation, the vessel is sealed off in a way that suggests a giant floating coffin.[12]

Judging from the images of the ark that appear on the Ark Encounter website, Petrusich has a point. More important, her description resonates with the theology and politics that appear to animate the Ark Encounter project. Rather than proclaim a message of hope and compassion, the plans for this project indicate a "terrifying-looking" boat that emphasizes a message of congratulations to the righteous, right-thinking Christians who have joined Ken Ham and his colleagues on board the "giant floating coffin," safe from what they imagine as the inevitable, horrific, and well-deserved eternal punishment awaiting billions of their fellow human beings once the door slams shut.

The Ark Encounter is expected to attract as many as two million visitors in its first year of operation. That is an impressive number given that the Creation Museum did not welcome its two millionth visitor until its sixth year. Moreover, according to an economic impact analysis conducted by Jerry Henry and Associates for AiG, on average each visitor will likely spend about $100 at the Ark Encounter and another $50 "in the market" per visit, which will result in nearly $245 million in total "direct spending impact for the first year." When indirect impacts are taken into account, that number rises to nearly $360 million. If capital expenditures are added, that number rises further to over $590 million. In addition, AiG anticipates directly employing two hundred people in the operations of Ark Encounter. Looking beyond the park, the impact study predicted that, "14,038 new jobs would be created in the area during the attraction's first year of operation."[13]

Not surprisingly, given these impressive projections, the state of Kentucky eagerly supported the building of Ark Encounter. By way of its Tourism Development Act, the state agreed initially to provide a sales tax rebate for the project. Once the ark was built and the park was operating, AiG would receive "up to 25 percent of their development costs over ten years from sales tax generated at the facility." That could amount to as much as $18 million for phase one.[14]

Questions soon emerged about the sales tax rebate among critics arguing that the state of Kentucky could not offer such a rebate to AiG without compromising the separation between church and state.[15] On December 10, 2014, the state of Kentucky's tourism board rejected AiG's application for the tax incentive. It said that an online job advertisement for a position at Ark Encounter (AiG later said the position was with AiG, not Ark Encounter) indicated that the successful applicant would have to, among other things, sign the Answers in Genesis faith statement, which states that the Bible is the inerrant word of God and that "the Great Flood of Genesis was an actual historic event, worldwide (global) in its extent and effect." According to the state, the job ad indicated that Ark Encounter intended to discriminate in its hiring practices on the basis of religious beliefs. The state also argued that the purpose of Ark Encounter had changed from the first application that AiG submitted (for a much larger project that included all phases) in which it was presented as a tourist attraction to the second application (just for phase one) in which it was identified as a tool of Christian evangelism. Answers in Genesis responded on February 5, 2015, by filing a forty-eight page federal lawsuit against Steven Beshear (governor of the state of Kentucky) and Bob Stewart (secretary of the Tourism, Arts, and Heritage Cabinet). That document argues, among other things, that the defendants "discriminated against [AiG] by wrongfully excluding them from participation in the Kentucky Tourism Development Program" because of their religious beliefs.[16]

Whatever happens with this case, the outcome is likely to have a significant impact on Ark Encounter. As of December 2015, it appears likely (but not certain) that AiG will raise the remaining funds needed to complete the Ark; it is less certain that AiG will secure the funding for the Tower of Babel, first-century city, and so forth. More generally, one cannot know how long the Creation Museum and AiG will remain afloat. The history of evangelical parachurch enterprises—with their dependence on a charismatic leader and with their lack of accountability structures and succession plans—is replete with examples of organizations that collapse due to scandal, internal conflict, and so forth. It is impossible to predict whether the Creation Museum and AiG will survive much beyond Ken Ham's lifetime, if that long.

This point is much less noteworthy than it might seem. The Creation Museum is not an inexplicable anomaly precariously existing on the fringes of American life. Instead, it is a mainstream Christian Right institution. Not only do its scientific, biblical, and political commitments represent and appeal to many of those on the right side of the American mainstream, but it joins with a host of other Christian Right organizations in seeking to shape, prepare, and arm conservative Christians to be aggressive and uncompromising culture warriors. And while history is always contingent, at the moment the Christian Right shows little indication of fading from American life. Perhaps some of the particular commitments will change. Perhaps the opposition to birth control will replace opposition to gay marriage as the hot-button issue. Perhaps geocentricity will replace young Earth creationism as the necessary corollary of reading the Bible literally. This said, there is no sign that the Christian Right is soon to find itself on the historical trash heap. So it seems likely that if/when the Creation Museum and AiG implode or fade away, similar evangelical leaders and similar parachurch ministries and similar Christian Right institutions will emerge, almost without notice, to take the place of Ken Ham's enterprise. Indeed, they may already be waiting in the wings.

ACKNOWLEDGMENTS

William Vance Trollinger, Sr., was a petroleum geologist who ran two small Denver-based companies that made use of aerial photographs and satellite imagery to produce geologic maps that would assist in the discovery of oil and gas. A fundamentalist in theology if not in behavior—he liked a good glass of wine, made prodigious use of remarkably colorful expletives, and won a good deal of money playing poker—he adamantly opposed evolution (never finding humorous his oldest son's observation that this was simply because he had not studied biology). But as a geologist, William Trollinger, Sr.—in keeping with most fundamentalists of his generation—held to the idea of an old Earth, believing that each Genesis day was a geologic "age." In the 1960s and 1970s, when ideas of flood geology and the notion of a young Earth began surfacing in his Baptist church and in other conservative Protestant churches in the Denver area, he was horrified. So he created a presentation—replete with charts and transparencies and slides that explained the geologic timetable and the law of stratigraphic succession (thanks to Paul Trollinger for locating and sending these materials)— that he gave to local churches, in the process making the case that the idea of an ancient Earth could be squared with a literal reading of Genesis. For all of his efforts, he was not able to stem the tide of young Earth creationism, something that troubled him until his death in 2002.

The Creation Museum opened in 2007, and that next year we made our first visit. While we had no intention of writing a book, we kept going and kept gathering material, in the process giving a few conference papers. In 2011 Sue accepted Ron Wells's invitation to give a presentation on the museum at his Maryville (TN) Symposium on Faith and the Liberal Arts. Historian Jay Green commented on the paper and afterward asked why we were not writing a book on the topic. That winter we pitched the topic to Greg Nicholl at Johns Hopkins University Press (who had served as editor for Sue's book, *Selling the Amish*); Greg was quite enthused, and he shepherded us through the first part of the project. When Greg left for another press, we were put in the able and experienced hands of Bob Brugger, who made crucial interventions along the way, particularly regarding the introduction. Thanks also to Andre Barnett, Greg Britton, Elizabeth Demers, Camille Hale, Becky Hornyak, Hilary Jacqmin, and Gene Taft.

Beyond Johns Hopkins University Press, thanks to our son-in-law, Dan Hatch (of Hatch Architecture), for his work on the museum map; to John Le-Comte and Patrick Thomas of the University of Dayton (UD) for their technical assistance; and to Heidi Gauder of the UD Roesch Library for assistance with databases. A number of PhD students in theology at UD provided important assistance, including Christine Dalessio, Dara Delgado, Sean Martin, Emily Mc-Gowin, Robert Parks, Anthony Rosselli, and especially, Julia Parks. We also had the benefit of frequent conversations with Jason Hentschel, whose 2015 disser-tation on inerrancy in post-1950 America provides invaluable theological con-text for understanding our work here. Thanks also to our religious studies col-leagues Meghan Henning, for her insights regarding the early Church's notions of hell, and Brad Kallenberg, for his patient assistance on multiple questions pertaining to biblical translation and theology. Thanks to Tiffany Walker for her assistance in the permission process and Answers in Genesis (answersingenesis. org) for granting us permission to publish photographs of the museum.

In the same vein, we are also grateful to our friend Frederick Schmidt of the Garrett-Evangelical Theological Seminary (IL). Rod Kennedy, senior pastor of the First Baptist Church of Dayton, helped us think more carefully about how the museum uses the Bible, while at the same time—in nine years of remarka-ble sermons—he challenged us to live the Gospel. Thanks to Jared Burkholder and Mark Norris of Grace College (IN) for background information on John C. Whitcomb; Noah Efron of Bar Ilan University for an advance look at his ac-count of rabbis visiting the Creation Museum; Jason Gerlach of Cambridge University for his insights concerning geochelone tortoises; Michael Hamilton of Seattle Pacific University for reading over our section on Calvin College (we look forward to his book on that institution); Suzanne Smailes of Wittenberg University (OH) for scanning and sending us the first chapter of Genesis from the 1909 Scofield Reference Bible; and Bill Trapani of Florida Atlantic Univer-sity for comments on an early paper on the museum. We are grateful to our anonymous informants who deepened our understanding of the controversies at Cedarville University. Thanks also go to Mike Trollinger, who took time away from watching obscure films to provide us with helpful information regarding the Beowulf-creationism connection. And many thanks go to Joseph and Peggy Huffman for the ongoing conversation that nurtures the soul.

We have had the great good fortune of participating in an ad hoc group of scholars who are writing on creationism and who have been willing to share their insights with us: James Bielo of Miami University of Ohio (who organized a two-day creationism "summit" at Miami, and who is writing a book on the Ark

Encounter), Steve Watkins of the University of Louisville (who has written a dissertation on the museum), and Carl Weinberg of Indiana University Bloomington (who is completing a book on creationism and politics).

Speaking of scholars on creationism, we do not have words to adequately convey our thanks to Ronald L. Numbers, the world's preeminent scholar of creationism. Ron was one of Bill's mentors at the University of Wisconsin and has remained so ever since. What is remarkable is that the number of people Ron has assisted, befriended, and inspired makes for a very long list, and it now includes Sue. This book could not have been written without Ron Numbers, and we are thrilled that it appears in his series at Johns Hopkins.

We are very fortunate to teach at the University of Dayton, which has been an extremely supportive environment for this project. Thanks to interim provost Paul Benson for his enormous assistance over the years, as well as dean Jason Pierce, associate dean Don Pair, and director of media relations Cilla Shindell. Thanks also to the department chairs who over the years have helped in various ways—Julius Amin, Sheila Hassell Hughes, Juan Santamarina, Andy Slade, Dan Thompson, and Sandra Yocum. There are too many UD colleagues and friends to mention here, but we can say that they have set a very high bar for both teaching and scholarship, and we are grateful for their example.

We do need to make specific mention of colleagues with whom we teach in the University of Dayton's Core Program. The first-year Core course is a yearlong, team-taught interdisciplinary class that includes faculty from English, history, philosophy, and religious studies and that traces the history of the West in its global context from the beginnings of civilization to the present. Those of us who have taught this course for a number of years find that we have been changed, by the course content, by the interdisciplinary collegiality, and by our remarkable students. This book is shaped by the fact that we teach in Core, as will surely be obvious to the faculty with whom we have taught over the past few years: David Darrow, Marilyn Fischer, John Inglis, Denise James, Alan Kimbrough, Tony Smith, and Una Cadegan.

As we worked on this book we were part of a group that met almost every Monday evening at our house. Brad and Jeannie Kallenberg, Johnelle and Rod Kennedy, and Sharon and Ed Wingham heard more than they probably ever wanted to hear about this book, but they were with us in the project and in our lives. They cared for us in multiple ways, and "thank you" is not nearly enough.

Finally, there are our four children, two in-laws, and one grandchild who have been extraordinarily supportive of our research and our writing. They are better human beings than we are, and they love us anyway. This book is for them.

NOTES

Introduction

1. "Membership with a Mission," Creation Museum, no date (accessed March 29, 2015), http://creationmuseum.org/members/mission; "Statement of Faith," *answersingenesis*, no date (accessed February 28, 2015), https://answersingene sis.org/about/faith.

2. While in this book we focus on young Earth creationism, the term "creationism" refers also to old Earth creationism, which has faded in popularity but has not disappeared. More than this, "intelligent design" is also a form of creationism, in that the similarities between intelligent design and young Earth creationism (even more so old Earth creationism) are much more salient than the differences. Nicholas J. Matzke, "The Evolution of Creationist Movements," *Evolution: Education and Outreach* 3 (June 2010): 145–162, http://link.springer .com/article/10.1007%2Fs12052-010-0233-1.

3. The last few decades have seen a remarkable flourishing of scholarship on the history of fundamentalism in the United States. This introduction—and the book as a whole—draws upon much of this scholarship. Please see the suggestions for further reading.

4. Regarding the latter and its role in precipitating the rise of the Christian Right, see Randall Balmer, *Redeemeer: The Life of Jimmy Carter* (New York: Basic Books, 2014), 102–108.

5. Luke Brinker, "Evolution and the GOP's 2016 Candidates: A Complete Guide," *Salon*, February 11, 2015, http://www.salon.com/2015/02/11/evolution _and_the_gops_2016_candidates_a_complet_guide; Heather Digby Parton, "Scott Walker Ably Demonstrates GOP Is Still the Party of Stupid," *Alternet*, February 18, 2015, http://www.alternet.org/scott-walker-ably-demonstrates-gop -still-party-stupid.

6. Ronald L. Numbers and Rennie Schoepflin, "Science and Medicine," in *Ellen Harmon White: American Prophet*, ed. Terrie Dopp Aamodt, Gary Land, and Ronald L. Numbers (Oxford: Oxford University Press, 2014), 214–216;

Ronald L. Numbers, *The Creationists: From Scientific Creationism to Intelligent Design*, expanded ed. (Cambridge, MA: Harvard University Press, 2006), 88–106.

7. Numbers, *Creationists*, 88–89, 116–117.

8. John C. Whitcomb and Henry M. Morris, *The Genesis Flood: The Biblical Record and Its Scientific Implications* (Phillipsburg, NJ: Presbyterian and Reformed, 1961), xx, 369; Numbers, *Creationists*, 116–119, 211–229.

9. Numbers, *Creationists*, 239–328.

10. "Ken A. Ham," http://www.creationmuseumnews.com/docs/bio_ham.pdf; Ken Ham, "Back to AiG Roots," *Ken Ham* (blog), November 28, 2012, http://blogs.answersingenesis.org/blogs/ken-ham/2012/11/28/back-to-aig-roots; Randall Stephens and Karl Giberson, *The Anointed: Evangelical Truth in a Secular Age* (Cambridge MA: Belknap Press of Harvard University Press, 2011), 39.

11. "The History of Answers in Genesis through February 2015," *answersin genesis*, https://answersingenesis.org/about/history; Margaret Buchanan, *Salem Revisited* (Brisbane, Australia: Margaret Buchanan, 1990), 10, 12, 15 (quotes); Numbers, *Creationists*, 365. For a skeptical but sympathetic look at Mackay, see Will Storr, *The Unpersuadables: Adventures with the Enemies of Science* (New York: Overlook, 2014), 1–20. For Ham and Mackay together again, see Ken Ham, "Creation Down Under," *Ken Ham* (blog), September 7, 2012, http://blogs.answersingenesis.org/blogs/ken-ham/2012/09/07/creation-down-under; Ken Ham, "Unexpected Message in an Unexpected Place," *Ken Ham*, August 27, 2013, http://blogs.answersingenesis.org/blogs/ken-ham/2013/08/27/unexpected-message-in-an-unexpected-place; Ken Ham, "Speechless and Convicted," *Ken Ham*, September 9, 2014, http://blogs.answersingenesis.org/blogs/ken-ham/2014/09/09/speechless-and-convicted.

12. Stephens and Giberson, *Anointed*, 41–43; Numbers, *Creationists*, 400–402; "History of Answers in Genesis." That the split was not amicable became obvious in 2007, when Creation Ministries International brought suit against Answers in Genesis. "Trouble in Paradise: Answers in Genesis Splinters," *Reports of the National Center for Science Education* 26 (2006), http://ncse.com/rncse/26/6/trouble-paradise; "Brisbane creationist group settles US legal dispute," *Brisbane Times*, April 29, 2009, http://www.brisbanetimes.com.au/queensland/brisbane-creationist-group-settles-us-legal-dispute-20090429-an0e.html.

13. "United States District Court Eastern District of Kentucky Central Division at Frankfort," 3; Corinne Ramey, "Meet the Creationist Group Building a Life-Size Noah's Ark," *Curbed*, July 8, 2015, http://curbed.com/archives/2015/07/08/answers-in-genesis-ark-encounter.php; "History of Answers in Genesis."

14. The following comes from a March 9, 2015, perusal of the website.

15. For more on the AiG enterprise, see chapter 5.

16. Ira S. Loucks [pseud.], "Fungi from the Biblical Perspective: Design and Purpose in the Original Creation," *Answers Research Journal* 2 (2009): 123–131, www.answersingenesis.org/contents/379/arj/v2/fungi_design_purpose.pdf; Anne Habermehl, "Where in the World Is the Tower of Babel?," *Answers Research Journal* 4 (2011): 25–53, www.answersingenesis.org/contents/379/arj/v4/Tower_Babel.pdf; Tom Hennigan, "An Initial Estimate toward Identifying and Numbering the Ark Turtle and Crocodile Kinds," *Answers Research Journal* 7 (2014): 1–10, www.answersingenesis.org/contents/379/arj/v7/ark_turtle_crocodile_kinds.pdf; Rod Martin, "A Proposed Bible-Science Perspective on Global Warming," *Answers Research Journal* 3 (2010): 91–106, https://answers ingenesis.org/environmental-science/climate-change/a-proposed-bible-sci ence-perspective-on-global-warming. Martin, who is identified as an "independent researcher," concludes his article with a list of "Global Warming Skeptics" materials, including a DVD produced by AiG, "Global Warming: A Scientific and Biblical Expose of Climate Change."

17. Callie Joubert, "The Unbeliever at War with God: Michael Ruse and the Creation-Evolution Controversy," *Answers Research Journal* 5 (2012): 125–139 (quote: 137), www.answersingenesis.org/contents/379/arj/v5/Michael_Ruse_creation_evolution_controversy.pdf; Danny Faulkner, "Interpreting Craters in Terms of the Day Four Cratering Hypothesis," *Answers Research Journal* 7 (2014): 11–25, www.answersingenesis.org/contents/379/arj/v7/craters_day_four.pdf.

18. "History of Answers in Genesis"; "Overturned! Creation Museum Finally Gets Approval from County!," *Creation Museum* (blog), http://creationmuseum .org/blog/1999/05/07/overturned-creation-museum-finally-gets-approval -from-county; "AiG Closes on Purchase of 47 Acres Near Cincinnati!," *Creation Museum* (blog), http://creationmuseum.org/blog/2000/05/04/aig-closes-on -purchase-of-47-acres-near-cincinnati; Stephens and Giberson, *Anointed*, 44.

19. This figure was reported in a lawsuit filed in early February 2015 by AiG (and other entities) against Kentucky state officials regarding AiG's most recent venture, the Ark Encounter, which we discuss in the epilogue. "United States District Court Eastern District of Kentucky Central Division at Frankfort" on behalf of Ark Encounter, LLC, Crosswater Canyon, Inc., and Answers in Genesis, Inc., against Bob Stewart (secretary of the Kentucky Tourism, Arts and Heritage Cabinet) and Steven Beshear (governor, Kentucky), 7, https://answersin genesis.org/religious-freedom/religious-discrimination-lawsuit-filed.

20. "Millionth Guest Visits Creation Museum," *answersingenesis*, April 26, 2010, https://answersingenesis.org/ministry-news/creation-museum/millionth -guest-visits-creation-museum; Ramey, "Meet the Creationist Group"; "United States District Court Eastern District of Kentucky," 15; "Return of Organization Exempt from Income Tax," Internal Revenue Service Form 990 (2012), 9.

21. For more on AiG's latest project, the Ark Encounter, see the epilogue.

Chapter 1. Museum

1. The claim about the trustworthiness of museums comes from the American Alliance of Museums (formerly, the American Association of Museums), which is an alliance of twenty-one thousand museums in the United States that, among other things, offers accreditation to qualified applicants such as the various Smithsonian Institutions in Washington, DC, along with hundreds of others. "Museum Facts," American Alliance of Museums, no date, accessed April 14, 2015, http://www.aam-us.org/about-museums/museums-facts. Sharon MacDonald, "Exhibitions of Power and Powers of Exhibitions: An Introduction to the Politics of Display," in *The Politics of Display: Museum, Science, Culture,* ed. Sharon MacDonald (New York: Routledge, 1998), 12; Andrew Barry, "On Interactivity: Consumers, Citizens, and Culture," *Politics of Display,* 100; Tony Bennett, *The Birth of the Museum: History, Theory, Politics* (New York: Routledge, 1995),47.

2. Randall J. Stephens and Karl W. Giberson make this point as well: "Passing under the watchful eyes of metallic stegosauruses atop the entrance gate, a tyrannosaur amid the trees, and sauropods lining the lobby, visitors have no idea they are entering anything other than a typical garden-variety museum of natural history." Randall J. Stephens and Karl W. Giberson, *The Anointed: Evangelical Truth in a Secular Age* (Cambridge, MA: Belknap Press of Harvard University Press, 2011), 22.

3. A. A. Gill, "Roll Over, Charles Darwin!," *Vanity Fair,* February 2010, http://www.vanityfair.com/culture/features/2010/02/creation-museum-201002; Matthew Wells, "Creation Museum Pushes 'True History,'" *BBC News,* December 11, 2006, news.bbc.co.uk/2/hi/Americas/6216788.stm; Lawrence Krauss, "Museum of Misinformation," *New Scientist,* May 26, 2007: 23, *Academic Search Complete,* EBSCO*host*; Daniel Phelps, "The Anti-Museum: An Overview and Review of the Answers in Genesis Creation 'Museum,'" *National Center for Science Education,* December 17, 2008, http://ncse.com/creationism/general/anti-museum-overview-review-genesis-creation-museum. Notably, Phelps draws his moniker for the Creation Museum from an article published on the AiG website that appeared in advance of the opening of the museum. In the closing paragraph of that article, Michael Matthews (identified by AiG as "Creation Museum scriptwriter") says: "By God's grace, AiG hopes to create the most dynamic, life-changing 'anti' museum ever built." Michael Matthews, "The Anti-Museum," *answersingenesis,* December 15, 2003, http://www.answersingenesis.org/museum/docs/031215_antimuseum.asp. Stephen T. Asma, "Solomon's House: The Deeper Agenda of the New Creation Museum in Kentucky," *Skeptic,* May 23, 2007, http://www.skeptic.com/eskeptic/07-05-23/; P. Z. Meyers, "The Creation 'Museum,'" *Pharyngula* (blog), *Science Blogs,* August 10, 2009, http://scienceblogs.com/pharyngula/2009/08/10/the-creation-museum-1; Stephen Bates,

"So What's with All the Dinosaurs?," *The Guardian*, November 12, 2006, http://www.guardian.co.uk/world/2006/nov/13/usa.religion.

4. Stephen Bates, "So What's with All the Dinosaurs?," *The Guardian*, November 12, 2006, http://www.guardian.co.uk/world/2006/nov/13/usa.religion; Edward Rothstein, "Adam and Eve in the Land of the Dinosaurs," *New York Times*, May 24, 2007, LexisNexis Academic, LexisNexis, http://www.nytimes.com/2007/05/24/arts/24crea.html?pagewanted=all&_r=0; Gordy Slack, "Inside the Creation Museum," *Salon*, May 31, 2007, http://www.salon.com/2007/05/31/creation_museum/; Daniel Radosh, *Rapture Ready! Adventures in the Parallel Universe of Christian Pop Culture* (New York: Soft Skull, 2010), 286.

5. Edward Rothstein, "Adam and Eve in the Land of the Dinosaurs," *New York Times*, May 24, 2007, LexisNexis Academic, LexisNexis. Julie Homchick rightly argues that the Creation Museum engages in a mimetic rhetoric that imitates or mimics natural history museums. This said, it matters a great deal whether the museum borrows from the forms characteristic of nineteenth-century museums or twentieth- and twenty-first-century museums. Julie Homchick, "Displaying Controversy: Evolution, Creation, and Museums," (PhD diss., University of Washington, 2009), UMI Microform (3377287).

6. In response to an article that appeared in the *Cincinnati Enquirer* that was critical of the Creation Museum's claim to be a museum, Ham refers to a description of museums from the American Alliance of Museums and a definition of museums from the International Council of Museums to argue that "The Creation Museum certainly fits under these definitions." Ken Ham, "Assistant Manager of Cincinnati Museum Center Derides Creation Museum," *Ken Ham* (blog), *answersingenesis*, June 7, 2013, http://blogs.answersingenesis.org/blogs/ken-ham/2013/06/07/assistant-manager-at-cincinnati-museum-center-derides-creation-museum/. In addition to those commentators mentioned above who contest the claim that the Creation Museum is a museum, mathematics professor Jason Rosenhouse argues that the Creation Museum is not a museum on the grounds that whereas "the appeal of a real museum is the opportunity to see actual physical exhibits and artifacts," the Creation Museum provides "very little that is comparable." Jason Rosenhouse, *Among the Creationists: Dispatches from the Anti-Evolutionist Front Line* (New York: Oxford University Press, 2012), 137.

7. Bruce W. Ferguson, "Exhibition Rhetorics: Material Speech and Utter Sense," in *Thinking about Exhibitions*, ed. Reesa Greenberg, Bruce W. Ferguson, and Sandy Nairne (New York: Routledge, 1996), 178–185.

8. Drawing on the work of Tony Bennett and others, scholarship in the field of rhetorical studies has paid particular attention to the stories that museums and memorials tell about the past that shape the way sense is made in the present and how those modes of making sense in turn shape their visitors. See esp., William V. Balthrop, Carole Blair, and Neil Michel, "The Presence of the Present: Hi-

jacking 'The Good War'?," *Western Journal of Communication* 74, no. 2 (2010): 170–210, doi:10.1080/10570311003614500; Barbara A. Biesecker, "Remembering World War II: The Rhetoric and Politics of National Commemoration at the Turn of the 21st Century," *Quarterly Journal of Speech* 88, no. 4 (2002): 393–409, *Communication and Mass Media Complete*, EBSCO*host*; Carole Blair, "Contemporary US Memorial Sites as Exemplars of Rhetoric's Materiality," in *Rhetorical Bodies*, ed. Jack Selzer and Sharon Crowley (Madison: University of Wisconsin Press, 1999): 16–57; Greg Dickinson, "Joe's Rhetoric: Finding Authenticity at Starbucks," *Rhetoric Society Quarterly* 32, no. 4 (2002): 5–27, http://www.jstor .org/stable/3886018; Greg Dickinson, "Memories for Sale: Nostalgia and the Construction of Identity in Old Pasadena," *Quarterly Journal of Speech* 83, no. 1 (1997): 1–27, *Communication and Mass Media Complete*, EBSCO*host*; Greg Dickinson, Brian Ott, and Eric Aoki, "Spaces of Remembering and Forgetting: The Reverent Eye/I at the Plains Indian Museum," *Communication and Critical/Cultural Studies* 3, no.1 (2006): 27–47, doi:10.1080/14791420500505619; Victoria J. Gallagher, "Remembering Together: Rhetorical Integration and the Case of the Martin Luther King, Jr. Memorial," *Southern Communication Journal* 60, no. 2 (1995): 109–119, doi:10.1080/10417949509372968; and Marouf Hasian, Jr., "Remembering and Forgetting the 'Final Solution': A Rhetorical Pilgrimage through the US Holocaust Memorial Museum," *Critical Studies in Media Communication* 21, no. 1 (2004): 64–92, doi:10.1080/073931804200 0184352. "Object" refers to anything—physical objects, processes, concepts, or whatever—that a museum displays as something one can know.

9. On the history and politics of museum display of objects and their relationship to the visitor and seeing, see Kevin Hetherington, "The Unsightly: Touching the Parthenon Frieze," *Theory, Culture, & Society* 19, no. 5/6 (2002): 187–205, *Communication and Mass Media Complete*, EBSO*host*. For a rhetorical analysis of the display of objects (in this case, guns) within a museum and its relationship to seeing, presence, and absence as well as its construction of the subject position of the museum visitor, see Brian L. Ott, Eric Aoki, and Greg Dickinson, "Ways of (Not) Seeing Guns: Presence and Absence at the Cody Firearms Museum," *Communication and Critical/Cultural Studies* 8, no. 3 (2011): 215–239, doi:10.1080/14791420.2011.594068. For a rhetorical analysis of a civil rights museum that contrasts traditional and nontraditional modes of displaying objects and considers their impact on how history is conceived as static (traditional) or fluid (nontraditional), see Victoria J. Gallagher, "Memory and Reconciliation in the Birmingham Civil Rights Institute," *Rhetoric and Public Affairs* 2, no. 2 (1999): 303–320, Communication and Mass Media Complete, EBSCO*host*.

10. For more on museums (and memorials) and their construction of subject positions, see Biesecker, "Remembering World War II"; Dickinson, Ott, and Aoki, "Spaces of Remembering"; Carole Blair and Neil Michel, "Reproducing

Civil Rights Tactics: The Rhetorical Performances of the Civil Rights Memorial," *Rhetoric Society Quarterly* 30, no. 2 (2000): 31–55, http://www.jstor.org/stable /3886159; and John Lynch, "'Prepare to Believe': The Creation Museum as Embodied Conversion Narrative," *Rhetoric and Public Affairs* 16, no. 1 (2013): 1–28, *Communication and Mass Media Complete*, EBSCO*host*.

11. Ferguson, "Exhibition Rhetorics," 181.

12. Bennett, *Birth of the Museum*, 126. In the field of rhetorical studies, Zagacki and Gallagher advance this kind of approach with their notion of "spaces of attention" within which visitors may consider new identities and relationships between themselves and nature (for instance) by way of their experience of moving through a multimodal space. Kenneth S. Zagacki and Victoria J. Gallagher, "Rhetoric and Materiality in the Museum Park at the North Carolina Museum of Art," *Quarterly Journal of Speech* 95, no. 2 (2009): 171–191, doi:10.1080/00335630902842087. Similarly, Ferguson argues that "Exhibitions are the material speech of what is essentially a political institution, one with legal and ethical responsibilities, constituencies and agents who act in relation to different sets of consequences and influences at any given historical moment." Ferguson, "Exhibition Rhetorics," 182. On the material rhetoric of memorials, see Blair, "Contemporary US Memorial Sites." Dickinson, Ott, and Aoki use the language of "experiential landscapes" to get at how the material speech of museums goes beyond the confines of the museum itself and includes the museum's surroundings as well as the "cognitive landscape" of its visitors. Dickinson, Ott, and Aoki, "Spaces of Remembering," 29.

13. Or, as Ferguson puts it, "The will to influence is at the core of any exhibition." Ferguson, "Exhibition Rhetorics," 179.

14. MacDonald, "Exhibitions of Power," 6–7; Bennett, *Birth of the Museum*, 95–96.

15. "Modern" here is meant to signify not "contemporary" but, instead, something that belongs to the cultural moment known as modernity.

16. Bennett, *Birth of the Museum*, 19–21, 41, 1–2. Bennett argues (borrowing from Michel Foucault) that, as the prisoner was being moved from the public scaffold, where his punishment could be witnessed as a lesson for all, to the inside of the penitentiary where he would modify his own behaviors by interiorizing the perspective and rule of the guard, what was once found in private curiosity cabinets was now to be found in the modern museum. As in the penitentiary, the contents were held indoors, but unlike the penitentiary the point of the museum was to make those contents visible to all for the general public's education and self-improvement. Ibid., 59–69.

17. Judith Barry discusses this relationship between the spectator and object under glass or situated in the diorama in "Dissenting Spaces," in *Thinking about Exhibitions*, ed. Reesa Greenberg, Bruce Ferguson, and Sandy Nairne (New York: Routledge, 1996), 308–309. Ott, Aoki, and Dickinson describe the kind

of looking that is encouraged by objects set behind glass as a distinctly Western style of looking that invites the visitor "to engage in surveillance, to see in a distanced, 'objective' way, to exercise a timeless, if occluded gaze." Ott, Aoki, and Dickinson, "Ways of (Not) Seeing Guns," 217.

18. For a discussion of this common strategy and its implications for the museumgoer, see Barbara Kirshenblatt-Gimblett, "Objects of Ethnography," in *Exhibiting Cultures: The Poetics and Politics of Museum Display*, ed. Ivan Karp and Steven D. Lavine (Washington DC: Smithsonian Institution Press, 1991), esp. 388–390.

19. Donna Haraway, "Teddy Bear Patriarchy: Taxidermy in the Garden of Eden, 1908–1936," *Social Text* 11 (Winter 1984–85), http://www.jstor.org/stable/466593, 24, 34–38.

20. Ibid., 24, 35–38.

21. Ibid., 20, 23–38.

22. Tracy Lang Teslow, "Reifying Race: Science and Art in *Races of Mankind* at the Field Museum of Natural History," in *The Politics of Display: Museums, Science, Culture*, ed. Sharon MacDonald (New York: Routledge, 1998), 55–58. Ott, Aoki, and Dickinson, "Ways of (Not) Seeing Guns," 217–218; Peter Markie, "Rationalism vs. Empiricism," *The Stanford Encyclopedia of Philosophy* (Summer 2013 Edition), ed. Edward N. Zalta, http://plato.stanford.edu/archives/sum2013/entries/rationalism-empiricism/.

23. For more on legibility, see Tony Bennett, "Speaking to the Eyes: Museums, Legibility and the Social Order," in *The Politics of Display: Museums, Science, Culture*, ed. Sharon MacDonald (New York: Routledge, 1998), 29. MacDonald, "Exhibitions of Power," 13. Importantly, the voice of those labels and other narratives in the modern museum was a disembodied male voice that at that time added to the sense that the stories told were universal. Bennett, *Birth of the Museum*, 32–33; Sharon MacDonald, "Afterword: From War to Debate," in *The Politics of Display: Museums, Science, Culture*, ed. Sharon MacDonald (New York: Routledge, 1998), 232–33; Teslow, "Reifying Race," 58. For more on specular dominance, see Bennett, *Birth of the Museum*, 66.

24. Bennet, *Birth of the Museum*, 96, 39.

25. MacDonald argues that objects were first organized in museums according to categories determined by similarities and differences in the seventeenth century. MacDonald, "Exhibitions of Power," 6–7. This new arrangement of objects was first made possible by the stratigraphical study of fossil remains according to which the location of a fossil in a particular layer of rock was taken as an indication of its place in the historical timeline. Bennett, *Birth of the Museum*, 77, 179–180. As both MacDonald and Bennett argue, much energy was dedicated in the design of modern museums and exhibits toward making both the objects on display and the underlying developmental logic that could

account for them "legible" to a mass public so that the narrative underlying it would be instantly recognizable. MacDonald, "Exhibitions of Power," 12; Bennett, "Speaking to the Eyes," 29–30.

26. Bennett, *Birth of the Museum*, 6.

27. Ibid., 43–46, 179–186. For a rhetorical analysis of the use of the path to structure the visitors' experience of a civil rights museum, see Gallagher, "Memory and Reconciliation."

28. Bennett, *Birth of the Museum*, 63–69. As Bennett puts it, "The exhibitionary complex [of the museum, arcade, department store, and the like] . . . perfected a self-monitoring system of looks in which the subject and object positions can be exchanged, in which the crowd comes to commune with and regulate itself through interiorizing the ideal and ordered view of itself as seen from the controlling vision of power—a site of sight accessible to all." Ibid., 69.

29. Ibid., 47. Bennett aptly terms this dynamic of watching and being watched "scopic reciprocity." Ibid., 51.

30. Ibid., 7, 98. MacDonald, "Exhibitions of Power," 9–12, Bennett, "Speaking to the Eyes," 30, and Barry, "On Interactivity," 100.

31. Although the exhibit was dismantled in 1966 after Malvina Hoffman (the artist who created the sculptures at the heart of the exhibit) died, many of the sculptures remain available for viewing at the Field Museum. "Malvina Hoffman," *The Field Museum*, no date, http://www.fieldmuseum.org/malvina -hoffman.

32. Teslow, "Reifying Race," 67.

33. Ibid., 55. As Teslow puts it, "the Field Museum offered a stable and familiar world order, frozen in bronze, at once humanitarian in its theme of unity and its preservation of endangered primitives, and reassuring in its reinforcement of the natural place of Europeans and Americans at the top of the evolutionary heap." Ibid., 61. For more on the role of modern museums in educating citizens of the nation-state in their racial superiority and the superiority of the nation compared to others especially in the context of imperialism, see MacDonald, "Exhibitions of Power," 9.

34. Barry, "On Interactivity," 103. For Oppenheimer's argument in his own words, see Frank Oppenheimer, "Rationale for a Science Museum," *Curator: The Museum Journal* 11, no. 3 (1968): 206–209, http://www.exploratorium.edu/ feiles/about/our_story/history/frank/pdfs/rationale.pdf. For his description of the interactive nature of the Exploratorium, see Frank Oppenheimer, "The Exploratorium: A Playful Museum Combines Perception and Art in Science Education," *American Journal of Physics* 40, no.7 (1972), http://www.explorato rium.edu/files/about/our_story/history/frank/pdfs/playful_museum.pdf.

35. MacDonald, "Exhibitions of Power," 13–16. Barry, "On Interactivity," 103.

36. Barry, "On Interactivity," 104. That said, the natural history museum has

held onto one of its signature forms—the diorama—for nearly a hundred years. Edward Rothstein, "Beyond Dioramas: Nature's New Story," *New York Times*, August 3, 2013. C1.

37. Barry, "On Interactivity," 101. Sharon MacDonald, "Supermarket Science? Consumers and 'the Public Understanding of Science,'" in *The Politics of Display: Museums, Science, Culture*, ed. Sharon MacDonald (New York: Routledge, 1998), 118–123. See also Davi Johnson, "Psychiatric Power: The Post-Museum as a Site of Rhetorical Alignment," *Communication and Critical/Cultural Studies* 5, no. 4 (2008): 344, doi:10.1080/14791420802412423

38. MacDonald, "Exhibitions of Power," 14–17. For an analysis of the supposedly democratizing effects of "new museums" (which have attempted to respond to these critiques through radical changes in "musealization") that also identifies some of their important limitations, see Jens Andermann and Silke Arnold-de Simine, "Introduction: Memory, Community and the New Museum," *Theory, Culture & Society* 29, no.1 (2012): 3–13, doi:10.1177/0263276411423041. For an analysis of the democratizing effects of postmodern architecture in public memorials, see Carole Blair, Marsha S. Jeppeson, and Enrico Pucci, Jr., "Public Memorializing in Postmodernity: The Vietnam Veterans Memorial as Prototype," *Quarterly Journal of Speech* 77, no. 3 (1991): 263–288, *Communication and Mass Media Complete*, EBSCO*host*.

39. At the Creation Museum website, readers can find a high-tech virtual tour in which they can "look" around most of the exhibits in the museum with some zoom capability (http://creationmuseum.org/whats-here/photo-preview/).

40. According to AiG, the exterior of the Creation Museum is not made of real stone but, instead, "lightweight hollow rock casts." Stacia McKeever, Gary Vaterlaus, and Diane King, *The Creation Museum: Behind the Scenes!* (Hebron, KY: Answers in Genesis, 2008), 27.

41. This film is discussed at length in chapter 4.

42. We note that the sign upon which these words appear was not present when the Museum opened but was added later.

43. The "Seven Cs" (and especially the first four) provide the narrative structure of the museum. They are Creation, Corruption, Catastrophe, Confusion, Christ, Cross, and Consummation.

44. This exhibit was added in May 2012 on the occasion of the Museum's fifth anniversary. "Stunning Exhibit Unveiled at Creation Museum: New High-Tech Display Opens on Museum's Fifth Anniversary," *answersingenesis*, May 28, 2012, https://answersingenesis.org/ministry-news/creation-museum/stunning -exhibit-unveiled-at-creation-museum/.

45. Answers in Genesis, *Journey through the Creation Museum: Prepare to Believe* (Green Forest, AR: Master Books, 2008), 39.

46. The implied claim of the scene with the *Tyrannosauras rex* is that prior

to human sin, all animals were vegetarian, and only after sin entered the world did some begin to eat one another. Ibid., 51.

47. Mark Looy, "$1.5 Million Dinosaur Exhibit Dedicated Today at the Creation Museum: Will Challenge Evolutionary Thinking," *answersingenesis*, May 23, 2014, https://answersingenesis.org/ministry-news/creation-museum/15-million -dinosaur-exhibit-dedicated-today/.

48. We use a masculine pronoun here since the museum refers to the Allosaurus as a male dinosaur named Ebenezer.

49. Answers in Genesis, *Journey*, 71.

50. Ibid., 73.

51. "Creation Museum Expands with Three World-Class Exhibits," *answers ingenesis*, May 24, 2013, https://answersingenesis.org/ministry-news/creation -museum/creation-museum-expands-with-three-world-class-exhibits/.

52. Answers in Genesis, *Journey*, 81.

53. ABC News, "Creation Museum Unites Adam, Eve, Dinos," *abcNews*, May 25, 2007, http://abcnews.go.com/GMA/Story?id=3211737&page=2; Jandos Rothstein, "Graphic Displays of Faith," *Print* 62, no. 1 (2008): 97, *Computer Source*, EBSCO*host*; Joseph Clark, "Specters of a Young Earth," *Triplecanopy*, December 1, 2008, www.canopycanopycanopy.com/4/specters_of_a_young_ earth; Lawrence Krauss, "Museum of Misinformation," *New Scientist*, May 26, 2007: 23, *Academic Search Complete*, EBSCO*host*.

54. The Creation Museum's home page can be found at creationmuseum .org. Answers in Genesis, *Journey*, 43.

55. Barbara Bradley Hagerty, "Creation Museum Promotes the Bible over Evolution," NPR, May 28, 2007, http://www.npr.org/templates/story/story.php?sto ryId=10498875; Bodie Hodge, "Is Christian Orthodoxy a Cult?," *answersingen esis*, June 11, 2010, http://www.answersingenesis.org/articles/2010/06/11/ feedback-is-christian-orthodoxy-a-cult; Stephen Bates, "So What's with All the Dinosaurs?," *The Guardian*, November 12, 2006, http://www.guardian.co.uk/ world/2006/nov/13/usa.religion.

56. Bruno Maddox, "Blinded by Science," *Discover*, February 2007, *Academic Search Complete*, EBSCO*host*.

57. Answers in Genesis, *Journey*, 13.

58. Ibid., 44.

59. As snapshots of the Creation story, these dioramas function in a manner similar to a photograph that also, as Barbie Zelizer argues, "catch[es] things in the middle." This is an important aspect of the dioramas in the Creation Museum not only for the ways that it gets visitors involved in a story that is unfolding right before their very eyes but also, as Zelizer argues, for the way that the subjunctive voice of the image invites visitors to think about each scene—how it might have turned out. Barbie Zelizer, "The Voice of the Visual in Memory,"

in *Framing Public Memory*, ed. Kendall R. Phillips (Tuscaloosa: University of Alabama Press, 2004), 158, 164.

60. We should note that the adaptations made here by the Creation Museum are reiterated in other contemporary natural history museums. That said, contemporary natural history museums also offer exhibitions that allow visitors to walk within a scene and some even invite visitors to touch elements of the scene. At the Field Museum in Chicago, for instance, an exhibit called the "Evolving Planet" included a stand of faux trees that visitors were free to walk among. For more on that exhibit, which opened on March 10, 2006, visit http://www.field museum.org/at-the-field/exhibitions/evolving-planet.

61. When the museum first opened, the Seven Cs were featured in a series of glass cases in the Portico. Thus, visitors encountered them as they entered the museum.

62. While the Last Adam Theater was a Creation Museum mainstay from its opening in 2007, sometime in 2015 the museum's website announced that the theater would be devoted to a new exhibit, with *The Last Adam* now to be shown in the Special Effects Theater.

63. Ken Ham and Stacia McKeever, *The Seven C's of History* (Hebron, KY: Answers in Genesis, 2004). A pdf of this booklet is available through the AiG online store at https://legacy-cdn-assets.answersingenesis.org/assets/pdf/radio/the7csofhistory.pdf.

64. A second edition appeared in 1917. More can be found on dispensational premillennialism in the introduction and on the *Scofield Reference Bible* in chapter 3.

65. Mark A. Noll, *The Scandal of the Evangelical Mind* (Grand Rapids, MI: Eerdmans, 1994), 195; Ronald L. Numbers, *The Creationists: From Scientific Creationism to Intelligent Design*, expanded ed. (Cambridge, MA: Harvard University Press, 2006), 371. Henry Morris's study Bible promotes dispensational premillennialism. See Henry M. Morris, ed., *The Henry Morris Study Bible* (King James Version), 1995, reprint (Green Forest, AR: Master Books, 2012), especially appendix 18, "Creation and Consummation—the Book of Revelation," 2127–2131.

66. According to Bodie Hodge of AiG, "Answers in Genesis does not take a position on this theological debate" between dispensational premillennialism and Covenant Theology, which offers another interpretive grid for a literal reading of the Bible, but sees both as biblical. Bodie Hodge, "Why Don't Christians Follow All the Old Testament Laws?," *answersingenesis*, May 6. 2013, https://answersingenesis.org/christianity/christian-life/why-dont-christians-follow-all-the-old-testament-laws/.

67. According to Frank, the power of dispensational premillennialism to explain every event in history and to predict the future has played a significant role in its popularity since the last decades of the nineteenth century. Douglas W.

Frank, *Less than Conquerors: How Evangelicals Entered the Twentieth Century* (Grand Rapids, MI: Eerdmans, 1986), 68–69, 73.

68. Robert O. Smith, *More Desired than Our Owne Salvation: The Roots of Christian Zionism* (New York: Oxford University Press, 2013), 151–154 (quote: 153); Frank, *Less than Conquerors*, 69–75 (quote: 71); Ernest R. Sandeen, *Roots of Fundamentalism: British & American Millenarianism, 1800–1930* (Chicago: Chicago University Press, 1970), 66–70.

69. Smith, *More Desired*, 154–158.

70. For more on the museum's efforts to link anti-Semitism and evolution, and particularly its promotion of the Darwin-to-Hitler trope, see chapter 4.

71. Noah J. Efron, *A Chosen Calling: Jews in Science in the Twentieth Century* (Baltimore: Johns Hopkins University Press, 2014), ix–xii.

72. Perhaps some of the museum's relative silence regarding the "church" has to do with the fact that AiG is a quintessentially evangelical "parachurch" organization with no formal ties or commitments to a particular ecclesial body.

73. AiG's preoccupation with the apostasy of Christian churches and leaders will be discussed in chapter 5.

74. As will be seen in chapter 4, the Creation Museum and AiG understand that it was Adam's, not Eve's, sin that brought God's judgment.

75. The one exception is the Facing the Allosaurus room. Visitors can enter that room either from the Flood Geology room or the Natural Selection room. There is only one exit.

76. Lynch argues that the Creation Museum moves its visitors, who are positioned as "seekers," through a "spatial sermon" that replicates the three-part structure of a conversion narrative: the creation of tension and dissonance, the resolution of that tension and dissonance through a symbolic death, and redemption through a new identity. According to Lynch, the museum creates the tension and dissonance early in the series of exhibits (from the Dig Site room through Culture in Crisis) but then resolves that tension and dissonance when the visitor enters the Six Days Theater, thereby enabling the visitor to experience a symbolic death of his/her prior self and a taking up of a new young Earth creationist self. If Lynch is right that the tension and dissonance early in the Walkthrough are resolved through the visitor's symbolic death of his or her former self, then why do the tension and dissonance reappear later in the museum (specifically, between scenes of the Fall and the Ark construction site)? Our reading of the narrative that drives the unidirectional path that is grounded in the Seven Cs better accounts for all of the rooms located along the Creation Museum's unidirectional path. John Lynch, "'Prepare to Believe.'"

77. Jackson offers an insightful rhetorical analysis of Hell Houses and their terrifying appeals to fear that borrow from the playbook of Jonathan Edwards but are even more powerful for the way they are made vivid and concrete via amateur dramaturgy. Brian Jackson, "Jonathan Edwards Goes to Hell (House):

Fear Appeals in American Evangelism," *Rhetoric Review* 26, no.1 (2007): 42–59, doi:10.1207/s15327981rr2601_3.

78. Ken Ham, "The Great Delusion: The Spiritual State of the Nation," Answers Mega Conference, Sevierville, Tennessee, July 24, 2013, http://www.answersin genesis.org/get-answers/features/the-great-delusion?utm_source=homepage &utm_medium=banner&utm_campaign=great-delusion-campaign.

79. Echoing classical dispensational premillennialism, the Ark serves the same function that the Rapture did. That is, it is the vehicle by which the righteous are spared the suffering of judgment and the End Times. On the Rapture's function and how its construction within classical dispensational premillennialism makes for a confident, self-satisfied, and self-righteous Christian, see Frank, *Less than Conquerors*, 80–83, 92–96.

80. In the Flood Geology room, this moment of double positioning is repeated. There a reproduction of a famous engraving by Gustave Dore, called *The Deluge*, appears on a large placard. Its title is "The Flood Drowns the Earth." The heading next to the image says, "The Flood was a global judgment," and the text below that says, "Human violence was worldwide, so judgment had to be worldwide. Animal violence was everywhere, so animals everywhere had to be destroyed. All land was cursed, so all land had to be restored." The miniature diorama in the Voyage room draws heavily from the design of Dore's image in that it too shows animals (adults and cubs) as well as men, women, and children about to perish on a rock being swallowed by rising floodwaters. The important innovation of the diorama is that it, unlike the engraving, includes the Ark.

81. Even when the Creation Museum does put technology in the hands of visitors (typically, in the form of touch screen computers), the purpose seems to be for visitors to answer preprogrammed questions that have right and wrong answers. Examples of this can be seen in the Ark Construction room and in the Christ, Cross, Consummation room (just outside the Last Adam Theater).

82. Molly Worthen, *Apostles of Reason: The Crisis of Authority in American Evangelicalism* (Oxford: Oxford University Press, 2014), 224–225, 258. Numbers' work adds nuance to Worthen's when he points out that not all young Earth creationists had the same epistemology even as they professed presuppositionalism. The authors of *The Genesis Flood* serve as one example of this. Whereas John C. Whitcomb believed that "the character and purposes of God could be found only through the Bible . . . rather than on the basis of external evidences," Henry M. Morris "believed that God had revealed himself in two books, nature and the Bible, which could be studied independently, though with priority given to the Scriptures." Numbers, *Creationists*, 225.

83. For a discussion of the concept of totalizing history, see Ernesto Laclau and Chantall Mouffe, *Hegemony and Social Strategy: Toward a Radical Democratic Politics*, 2nd ed. New York: Verso, 2001. Others have noted that while the use of advanced technologies may enable museum visitors to engage exhibits

in more active ways that may encourage democratic politics, this is not always the case. See, for instance, Dickinson, Ott, and Aoki, "Spaces of Remembering."

84. The fact that the Creation Museum is located in an isolated spot and, thus (unlike so many museums in America) does not make its claims amidst the claims of other museums and the fact that it resembles an exurban gated community is not surprising, given this reading. Like the one totalizing history that it tells, the museum itself is situated as the one-and-only voice and story. For more on the location and architectural appearance of the Creation Museum, see Lynch, "'Prepare to Believe,'" and Clark, "Specters of a Young Earth," *Triplecanopy*, December 1, 2008, www.canopycanopycanopy.com/4/specters_of_a_ young_earth.

85. Casey Ryan Kelly and Kristen E. Hoerl, "Genesis in Hyperreality: Legitimizing Disingenuous Controversy at the Creation Museum," *Argumentation and Advocacy* 48, no. 3 (2012): 125. *Communication and Mass Media Complete*, EBSCO*host*.

86. Answers in Genesis, *Journey*, 43.

87. Kelly and Hoerl, "Genesis in Hyperreality," 125–126, 132–133, 135. Kelly and Hoerl take this point further when they claim, "All of the fossils in the Creation Museum are replicas donated by private individuals." Ibid., 132. They offer no evidence for this claim, and other commentators suggest otherwise.

88. Ken Ham, "'Not What I Expected!,'" *answersingenesis*, July 18, 2011, https://answersingenesis.org/ministry-news/core-ministry/not-what-i-expected/.

89. The construction of the Garden as a real historical location enhances the visitor's experience of the underlying narrative of the Creation Museum as a stabilizing force for their lives and, especially, their identities in a time of perceived confusion and chaos. For an exploration of the relationships among constructed space, memory, and identity, see Dickinson, "Memories for Sale."

Chapter 2. Science

1. "'Science Guys' Garner Huge Attention for Evolution/Creation Debate at the Creation Museum," *answersingenesis*, February 3, 2014, https://answersin genesis.org/creation-vs-evolution/science-guys-garner-huge-attention-for-evo lutioncreation-debate-at-the-creation-museum/.

2. Jerry A. Coyne, "Bill Nye Talks about His Upcoming Debate with Ken Ham," *Evolution Is True* (blog), January 8, 2014, http://whyevolutionistrue.word press.com/2014/01/08/bill-nye-talks-about-his-upcoming-debate-with-ken -ham/; Dan Arel, "Why Bill Nye Shouldn't Debate Ken Ham," Richard Dawkins Foundation, January 16, 2014, https://richarddawkins.net/2014/01/why-bill -nye-shouldnt-debate-ken-ham/ (Ham quoted this claim by Arel in his opening remarks in the debate); Josh Rosenau, "Getting Ready for the Nye-Ham Debate," *National Center for Science Education*, February 4, 2014, http://ncse .com/blog/2014/02/getting-ready-nye-ham-debate-0015367. Notably, there is

disagreement regarding this claim that evolutionists should not debate (or otherwise seriously engage creationists) even among evolutionists. For more, see Paul Nelson, "Jerry, PZ, Ron, Faitheism, Templeton, Bloggingheads, and All That—Some Follow-Up Comments," *Uncommon Descent*, July 26, 2009, http:// www.uncommondescent.com/intelligent-design/jerry-pz-ron-faitheism-tempeton-bloggingheads-and-all-that-some-follow-up-comments/.

3. Greg Laden, "Bill Nye on the Inside Story of the Nye-Ham Debate," *Greg Laden's Blog* (blog) April 17, 2014, http://scienceblogs.com/gregladen/2014/04/17/ bill-nye-on-the-inside-story-of-the-nye-ham-debate/. Dan Arel, "Did Bill Nye Hurt Science?," Richard Dawkins Foundation, February 5, 2014, https://richard dawkins.net/2014/02/did-bill-nye-hurt-science-2/. According to the *Christian Science Monitor*, Jerry A. Coyne stated, "The debate was Ham's to win and he lost. And the debate was Nye's to lose and he won." Sudeshna Chowdhury, "Bill Nye versus Ken Ham: Who Won?," *Christian Science Monitor*, February 5, 2014, http://www.csmonitor.com/Science/2014/0205/Bill-Nye-versus-Ken-Ham -Who-won-video.

4. Don Boys, "Ham Won the Debate but No Grand Slam!," *Don Boys: Common Sense for Today* (blog), donboys.cstnews.com/ham-won-the-debate-but-no -grand-slam. See, for instance, Tim Gilleand, "Why Ken Ham Won the Debate Whether You Agree with His Position or Not," *Grace with Salt*, gracesalt/word press.com/tag/evolution/page/4/ and Shari Abbott, "Who Won the Bill Nye/ Ken Ham Debate?," *Reasonsforhope*, February 22, 2014, http://www.rforh.com/ blog/nye-ham-debate/. Ezra Byer, another conservative Christian blogger argued that Ham won by losing since, although Nye technically was the stronger debater, Ham *"raised huge awareness about Biblical Creationism on a mammoth scale," "connected with Christians who did not think the age of the earth was a big deal,"* and *"preached the gospel over four times to hundreds of thousands of people!"* (emphasis in original). Ezra Byer, "Why I Believe Ken Ham Won by Losing the Debate," *Powerline: Kingdom Network* (blog) February 4, 2014, http:// www.powerlinekingdom.com/kingdom-news/why-i-believe-ken-ham-won-by -losing-the-debate. AiG said, "Both men conducted themselves respectfully and calmly. Their arguments were clear, pointed, and demonstrated a passion for their views." As to who won, AiG concluded that "the overarching victory for Christians in this debate is that the gospel of Jesus Christ was shared with the millions of people watching the debate." "Debate Reactions Mixed," Answers in Genesis, February 6, 2014, https://answersingenesis.org/creation-vs-evolution/ debate-reactions-mixed/.

5. Michael Schulson, "The Bill Nye-Ken Ham Debate Was a Nightmare for Science," *The Daily Beast*, February 5, 2014, http://www.thedailybeast.com/ar ticles/2014/02/05/the-bill-nye-ken-ham-debate-was-a-nightmare-for-science .html. Of course, in a debate like this one wherein the debaters give long speeches and rarely addressed one another directly, it is extremely difficult to assess who

won. In academic debate, judges have a clear sense of the standard issues that should be addressed and assess one side's arguments against the other's. In this exchange, the criteria for who won ranged broadly from credible scientific evidence to whether the Gospel was preached.

6. Douglas Quenqua, "Debating Evolution and Dealing with Climate Change," *New York Times*, February 10, 2014, http://www.nytimes.com/2014/02/11/sci ence/debating-evolution-and-dealing-with-climate-change.html?_r=0. Nate Anderson confirmed the *New York Times* number. Nate Anderson, "Ham on Nye: The High Cost of 'Winning' an Evolution/Creation Debate," February 7, 2014, http://arstechnica.com/science/2014/02/ham-on-nye-the-high-cost-of-win ning-an-evolutioncreation-debate/. "Did Bill Nye Win the Creation Debate?," *MSNBC Live*, last updated February 5, 2014, http://www.msnbc.com/msnbc -live/watch-now-bill-nye-debates-ken-ham. Of course, if eight hundred thousand were registered, likely more actually watched since one person may have registered for multiple viewers. Sudeshna Chowdhury, "Bill Nye versus Ken Ham: Who Won?," *Christian Science Monitor*, February 5, 2014, http://www.cs monitor.com/Science/2014/0205/Bill-Nye-versus-Ken-Ham-Who-won-video. "Over 3 Million Tuned in Live for Historic Bill Nye and Ken Ham Evolution/ Creation Debate," *answersingenesis*, February 5, 2014, https://answersingenesis .org/creation-vs-evolution/over-3-million-tuned-in-live-for-historic-bill-nye -and-ken-ham-evolutioncreation-debate. "Bill Nye Debates Ken Ham-HD (Official)," *YouTube*, accessed October 30, 2014, https://www.youtube.com/watch? v=z6kgvhG3AkI.

7. "Over 3 Million Tuned in Live for Historic Bill Nye and Ken Ham Evolution/ Creation Debate," *answersingenesis*, February 5, 2014, https://answersingenesis .org/creation-vs-evolution/over-3-million-tuned-in-live-for-historic-bill-nye -and-ken-ham-evolutioncreation-debate.

8. James Hoskins, "Bill Nye vs. Ken Ham: Continuing Our Long American Tradition of Spectacle and Culture War," *Christ and Pop Culture*, January 7, 2014, http://www.patheos.com/blogs/christandpopculture/2014/01/bill-nye -vs-ken-ham-continuing-the-american-tradition-of-spectacle-and-culture -war/. "Scopes II" showed up in lots of other places on the Internet typically in reference to other unnamed people calling it that. AiG picked it up too and reiterated it in Ken Ham, "History: The History of Answers in Genesis through May 2014," *answersingenesis*, accessed July 10, 2014, https://answersingenesis .org/about/history/. Not surprisingly, other trials have been given this moniker. See, for instance, Jeffrey Katz, "Scopes 2: Arkansas' Creationism Trial," NPR, July 5, 2005, www.npr.org/templates/story/story.php?storyId=4726786. Of course, Ham is not the first creationist to debate a scientist. For instance, Henry Rimmer debated retired biologist Christian Schmucker in 1930. See Ronald L. Numbers, *The Creationists: From Scientific Creationism to Intelligent Design*, expanded ed. (Cambridge, MA: Harvard University Press, 2006), 82.

9. Although some commentators at the time thought that creationism had been put to rest at the Scopes trial, just the opposite was underway as conservative Christian leaders became convinced that a fight against evolution should be their top priority. George M. Marsden, *Understanding Fundamentalism and Evangelicalism* (Grand Rapids, MI: Eerdmans, 1991), 148.

10. In the immediate aftermath of the trial there were few fundamentalists who would join Seventh-day Adventist George McCready Price (who already in the 1920s argued that "day" in Genesis 1 means a literal twenty-four-hour day) in objecting to William Jennings Bryan's performance on the stand. By the 1980s, however, prominent fundamentalists such as Henry Morris, Jerry Falwell, and their many followers lamented Bryan's testimony on the stand that indicated he did not subscribe to a literal six-day, twenty-four-hour creation. Ronald L. Numbers, *Darwinism Comes to America* (Cambridge, MA: Harvard University Press, 1998), 82–89. For an insightful analysis of the Scopes trial and especially William Jennings Bryan's performance as a template for subsequent young Earth creationist arguments, see Edward Caudill, *Intelligently Designed: How Creationists Built the Campaign against Evolution* (Urbana: University of Illinois Press, 2013).

11. Marsden, *Understanding Fundamentalism*, 162–163. Other young Earth creationists, specifically members of the Institute for Creation Research, had debated evolutionists in the past, and although those debates often occurred on university campuses with significant audiences, none garnered the kind of national audience that the Ham–Nye debate did. Numbers, *Creationists*, 316.

12. Of course, Ken Ham is not the first conservative Christian to adopt this posture. Other antievolutionists, like George McCready Price, embraced science as far back as the mid-1920s. Numbers, *Creationists*, 88–119.

13. All direct references to the content of the debate are based on a transcript produced by the authors of this book on the basis of the DVD produced by AiG. *Uncensored Science: Bill Nye Debates Ken Ham*, aired February 4, 2014 (Hebron, KY: Answers in Genesis-USA, 2014), DVD.

14. On our count, Ham spoke 5,039 words in the course of the debate and devoted 1,813 of them to arguments that either focused on the importance of making the distinction between historical science and observational science or that mobilized that distinction explicitly to make another, related argument.

15. We are not the first to make this claim. See Casey Ryan Kelly and Kristen E. Hoerl, "Genesis in Hyperreality: Legitimizing Disingenuous Controversy at the Creation Museum," *Argumentation and Advocacy* 48 (2012): 123–141, *Communication and Mass Media Complete*, EBSCOhost.

16. *Uncensored Science*. This definition similarly appears in Ham's books. See Ken Ham, *The Lie: Evolution/Millions of Years*, rev. ed. (Green Forest, AR: Master Books, 2012), 35, and Ken Ham, *Six Days: The Age of the Earth and the Decline of the Church* (Green Forest, AR: Master Books, 2013), 58.

17. Ham, *Six Days*, 58. Ham's colleague, Roger Patterson (who has written books on Earth science and biology for AiG) defines it similarly: "A systematic approach to understanding that uses observable, testable, repeatable, and falsifiable experimentation to understand how nature commonly behaves." Roger Patterson, "What Is Science?," *answersingenesis*, February 22, 2007, https://answers ingenesis.org/what-is-science/what-is-science/. Ham's definition of observational science echoes Henry Morris's definition of science, according to which science is "the organized body of observation and experiment on present processes." Quoted in Numbers, *Creationists*, 268. John S. Park made a similar distinction between "'actually known and observed scientific facts," and "speculat[ion] on philosophical subjects such as organic evolution" back in the mid-1920s. Quoted in Numbers, *Creationists*, 64–65.

18. Ham, *Six Days*, 58.

19. Ibid., 59; Ham, *Lie*, 35.

20. Patterson, of AiG, defines historical science as "interpreting evidence from past events based on a presupposed philosophical point of view." Roger Patterson, "What Is Science?," *answersingenesis*, February 22, 2007, https://an swersingenesis.org/what-is-science/what-is-science/.

21. Ham, *Six Days*, 59.

22. Ham explains the relationship this way: "Genesis tells us that because of man's wickedness, God judged the world with a worldwide Flood. If this is true, what sort of evidence would we find? We could expect that we would find billions of dead things (fossils) buried in rock layers, laid down by water and catastrophic processes over most of the earth. This is exactly what we observe. Observational science confirms the Bible's historical science." Ham, *Lie*, 59.

23. Ham is joined by his colleagues at AiG in making these arguments. See, for instance, Roger Patterson, "What Is Science?," *answersingenesis*, February 22, 2007, https://answersingenesis.org/what-is-science/what-is-science. See also Troy Lacey, "Deceitful or Distinguishable Terms—Historical and Observational Science," *answersingenesis*, June 10, 2011, https://answersingenesis.org/what-is -science. These arguments also appear on the AiG website without authorial attribution, indicating that they are the official position of the organization. See "Two Kinds of Science," *answersingenesis*, accessed July 6, 2014, https://answers ingenesis.org/what-is-science/two-kinds-of-science/.

24. Ham, *Lie*, 35.

25. Ham, *Six Days*, 59. It may be more accurate to say that Ham objects to what Thomas M. Lessl, borrowing from Michael Ruse, calls "evolutionism" rather than "evolution." Evolutionism, according to Lessl, is a worldview whereas evolution is a science. For a rhetorical analysis of the difference and its impact on debates about creationism and evolution, see Thomas M. Lessl, *Rhetorical Darwinism: Religion, Evolution, and the Scientific Identity* (Waco, TX: Baylor University Press, 2012).

26. Efforts to move the boundaries of what is to be considered science are by no means new to young Earth creationists like Ham. Since at least the mid-nineteenth century, scientists have argued over the proper boundaries for science with the results of those arguments having profound effects on, among other things, whose work is considered legitimate, who is considered a real scientists or a pseudoscientist, and what responsibilities science and scientists rightly carry as a result of their work. For more on the history of those boundary struggles and the rhetorical strategies used therein, see Thomas F. Gieryn, "Boundary Work and the Demarcation of Science from Non-science: Strains and Interests in Professional Ideologies of Scientists," *American Sociological Review* 48, no. 6 (1983): 781–795. Ella Butler makes the additional argument that this line of argument (that evolution is not science) as it appears in the Creation Museum powerfully draws on the discourse of conspiracy theory as it positions the creationist as the subject uniquely positioned to unmask the sham of mainstream science. Ella Butler, "God Is in the Data: Epistemologies of Knowledge at the Creation Museum," *Ethnos* 75, no. 3 (2010): 237.

27. George McCready Price, the Seventh-day Adventist who wrote *The New Geology* (1923), which argued that the catastrophe of the Flood could account for the scientific evidence that seemed to indicate an old Earth, and who "more than any other creationist of his time . . . concerned himself with questions of the scientific method," not only argued that facts do not speak for themselves but that they must be interpreted. He also "use[d] this knowledge to attempt to place creation on an equal footing with evolution." Numbers, *Creationists*, 107–108.

28. On the emergence of creation science, see Numbers, *Creationists*, 268–285. As Numbers argues: "By narrowly drawing the boundaries of science and emphasizing its empirical nature, creationists could at the same time label evolution as false science, claim equality with scientific authorities in comprehending facts, and deny the charge of being antiscience." Numbers, *Creationists*, 65.

29. This argument dates back to the emergence of creation science in the mid-1970s when "scientific creationists granted creation and evolution equal scientific standing. Instead of trying to bar evolution from the classroom, as their predecessors had done in the 1920s, they fought to bring creation into the schoolhouse and repudiated the epithet, 'antievolutionist.'" Numbers, *Creationists*, 269.

30. In Butler's analysis of the Creation Museum's mobilization of evidence such as this, she argues that scientific evidence is both denied the power to support a claim and mobilized in order to show that the Bible is literally true. Butler, "God Is in the Data," 241.

31. As quoted in Jason Lisle, "Just Show the 'Evidence,'" *answersingenesis*, May 16, 2008, https://answersingenesis.org/presuppositions/just-show-the-evidence/. Since AiG did not set this quote off in quotation marks (they used a gray

vertical bar in the margin instead), we have set it off in regular quotation marks rather than single quotation marks.

32. Lisle, "Just Show the 'Evidence.'"

33. Stacia McKeever,Gary Vaterlaus, and Diane King, *The Creation Museum: Behind the Scenes!* (Hebron, KY: Answers in Genesis, 2008), 72.

34. *Uncensored Science.* What Ken Ham refers to here as "the scientific method," is not as straightforward as it might seem given that the term has meant many different ways of doing science even just within the twentieth century. For a history of the term from the nineteenth century to the twenty-first, see Daniel P. Thurs, "Scientific Methods," in *Wrestling with Nature: From Omens to Science*, ed. Peter Harrison, Ronald L. Numbers, and Michael H. Shank (Chicago: University of Chicago Press, 2011): 307–335.

35. Henceforth whenever "real science" appears in quotation marks it refers to AiG's definition of what constitutes true science.

36. The other film shown there every day is *Fires in the Sky: The Sun-Grazing Comets*, which was added in 2013. Our analysis focuses on *Created Cosmos* since it has been shown in the planetarium since the Creation Museum opened. "Stargazer's Planetarium," Creation Museum, n.d., accessed August 3, 2014, http://creationmuseum.org/whats-here/exhibits/planetarium/.

37. *Created Cosmos*, special edition (Hebron, KY: Answers in Genesis-USA, 2011), DVD. All quotations from the film that appear in this chapter were taken directly from the DVD.

38. Ibid.

39. For a history of intelligent design and its arguments, see Numbers, *Creationists*, 373–398. Importantly, AiG does on occasion use the language of intelligent design when, for instance, it argues in a description of the Wonders of Creation room that "there has to be a powerful Intelligence behind the universe." Answers in Genesis, *Journey through the Creation Museum: Prepare to Believe* (Green Forest, AR: Master Books, 2008), 39. However, in general AiG seems to resist the intelligent design movement on the grounds that it "fails to reference the God of the Bible and the Curse's impact on a once-perfect world." "ID'ed for a [*sic*] Imperfect Argument," *answersingenesis*, October 1, 2010, https://answersingenesis.org/intelligent-design/ided-for-a-imperfect-argument/. Given AiG's apparent discomfort with the intelligent design movement, we refer here to a "divine designer" rather than an "intelligent designer" in order to signal that for AiG the designer must be the God of the Bible.

40. *Created Cosmos.*

41. The one that does not have a voiceover is "Creator Clearly Seen." "Creator Clearly Seen," *Life* (Petersburg, KY: Answers in Genesis, 2007), DVD.

42. To calculate this percentage, we divided the amount of time dedicated in the video to the presentation of information on observational science (111 seconds) by the total time of the video (from the moment the voiceover begins until

it ends—152 seconds). We used the time indicated on our DVD player to calculate these durations. The duration of each video as indicated by our DVD player differed from the duration of the videos reported on the AiG DVD case, so we checked the timer on our DVD player against a stopwatch and found it to be correct. According to our DVD player, the "Solar System" video is one of the three longest videos with the "Creator Clearly Seen" video (which has no voiceover) being the longest at about four minutes and thirty-five seconds. "Solar System," *Heaven and Earth* (Petersburg, KY: Answers in Genesis, 2007), DVD.

43. We calculated that the video lasts seventy-five seconds (from the beginning of the voiceover to the closing scriptural quotation) and that the presentation of descriptive information takes fifty-two of those seventy-five seconds (68 percent). "Stars," *Heaven and Earth* (Petersburg, KY: Answers in Genesis, 2007), DVD.

44. "Designed for Flight," *Life* (Petersburg, KY: Answers in Genesis, 2007), DVD.

45. "DNA," *Heaven & Earth* (Petersburg, KY: Answers in Genesis, 2007), DVD.

46. These arguments are found in the following videos: "Eyes," "Common Designer," "Kinds," and "Communities." All four of these videos can be found on *Life* (Petersburg, KY: Answers in Genesis, 2007), DVD.

47. "Eyes," *Life* (Petersburg, KY: Answers in Genesis, 2007), DVD.

48. "Designed for Flight." Hebrew text also appears upside down in the Creation Museum.

49. "Communities." The full text of these verses is the following: "There is one body and one Spirit, just as you were called to the one hope of your calling, one Lord, one faith, one baptism, one God and Father of all, who is above all and through all and in all." "Ephesians 4:4–6," *New Revised Standard Version (NRSV)*, BibleGateway.com, https://www.biblegateway.com/passage/?search=Ephesians +4&version=NRSV.

50. "Solar System."

51. An exhibit that was added in May 2013, Dr. Crawley's Insectorium, features lighted cases displaying lots of insects, informative placards, an animatronic scientist, a video of an insect metamorphosis, and more. Not surprisingly, several of the placards argue much like the videos in the Wonders of Creation room that the beauty, complexity, or apparent purposefulness that can be seen in insects confirms a special creation insofar as it suggests a divine designer. A few placards make the additional move, again as in the Wonders of Creation room, of quoting scripture to suggest that the divine designer is the God of the Bible.

52. Whereas in the Wonders of Creation room all of the flat screens show all fifteen videos in the same continuous loop (so that the same video is shown on all the screens at once), in the Flood Geology room, each flat screen only plays the one video associated with the placards nearby. It seems that with the

addition of the Allosaurus exhibit, the number of videos on display in the Flood Geology room decreased from ten to seven.

53. As Ronald L. Numbers tells the story, "In bringing *The Genesis Flood* to a climax, Morris stressed the central role of flood geology in adjudicating the conflict between creation and evolution. If it could be established that the fossil-bearing strata had been deposited during the brief period of Noah's flood, he argued passionately, then 'the last refuge of the case for evolution immediately vanishes away, and the record of the rocks becomes a tremendous witness . . . to the holiness and justice and power of the living God of Creation!' Clearly, Christians could ignore the issue only at their peril" (ellipses in original). Numbers, *Creationists*, 229. For more on Price and *The New Geology*, see Numbers, *Creationists*, 88–119.

54. Roger Patterson, "Chapter 1: What Is Science?," *answersingenesis*, February 22, 2007, https://answersingenesis.org/what-is-science/what-is-science/. Notably, Patterson's definition of historical science in this quote says, "Historical (origins) science involves interpreting evidence *from the past* and includes the models of evolution and special creation" (emphasis added). According to that definition even observations made in the present of objects *from the past* would not count as "real science." Applied seriously, this definition would appear to render all observations of rocks, sediment, fossils, or anything "from the past" as not real science. On this logic, the whole field of geology (along with any other field that studies objects from the past) would have to be considered not scientific.

55. Other placards that use evidence from the Mount Saint Helens eruption that cannot be considered observational science include "County Scale" in the Flood Geology room and "The Land Recovers" in the Post-Flood room.

56. One might argue that footage from the eruption can be observed in the present. But any footage would provide only a certain perspective on the eruption since the footage would have been taken from a particular location at a particular time. No footage could count as a full repetition of the event that would allow a scientist to conduct whatever measurements were desired and so forth.

57. "Volcano Hazards Program: Crater Lake," USGS, last updated December 5, 2013, http://volcanoes.usgs.gov/volcanoes/crater_lake/; "Volcano Hazards Program: Yellowstone Volcano Observatory," USGS, last updated May 2, 2012, http://volcanoes.usgs.gov/volcanoes/yellowstone/yellowstone_geo_hist_88.html.

58. *Uncensored Science.*

59. Mark A. Noll, *The Scandal of the Evangelical Mind* (Grand Rapids, MI: Eerdmans, 1994), 178, 197. George Marsden dates the privileging of this approach to science all the way back to the eighteenth century. See Marsden, *Understanding Fundamentalism*, 162. For another account of the evangelical commitment to this kind of science, see Christopher M. Rios, *After the Monkey Trial: Evangelical Scientists and a New Creationism* (New York: Fordham University Press, 2014), 20–21. For a history of Presbyterian uses of this sort of

science to protect the Bible from threats of secular science, see Theodore Dwight
Bozeman, *Protestants in an Age of Science: The Baconian Ideal and Antebellum
American Religious Thought* (Chapel Hill: University of North Carolina Press,
1977), especially 101–131.

60. Numbers, *Creationists*, 274. Butler argues that the Creation Museum
adopts the discourse of science and at the same time "destabilises scientific
claims to objectivity" by arguing that "scientific practice is always influenced by
the starting assumptions of its practitioners." Butler, "God Is in the Data," 231.

61. As stated here, seven of the thirteen remaining placards appear to display observational science and ground arguments for a biblical creation in them.
The other six placards (of the thirteen) also appear to display observational science but do not ground arguments for a biblical creation in that observational
science. The reasoning at work in those six placards is analyzed below in a discussion of Ken Ham's scientific method.

62. For AiG's explanation of the biblical term "kind," see Bodie Hodge and
Georgia Purdom, "What Are 'Kinds' in Genesis," *answersingenesis*, April 16, 2013,
https://answersingenesis.org/creation-science/baraminology/what-are-kinds
-in-genesis/. For a history of the emergence of the notion of kinds and "bar-
aminology"—the study of biblical kinds—within young Earth creationism, see
Numbers, *Creationists*, 100, 150, and Ronald L. Numbers, "Ironic Heresy: How
Young-Earth Creationists Came to Embrace Rapid Microevolution by Means of
Natural Selection," in *Darwin Heresies*, ed. Abigail Lustig, Robert J. Richards,
and Michael Ruse (Cambridge: Cambridge University Press, 2004), 84–100.

63. It should be pointed out that at least one piece of evidence—"sand grains
in the Coconino Sandstone come from the Appalachian Mountains"—is not observational since the "coming" of the Sandstone occurred in the past and, therefore, cannot be observed in the present.

64. "Rock Layers," *Flood Geology* (Hebron, KY: Answers in Genesis, 2007)
DVD.

65. Butler describes the rhetorical strategy of placards like this one in the
following manner: "The text displays the typical rhetorical style used throughout the museum: it begins with what might be termed a scientifically neutral
descriptive statement about the specimen under discussion, then it poses a
problem that science is apparently puzzled by, suggesting that it is not 'not fully
understood,' and then the text forecloses that problem with the key explanatory
evidence provided by the Bible. The fossil thus can be read by the viewer as the
material evidence of Genesis; the fossil is an indexical trace of Noah's Flood, and
the particular conditions under which the fossilisation occurred attest to the
infallibility of the Bible" (239). Although this line of argument appears often in
certain areas of the museum, it is not "the typical rhetorical style used throughout the museum." Many other argumentative strategies are on display, including the move from phenomena in the present to their construction as beautiful,

complex, or purposeful, to the claim that a divine designer must have created them; the move from historical claim to model to supposed observational science that confirms the model; or the move from a biblical claim to an interpretation of supposed observational science; or the construction of an instance in the present (or past) as an analogy for a biblical event (such as the Flood).

66. Importantly, the definition of observational science, which may at first seem straightforward, seems to collapse when scrutinized. We have counted the information on this placard as observational science because river floods happen again and again and, thus, presumably can be observed in the present. They are not events that only happened in the past. That said, any particular river flood happens in one moment in time that cannot be repeated ever again as it happened that one time. Like a Mount Saint Helens eruption, any particular flood recedes into the past and cannot be observed or measured in the present. Does the data on this placard qualify as observational science? For the purposes of this analysis, we are counting this as observational science. But it just may be that, in the end, the whole notion of observational science by which evolution is dismissed as not real science fails.

67. "Canyons," *Flood Geology* (Hebron, KY: Answers in Genesis, 2007) DVD.

68. Apparently seeking to rescue by some means the ability to reason by analogy, the placard states at the bottom, "At best, modern catastrophes provide only clues about those times." But, again, if Noah's Flood was so huge as to be qualitatively different from anything observable in the present, then in what sense can any phenomenon observed in the present provide even a "clue"? And what exactly is the difference between an analogy and a "clue"?

69. The only clue is a footnote at the point where the placard mentions the "supercontinent" that says, "As interpreted from Genesis 1:9." Although this footnote does not explain much, it is an unusual and perhaps notable admission that the Bible is being *interpreted*, not just read literally, at the Creation Museum.

70. In 1970, Henry Morris and Tim LaHaye established a center for creation research designed to produce genuine science on behalf of a biblical creation. It split in 1972, with Morris forming the Institute for Creation Research (ICR). But the promised genuine science never materialized, so by the late 1980s, the ICR had turned its focus to "old-fashioned biblical creationism." It appears that producing genuine science on behalf of a biblical creation remains a challenge. For a history of the ICR, see Numbers, *Creationists*, 315–319.

71. *Uncensored Science*.

72. Ham, *Lie*, 58.

73. Ibid., 59. Putting this form of argumentation in the language of epistemology, Butler says that "the Creation Museum combines the axiom with objectivity by simultaneously deploying and dismantling the truth effects of material objects. The material world is portrayed as grounds for proof—but only if it accords with the 'factual data' of the Bible." Butler, "God Is in the Data," 243.

74. *Uncensored Science.* This logic is confirmed by a placard that hung in the Flood Geology room prior to the construction of the Ebenezer exhibit. The title of the placard is, "God's Word: The Key to Understanding God's World." Below the title, the placard says, "Starting with the facts of God's Word and world, we fashion models to know God and see His truth."

75. "Statement of Faith," *answersingensis,* https://answersingenesis.org/about/faith, n.d., accessed October 25, 2014.

76. Butler also notes this mode of scientific reasoning in the Creation Museum: "The facts cannot change the interpretation because, at the level of the Bible, fact and interpretation are one." According to Butler, whereas the Creation Museum constructs facts like fossils as material support for claims and also as incapable of providing such support since they cannot speak for themselves, she says that the Bible functions differently. Butler argues that the Creation Museum constructs the Bible as a fact (by putting it in cases and so forth) that can speak for itself since it is a text. Thus, it serves as fact and interpretation simultaneously. Therefore, it trumps all other facts to become the single source for indisputable truth beyond challenge. While interesting, Butler's argument does not notice that the most important "fact" constructed by the Creation Museum is the inerrant word. Also, Butler argues that the Bible is "a fact, only better: it is also an interpretation" (247). Butler's claim misses the point (which will be made clear in the next chapter), that the Bible-as-text is the problem, not the solution.

77. While it may be possible to have clarity about Ken Ham's scientific method, it is much more challenging to determine a single definition for the scientific method more generally. For a history of its varied meanings, see Thurs, "Scientific Methods."

78. Two of the three that do not identify models are in the Facing the Allosaurus room. Although they do not name a model, both discuss effects of the Flood and then suggest that Ebenezer confirms those effects. Likewise, the third assumes the Flood, works through a post-Flood event (the migration of mammals), and then argues that evidence from those mammals is consistent with the migration.

79. The figure for the number of "kinds" that were on the Ark comes from Marcus Ross, "No Kind Left Behind: Recounting the Animals on the Ark," *Answers Magazine,* December 11, 2012, https://answersingenesis.org/noahs-ark/no-kind-left-behind/. According to *ScienceDaily,* there are about 8.7 million species of plant and animal life on Earth. If fungi were not on the Ark (as Ross argues), then the number of relevant species here becomes about 8 million. Of those, 1,116,078 (again, not counting fungi) have been described and catalogued. Census of Marine Life, "How Many Species on Earth? About 8.7 Million, New Estimate Says," *ScienceDaily,* Science Daily, August 24, 2011, http://www.sciencedaily.com/releases/2011/08/110823180459.htm.

80. To clarify, by "landing sites" the placard appears to mean locations where

animals (and perhaps plants) landed after having migrated by way of ocean currents from the original landing site of the Ark.

81. These three highlighted areas are not labeled on the map. Given their location and the fact that each area is home to large tortoises, we take these to be the places identified on the map.

82. According to Gerlach, "there is Geochelone carbonaria of the Amazon which is quite big, but it's not a giant." Personal communication via e-mail, August 28, 2014.

83. For more on that controversy, see Justin Gerlach et al., "Giant Tortoise Distribution and Abundance in the Seychelles Islands: Past, Present, and Future," *Chelonian Conservation and Biology* 12, no. 1 (2013): 70–83.

84. "Galapagos Tortoise: Geochelone nigra," *Speed of Animals*, n.d., accessed August, 28, 2014, http://speedofanimals.com/animals/galapagos_tortoise.

85. According to Gerlach, "South America to Galapagos is plausible as there is an extremely strong current that flows in exactly the right direction and so could move them the required distance in a few weeks." Personal e-mail communication, August 28, 2014.

86. Gerlach, personal e-mail communication, August 28, 2014.

87. For a history of these events, see Maurice A. Finocchiaro, "That Galileo Was Imprisoned and Tortured for Advocating Copernicanism," in *Galileo Goes to Jail: And Other Myths about Science and Religion*, ed. Ronald L. Numbers (Cambridge, MA: Harvard University Press, 2009), 68–71.

88. For a history of the struggle between Galileo and the Church, see David C. Lindberg, "Galileo, the Church, and the Cosmos," in *When Science and Christianity Meet*, ed. David C. Lindberg and Ronald L. Numbers (Chicago: University of Chicago Press, 2003), 33–60.

89. For more on biblical cosmology, see John H. Walton, *Ancient Near Eastern Thought and the Old Testament: Introducing the Conceptual World of the Hebrew Bible* (Grand Rapids, MI: Baker Academic, 2006), and John H. Walton, *The Lost World of Genesis One: Ancient Cosmology and the Origins Debate* (Downers Grove, IL: InterVarsity, 2009).

90. Throughout this section, we draw upon Lloyd R. Bailey, *Genesis, Creation, and Creationism* (New York: Paulist, 1993), and Denis O. Lamoureux, *I Love Jesus & I Accept Evolution* (Eugene, OR: Wipf & Stock, 2009). The translation used throughout this paragraph is the NRSV unless otherwise indicated.

91. The NRSV translation suggests a similar movement across the sky: "Its rising is from the end of the heavens, and its circuit to the end of them."

92. Lamoureux, *I Love Jesus*, 53. Driving the point home, Lamoureux notes that while the word for Earth in the Old Testament (*'eres*) appears in the Bible about 2,500 times, and the word for Earth in the New Testament (*ge*) appears 250 times, "never once" in these 2,750 instances "is the earth referred to as spherical. Nor is a spherical shape implied by the context of any passage." Ibid., 46.

93. For commentary on the implications of taking seriously the fact that the Bible echoes an ancient cosmology, see Walton, *Ancient Near Eastern Thought*, and Lamoureux, *I Love Jesus*.

94. *Uncensored Science*.

95. Notably, the script for *Six Days*, which comes from the first chapter of Genesis, excises seven sets of verses. Two of those seven include language that vividly depicts an ancient cosmology. They include Genesis 1:7, which talks about the waters that were under the dome and the waters that were above the dome, and Genesis 1:17, which talks about how the two lights and the stars were "set . . . in the dome."

96. Butler, "God Is in the Data."

97. Walton goes on in that paragraph to say that imposing a modern universe on the Bible "is not just a case of adding meaning (as more information has become available) it is a case of changing meaning. Since we view the text as authoritative, it is a dangerous thing to change the meaning of the text into something it never intended to say." Walton, *Lost World*, 15.

Chapter 3. Bible

Portions of this chapter draw from Susan Trollinger, "From Reading to Revering the Good Book: How the Bible Became Fossil at the Creation Museum," *The Maryville Symposium: Conversations of Faith and the Liberal Arts: The Proceedings* (Maryville, TN: Maryville College, 2011), 29–49.

1. Plato, *Phaedrus*, in *Plato on Rhetoric and Language*, ed. Jean Nienkamp, trans. Alexander Nehamas and Paul Woodruff (New York: Routledge, 1999), 210 (275e).

2. For an insightful reading of Plato's *Phaedrus* that contrasts Plato's "dream of communication" (wherein only the proper vessel ever receives the word) with Jesus's broadcast word (wherein all are invited to hear it), see John Durham Peters, *Speaking into the Air: A History of the Idea of Communication* (Chicago: University of Chicago Press, 1999).

3. Nicholas Carr, *The Shallows: What the Internet Is Doing to Our Brains* (New York: W. W. Norton, 2010), 67.

4. Alister E. McGrath, *Christianity's Dangerous Idea: The Protestant Revolution—a History from the Sixteenth Century to the Twenty-First* (New York: HarperOne, 2007), 2–3, 53.

5. McGrath, *Dangerous Idea*, 3, 53.

6. We are here referring to, respectively, the Anabaptists, the Jehovah's Witnesses, and the Weigh-Down movement.

7. "Statement of Faith," https://answersingenesis.org/about/faith; Don Landis, "Not Just a Container of Truth: The Bible and Existentialism," Biblical Authority, *Answers* 5 (January–March 2010), http://www.answersingenesis.org/articles/am/v5/n1/container-truth; Paul Taylor, "Isn't the Bible Full of Errors?,"

Answers 2 (October–December 2007), https://answersingenesis.org/is-the-bible -true/isnt-the-bible-full-of-errors; Don Landis, "Science in the Balance," Biblical Authority, *Answers* 5 (July–September 2010), http://www.answersingenesis .org/articles/am/v5/n3/science-balance.

8. Tim Chaffey, "How Should We Interpret the Bible, Part 1: Principles for Understanding God's Word," *answersingenesis*, February 22, 2011, https:// answersingenesis.org/hermeneutics/how-we-interpret-the-bible-principles -for-understanding; Brian H. Edwards, "Unlocking the Truth of Scripture," *an swersingenesis*, September 18, 2007, https://answersingenesis.org/hermeneu tics/unlocking-the-truth-of-scripture; Henry M. Morris, ed., *The Henry Morris Study Bible* (King James Version) (1995; repr., Green Forest, AR: Master Books, 2012), 2045–2046. Morris's comments are part of his commentary on Revelation 22:18: "For I testify unto every man that heareth the words of the prophecy of this book, 'If any man shall add unto these things, God shall add unto him the plagues that are written in this book.'"

9. Ken Ham, "Christian Leader Agrees There Isn't 'One True Christian View on Evolution,'" *Ken Ham* (blog), *answersingenesis*, July 19, 2011, http://blogs .answersingenesis.org/blogs/ken-ham/2011/07/19/christian-leader-agrees -there-isnt-one-true-christian-view-on-evolution; Terry Mortenson, "Embracing Christ's View of Scripture," *Answers* 8 (July–September 2013), https://an swersingenesis.org/is-the-bible-true/embracing-christs-view-of-scripture; "Is Six Literal Days Really So Important?," Readers Respond, *Answers* 5 (January–March 2010), http://www.answersingenesis.org/articles/am/v5/n1/readers-respond; Ken Ham, "Maturing the Message," *Answers* 5 (January–March 2010), https://an swersingenesis.org/apologetics/maturing-the-message/; John UpChurch, "The Danger of BioLogos," *Answers* 6 (October–December 2011), https://answersin genesis.org/theistic-evolution/the-danger-of-biologos.

10. Mike Matthews, "Part One: The Ultimate Proof," *Answers* 6 (April–June 2011), https://answersingenesis.org/is-the-bible-true/part-one-the-ultimate -proof; Todd Charles Wood and Andrew A. Snelling, "Looking Back and Moving Forward," *Answers* 3 (October–December 2008), https://answersingenesis.org/ hermeneutics/looking-back-and-moving-forward.

11. Terry Mortenson, "Exposing a Fundamental Compromise," *Answers* 5 (July–September 2010), https://answersingenesis.org/theistic-evolution/expos ing-a-fundamental-compromise; Don Landis, "Jesus . . . Plus," *Answers* 5 (April–June 2010), http://www.answersingenesis.org/articles/am/v5/n2/jesus-plus; Ken Ham, "Does the Gospel Depend on a Young Earth?," *Answers* 6 (January–March 2011), https://answersingenesis.org/creationism/young-earth/does-the-gospel -depend-on-a-young-earth; Ken Ham, "Millions of Years—Are Souls at Stake?," *Answers* 9 (January–March 2014), https://answersingenesis.org/theory-of-evo lution/millions-of-years/are-souls-at-stake.

12. There are also Bibles—inaccessible for visitor perusal—in the Verbum

Domini exhibit in the museum, which consists of an ever-changing collection of items from the Green collection; owners of Hobby Lobby, the Greens will eventually put their collection on display in the Museum of the Bible in Washington. Of course, there are Bibles for sale in the Dragon Hall Bookstore, including the English Standard Version, Gift & Award Bibles, and the *Henry Morris Study Bible.*

13. AiG, *Journey through the Creation Museum: Prepare to Believe* (Green Forest, AR: Master Books, 2008), 39.

14. This verse seems to be drawn from the King James Version, but with slight editorial adjustments. The full text from Romans 1:20 (KJV) is as follows: "For the invisible things of him from the creation of the world are clearly seen, being understood by the things that are made, even his eternal power and Godhead; so that they are without excuse."

15. Carr's book is a great resource for the many scientific studies that have been published on the topic. Also see Maryanne Wolf, "Our 'Deep Reading' Brain: Its Digital Evolution Poses Questions," *Nieman Reports* 64 (Summer 2010): 7–8.

16. The material that is actually on the placard represents an eclectic mishmash of biblical translations: verse 2 ("It came to pass . . . ") is from the King James Version, but the middle clause is missing; verse 4 ("And they said . . . ") is a combination of KJV and the New King James Version (NKJV), with some additional editing; verse 6a ("And the Lord said . . . ") is essentially KJV; verse 7 ("Let us go down . . . ") is NKJV; verse 8 is closest to the New Revised Standard Version.

17. The video's editing of Genesis 11:5 would seem to be unnecessary: not only does God's descent presume knowledge of happenings in Babel, but it could humorously suggest that "the tower is so puny that God has to come down to see it." Bruce Waltke with Cathi J. Fredricks, *Genesis: A Commentary* (Grand Rapids, MI: Zondervan, 2001), 180.

18. Nahum Sarna notes that "some rabbinic sages" viewed Noah unfavorably in comparison to Abraham, given that, "unlike Abraham's response to the case of Sodom and Gomorrah in [Genesis] 18:23–52, Noah does not plead for mercy for his contemporaries. Sensitive to this moral issue, rabbinic lore supplements the text by having Noah warn his fellow men of impending disaster and call them to repentance." Nahum M. Sarna, *Genesis = Be-reshit: The Traditional Hebrew Text with New JPS Translation*, commentary by Nahum M. Sarna (Philadelphia: Jewish Publication Society, 1989), 50.

19. We are using the KJV here, as this display matches most closely this version. The bracketed material was added by the museum. As is often the case at the museum, no ellipses were used to show where text has been removed.

20. At some point in 2013 or 2014 this passage disappeared, or at least we could not find it, although as of July 3, 2014, it was still on the website's virtual

tour of the Flood Geology room. The full passage from Job 12:7–9 (KJV, which seems to come closest to this display) reads as follows: "But ask now the beasts, and they shall teach thee; and the fowls of the air, and they shall tell thee: Or speak to the earth, and it shall teach thee: and the fishes of the sea shall declare unto thee. Who knoweth not in all these that the hand of the LORD hath wrought this?" Much text was removed from the museum's display, with no ellipses provided, and no indication that the Hebrew text is referring to YHWH.

21. Carol A. Newsom, "The Book of Job: Introduction, Commentary, and Reflections," in *The New Interpreter's Bible: A Commentary in Twelve Volumes*, ed. Leander Keck et al., vol. 4 (Nashville, TN: Abingdon, 1996), 429.

22. The placard entitled "The Waters Rise" provides a particularly telling example. According to the placard the text included here is from Genesis 7:11, 17, and 22. The text does include ellipses, but only to indicate the gaps between verses, and not to indicate the text that has been omitted from within these verses. Moreover, the text actually comes from Genesis 7:11, 17, and 21 (not 22).

23. Troy Lacey and Lee Anderson, "The Genesis Flood—Not Just Another Legend," *Answers* 8 (October–December 2013), https://answersingenesis.org/the-flood/flood-legends/the-genesis-floodnot-just-another-legend; Gordon J. Wenham, *Genesis 1–15* (Waco, TX: Word, 1987), 139. See also Terence E. Fretheim, "The Book of Genesis: Introduction, Commentary, and Reflections," in *The New Interpreter's Bible: A Commentary in Twelve Volumes*, ed. Leander Keck et al., vol. 1 (Nashville, TN: Abingdon, 1994), 382–383. Given the AiG emphasis on the perspicuity of scripture, it is interesting that the organization "does not hold an official position on these verses"; according to AiG commentator Tim Chaffey, while "studying the nature of the 'sons of God' is interesting and profitable," it is "certainly not as crucial as the clear, unequivocal revelation about six days of creation." "Battle over the Nephilim," *Answers* 7 (January–March 2012), https://answersingenesis.org/bible-characters/battle-over-the-nephilim.

24. Wenham, *Genesis*, 140; Peter Enns, *The Evolution of Adam: What the Bible Does and Doesn't Say about Human Origins* (Grand Rapids, MI: Brazos, 2012), 48–49; Donald Harman Akenson, *Surpassing Wonder: The Invention of the Bible and the Talmuds* (New York: Harcourt Brace, 1998), 258. Akenson goes on to argue that God's destruction of humankind in the Flood confirms that in the Torah "the idea that [quoting Ulrich Luz] 'God and human beings can sexually interact is the pinnacle of sacrilege.'"

25. See, for example, Victor P. Hamilton, *The Book of Genesis: Chapters 1–17* (Grand Rapids, MI: Eerdmans, 1990), 287.

26. For a succinct description of the history and basic outlines of the documentary hypothesis, which has its origins in the nineteenth century (particularly with the work of Julius Wellhausen), see Richard Elliott Friedman, *Who Wrote the Bible?* (New York: Summit Books, 1987), 15–32. For Friedman's de-

tailed response to scholarly critics of the documentary hypothesis, see his *The Hidden Book in the Bible* (San Francisco: HarperSanFrancisco, 1998), 350–378. For Friedman's summary of evidence in behalf of the documentary hypothesis, see his *The Bible with Sources Revealed: A New View into the Five Books of the Bible* (San Francisco: HarperSanFrancisco, 2003), 7–31. The latter book also contains the first five books of the Bible, with the different sources indicated by differently colored and styled fonts: 32–368.

27. Friedman, *Who Wrote the Bible?*, 53–60. Friedman argues that the two different accounts of Noah and the Flood are to be found as follows (for Genesis 6:1–4 and 9:1–17 please refer to *Bible with Sources*, 42, 46–47):

J: 6:1–8; 7:1–5, 7, 10, 12, 16b–20, 22–23; 8:2b–3a, 6, 8–12, 13b, 20–22.

P: 6:9–22; 7:6, 8–9, 11, 13–16a, 21, 24; 8:1–2a, 3b–5, 7, 13a. 14–19; 9:1–17.

28. Akenson, *Surpassing Wonder*, 53; Walter Brueggemann, *Genesis* (Atlanta: John Knox, 1982), 75; Ken Ham, "The Enns Justifies the Means? A Review of the New Book by Peter Enns, *The Evolution of Adam,*" *answersingenesis*, January 19, 2012, https://www.answersingenesis.org/articles/2012/01/19/evolution-of-adam-review. As for who has "thoroughly debunked" the documentary hypothesis, Ham points to Bodie Hodge and Terry Mortenson, "Did Moses Write Genesis?," *answersingenesis*, June 28, 2011, https://answersingenesis.org/bible-characters/moses/did-moses-write-genesis.

29. Ham, "Wolves in Sheep's Clothing?"; Hodge and Mortenson, "Did Moses Write Genesis?"; Tim Chaffey, "Does It Really Matter If Moses Wrote Genesis?," Get Answers, *Answers* 7 (April–June 2012) (online only), www.answersingenesis.org/articles/am/v7/n2/get-answers (quote). Henry Morris agrees that "Moses was the compiler and editor of a number of earlier documents," as "it is reasonable that Adam and his descendents all knew how to write and, therefore, kept records of their own times." *Henry Morris Study Bible*, 6.

30. Enns, *Evolution of Adam*, 50–52 (quote: 52).

31. On one occasion the verse is misidentified: on the signboard labeled "God's Word Is questioned," the text that is said to be Genesis 3:4 is actually from Genesis 3:1.

32. Some of these placards have words missing from the Bible passages, without ellipses provided. There are also placards without Bible verses that describe Eden as having been a place without aging, scavengers, and weeds. While these placards lack biblical support, they do include detailed explanations of these claims, for example, "Once animals began overproducing to replace the dead and diseased, they would eat too many plants. So God introduced overproduction of plants to replace the plants that would be lost. As a result: Plants struggle against other plants for survival [and] Plants grow where they are not wanted (weeds)."

33. See also the part of Revelation 21:4 omitted from the Creation Museum placard: "the former things have passed away."

34. *Henry Morris Study Bible*, 8–9.

35. John H. Walton, *The Lost World of Genesis One: Ancient Cosmology and the Origins Debate* (Downers Grove, IL: InterVarsity, 2009), 90–91, 94, 161–162. Walton is at pains to make clear that his argument is not an argument against God as the Creator of all: *"Viewing Genesis 1 as an account of functional origins of the cosmos as temple does not in any way suggest or imply that God was un- involved in material origins—it only contends that Genesis 1 is not that story"* (95, emphasis Walton's). See also William P. Brown, *The Seven Pillars of Crea- tion: The Bible, Science and the Ecology of Wonder* (New York: Oxford University Press, 2010), 37–41; Michael W. Pahl, *The Beginning and the End: Rereading Genesis's Stories and Revelation's Visions* (Eugene, OR: Cascade Books, 2011), 11–12, 19.

36. Walton, *Lost World*, 52, 169; Enns, *Evolution of Adam*, 45; Morris, *Study Bible*, 7.

37. Enns, *Evolution of Adam*, 154n3; Gerhard May, *Creatio ex Nihilo: The Doctrine of "Creation out of Nothing" in Early Christian Thought*, trans. A. S. Wor- rall (Edinburgh: T & T Clark, 1994), xiii; Christian Smith, *The Bible Made Im- possible: Why Biblicism Is Not a Truly Evangelical Reading of Scripture* (Grand Rapids, MI: Brazos, 2011), 84. In an article published by AiG, "Have We Misun- derstood Genesis 1:1?," Josh Wilson rejects what he calls "the dependent-clause understanding of Genesis 1:1," arguing that it is grammatically "difficult and awkward." Instead, "the traditional understanding of Genesis 1:1 is trustworthy," and "in the absolute beginning God did indeed create the heavens and earth out of nothing." *Answers in Depth* 8 (2013), https://answersingenesis.org/herme neutics/have-we-misunderstood-genesis-11.

38. Terry Mortenson, "Evangelical Popes," *Answers* 7(April–June 2012), https://answersingenesis.org/christianity/evangelical-popes.

39. Ham, "Compromise with Millions of Years"; Ken Ham, "Wheaton College and False Teaching in Tennessee," *Ken Ham* (blog), *answersingenesis*, February 18, 2011, http://blog.answersingenesis.org/blogs/ken-ham/2011/02/18/wheaton -college-and-false-teaching-in-Tennessee; Ken Ham, *"The Genesis Flood*—The Battle Still Rages!,"* Ken Ham* (blog), *answersingenesis*, February 20, 2011, http://blogs.answersingenesis.org/blogs/ken-ham/2011/02/20/the-genesis -flood-the-battle-still-rages.

40. Smith, *Bible Made Impossible*, 17.

41. Kathleen C. Boone, *The Bible Tells Them So: The Discourse of Protestant Fundamentalism* (Albany: State University of New York Press, 1989), 73, 79.

42. Ham, "The Enns Justifies the Means?"; Peter Enns, " 'Ken Ham Clubs Baby Seals' (or it may be time for him to rethink his ministry strategy)," *Re- thinking Biblical Christianity* (blog), patheos, September 21, 2012, http:// www.patheos.com/blogs/peterenns/2012/09/ken-ham-clubs-baby-seals-or -it-may-be-time-for-him-to-rethink-his-ministry-strategy; Ken Ham, "Peter Enns

Responds with 'Ken Ham Clubs Baby Seals," *Ken Ham* (blog), answersingen esis, September 24, 2012, http://blogs.answersingenesis.org/blogs/ken-ham/ 2012/09/24/peter-enns-responds-with-ken-ham-clubs-baby-seals. See also Ken Ham, "Peter Enns Wants Children to Reject Genesis," *Ken Ham* (blog), answers ingenesis, September 20, 2012, http://blogs.answersingenesis.org/blogs/ken -ham/2012/09/20/peter-enns-wants-children-to-reject-genesis; Ken Ham, "Peter Enns—Mutilating God's Word," *Ken Ham* (blog), *answersingenesis*, December 14, 2012, http://blogs.answersingenesis.org/blogs/ken-ham/2012/12/14/peter -enns-mutilating-gods-word.

43. James Barr, *Escaping from Fundamentalism* (London: SCM, 1984), 149; Harriet A. Harris, *Fundamentalism and Evangelicals* (Oxford: Oxford University Press, 1998), 183–185.

44. Gordon Campbell, *Bible: The Story of the King James Version, 1611–2011* (Oxford: Oxford University Press, 2010), 243–244; William Vance Trollinger, Jr., "An Outpouring of 'Faithful' Words: Protestant Publishing in the United States," in *Print in Motion: The Expansion of Publishing and Reading in the United States, 1880–1940*, ed. Carl Kaestle and Janice Radway, vol. 4: *History of the Book in America* (Chapel Hill: University of North Carolina Press, 2009), 367–368.

45. C. I. Scofield, ed., *The Scofield Reference Bible* (King James Version) (New York: Oxford University Press, American Branch, 1909), 3.

46. Dewey M. Beegle, *Scripture, Tradition, and Infallibility* (Grand Rapids, MI: Eerdmans, 1973), 110, as quoted in Boone, *The Bible Tells Them So*, 79.

47. Elizabeth Rundle Charles, *Chronicles of the Schonberg-Cotta Family by Two of Themselves* (New York: M. W. Dodd, 1864), 321; Bob Caldwell, "'If I Profess': A Spurious, If Consistent, Luther Quote?," *Concordia Journal* 35 (Fall 2009): 356–359. The museum has made a few small changes to the quote from the novel, including replacing "truth of God" with "Word of God," and substituting "professing Him" for "professing Christianity."

48. One wonders if the museum's designers have some subconscious awareness that it is "sola scriptura" itself that produces the proliferation of competing interpretations of the Bible.

49. Michael B. Roberts, "Genesis Chapter 1 and Geological Time from Hugo Grotius and Marin Mersenne to William Conybeare and Thomas Chalmers (1620–1825)," in *Myth and Geology*, ed. Luigi Piccardi and W. Bruce Masse (London: Geological Society of London, 2007), 45–47; Ronald L. Numbers, *Darwinism Comes to America* (Cambridge, MA: Harvard University Press, 1998), 2. Interestingly, Whitcomb and Morris wrote *The Genesis Flood* specifically to counter the gap and day-age theories. Ronald L. Numbers, *The Creationists: From Scientific Creationism to Intelligent Design*, expanded ed. (Cambridge, MA: Harvard University Press, 2006), 230.

50. For more on how contemporary fundamentalists understood the Scopes

trial, and on post-Scopes antievolutionist activity, see Numbers, *Darwinism*, 76–91.

51. Here Darrow was clearly referring to the notion that the earth was created in 4004 BC, which would have made the world six thousand years old.

52. Susan Friend Harding, *The Book of Jerry Falwell: Fundamentalist Language and Politics* (Princeton, NJ: Princeton University Press, 2000), 70–71; *The World's Most Famous Court Trial: Tennessee Evolution Case*, 2nd rpt. ed. (Dayton, TN: Bryan College, 1990), 296, 298–299, 302.

53. Harding, *Jerry Falwell*, 73. For the possibility that the Scopes trial was a catalyst for the increased popularity of young Earth creationism, see Adam R. Shapiro, *Trying Biology: The Scopes Trial, Textbooks, and the Antievolution Movement in American Schools* (Chicago: University of Chicago Press, 2013), 106–107.

54. Harding, *Jerry Falwell*, 214.

55. Galileo Galilei, "Letter to Madame Christina of Lorraine, Grand Duchess of Tuscany, Concerning the Use of Biblical Quotations in Matters of Science [1615]," in *Discoveries and Opinions of Galileo*, ed. and trans. Stillman Drake (Garden City, NY: Doubleday Anchor Books, 1957), 182–183.

56. Galileo, "Letter to Madame Christina," 181; Maurice A. Finocchiaro, "That Galileo was Imprisoned and Tortured for Advocating Copernicanism," in *Galileo Goes to Jail: And Other Myths about Science and Religion*, ed. Ronald L. Numbers (Cambridge, MA: Harvard University Press, 2009), 70; Maurice A. Finocchiaro, ed. and trans., *The Galileo Affair: A Documentary History* (Berkeley: University of California Press, 1989), 291.

57. Wood and Snelling, "Looking Back and Moving Forward"; Martin Luther, *Luther's Works*, ed. Helmut T. Lehmann, vol. 54, *Table Talk*, ed. and trans. Theodore G. Tappert (Philadelphia: Fortress, 1967), 358–359; Martin Luther, *Luther's Works*, ed. Jaroslav Pelikan, vol. 1, *Lectures on Genesis: Chapters 1–5*, trans. George V. Schick (St. Louis, MO: Concordia, 1958), 44.

58. Danny Faulkner, "Should You Follow *The* (Copernican) *Principle?*," *answersingenesis*, April 23, 2015, https://answersingenesis.org/astronomy/should-you-follow-the-copernican-principle.

59. Jason Lisle and Tim Chaffey, "Defense—Poor Reasoning," *answersin genesis*, January 12, 2012, https://answersingenesis.org/creationism/old-earth/defense-poor-reasoning. This article originally appeared as chapter 4 in their book, *Old-Earth Creationism on Trial: The Verdict Is In* (Green Forest, AR: Master Books, 2008).

60. Gerardus D. Bouw, *Geocentricity* (Cleveland, OH: Association for Biblical Astronomy, 1992). See www.geocentricity.com for more from Bouw.

61. Danny Faulkner, "Geocentrism and Creation," *TJ* 15 (August 2001): 110–121; posted on AiG website: https://answersingenesis.org/creationism/arguments-to-avoid/geocentrism-and-creation. As elsewhere at AiG, and as

discussed in chapter 2, the unstated presumption—without accompanying evidence—is that the ancient Hebrews held to a heliocentric cosmology.

62. Gerardus D. Bouw, "Geocentricity: A Fable for Educated Man?," http://www.geocentricity.com/ba1/fresp. Bouw told journalist Daniel Radosh that his scientific work needed to be distinguished from "*those other, really crazy geocentrists*" (italics in original). *Rapture Ready! Adventures in the Parallel Universe of Christian Pop Culture* (New York: Soft Skull, 2010), 294.

63. This figure comes from a 1999 Gallup Poll: www.gallup.com/poll/3742/new-poll-gauges-americans-general-knowledge-levels.aspx.

Chapter 4. Politics

1. The Creation Museum's interpretation of Proverbs 1:7 as instructing humans to use the Bible as the "starting point" is not obvious from the verse itself, which reads: "The fear of the LORD is the beginning of knowledge: fools despise wisdom and instruction" (NRSV).

2. Ernesto Laclau and Chantal Mouffe, *Hegemony and Socialist Strategy: Towards a Radical Democratic Politics*, 2nd ed. (New York: Verso, 2001), 93–148.

3. *Men in White*, directed by John Grooters (Hebron, KY: Answers in Genesis in association with Grooters Productions, 2007), DVD. The actors with Hollywood credits include John Allsopp, Jeff Charlton, Jim Custer, Michelle Dunker, Suzanne Friedline, Kelly Keaton, and Scott Ward.

4. Ken Ham, "So Obvious, Even a Child Gets It!," *answersingenesis*, April 29, 2008, http://www.answersingenesis.org/articles/au/so-obvious-child-gets -it; "Evolution in Kindergarten," *Answers* 2 (January–March 2007), http://www.answersingenesis.org/articles/am/v2/n1/evolution-in-kindergarten; Terry Mortenson, "Slam Dunk for Science and the Constitution?," *answersingenesis*, February 23, 2009, http://www.answersingenesis.org/articles/2009/02/23/ slam-dunk-science-constitution; Ken Ham and Mark Looy, "Petitioning President Obama to Ban Creation," *answersingenesis*, June 25, 2013, http://www .answersingenesis.org/articles/2013/06/25/petitioning-president-obama -to-ban-creation; Elizabeth Mitchell, "Mother Jones: 'Insist That People Co-existed with Dinosaurs . . . and Get an A in Science Class!,'" News to Note, *answersingenesis*, March 2, 2013, http://www.answersingenesis.org/articles/2013 /03/02/news-to-note-03022013.

5. Ham, "So Obvious"; Ken Ham, "Australian Students to Be Taught There Is No God," *Ken Ham* (blog), *answersingenesis*, December 18, 2008, http://blogs .answersingenesis.org/blogs/ken-ham/2008/12/18/australian-students-to-be -taught-there-is-no-god; Ken Ham, "Public Schools Promoting a Religion," *Ken Ham* (blog), *answersingenesis*, January 21, 2012, http://blogs.answersingene sis.org/blogs/ken-ham/2012/01/21/public-schools-promoting-a-religion.

6. Georgia Purdom, "Public Schools and Parenting," *answersingenesis*, November 6, 2007, http://www.answersingenesis.org/articles/2007/11/06/public

-schools-and-parenting; Ken Ham, "Is Our Moral Code Out of Date?," *Ken Ham* (blog), *answersingenesis*, September 21, 2010, http://blogs.answersingenesis .org/blogs/ken-ham/2010/09/21/is-our-moral-code-out-of-date.

7. Ken Ham, "Evolution Connection to School Violence," *Ken Ham* (blog), *answersingenesis*, November 8, 2007, http://blogs.answersingenesis.org/blogs/ ken-ham/2007/11/08/evolution-connection-to-school-violence; Bodie Hodge, "Finland School Shootings: The Sad Evolution Connection," *answersingenesis*, November 8, 2007, http://www.answersingenesis.org/articles/2007/11/08/ finland-fruits-of-humanism; "BBC News: 'Finnish College Gunman Kills 10,'" News to Note, *answersingenesis*, September 27, 2008, http://www.answersin genesis.org/articles/2008/09/27/news-to-note-09272008; Ken Ham, "If You Don't Matter to God, You Don't Matter to Anyone," *answersingenesis*, April 20, 2009, http://www.answersingenesis.org/articles/2009/04/20/if-you-dont-mat ter-to-god-you-dont-matter-to-anyone; Ralph W. Larkin, *Comprehending Columbine* (Philadelphia: Temple University Press, 2007), 200; Dave Cullen, *Columbine* (New York: Twelve, 2009), esp. 222–233.

8. "Unnaturally Selected," Culture News, *Answers* 3(April–June 2008), https://answersingenesis.org/sanctity-of-life/mass-shootings/unnaturally -selected; Hodge, "Finland School Shootings."

9. Donald B. Kraybill, Steven M. Nolt, and David Weaver-Zercher, *Amish Grace: How Forgiveness Transcended Tragedy* (San Francisco: Jossey-Bass, 2007), 52.

10. "Reuters: Gunman Kills 3 girls at Pennsylvania Amish School," News to Note, *answersingenesis*, October 7, 2006, http://www.answersingenesis.org /articles/2006/10/07/news-to-note-1072006.

11. Kraybill, Nolt, and Weaver-Zercher, *Amish Grace*, 63.

12. The following discussion is based on our photos taken of Graffiti Alley on May 18, 2013. The photos taken in 2010 reveal that the earlier Graffiti Alley had fewer clippings. But a close examination of the 2013 photos reveals that many or most of the older clippings remain on the walls, with some of these partially or almost completely covered by the newer clippings. As far as general topics (e.g., homosexuality, school shootings) that are included in the newspaper and magazine fragments, there is very little difference between the 2010 and the 2013 Graffiti Alley.

13. "Be Worried. Be Very Worried" came from the April 3, 2006, *Time* cover on the dangers of global warming, an odd headline given that the folks at AiG reject the idea of human-caused climate change. Apparently, this headline was just too frightening to pass up.

14. Richard T. Hughes, *Christian America and the Kingdom of God* (Urbana: University of Illinois Press, 2009), 35, 51.

15. This could also be a photo of a woman protesting outside Terri Schiavo's hospice.

16. Charles P. Pierce, *Idiot America: How Stupidity Became a Virtue in the Land of the Free* (New York: Anchor Books, 2009), 189 (DeLay quote); Michelle Goldberg, *Kingdom Coming: The Rise of Christian Nationalism* (New York: W. W. Norton, 2006), 158–159; Timothy Williams, "Schiavo's Brain Was Severely Deteriorated, Autopsy Says," *New York Times*, June 15, 2005, http://www .nytimes.com/2005/06/15/national/15cnd-schiavo.html. Pierce's chapter on the Schiavo case, "A Woman Dies on Beech Street," is a vivid description of how the Christian Right made use of the Schiavo case.

17. Ken Ham has memorably summarized the theological problem regarding alien beings: "If there are intelligent beings on other planets, then they would have been affected by the fall of Adam because the whole creation was affected. So these beings would have to die because death was the penalty for sin . . . [But] Jesus didn't become a 'God-Klingon,' a 'God-Vulcan,' or a 'God-Cardassian'—He became the God-man. It wouldn't make sense theologically for there to be other intelligent physical beings who suffer because of Adam's sin but cannot be saved." Ken Ham, "Do I Believe in UFOs? Absolutely!," *Answers* 3 (January–March 2008), http://www.answersingenesis.org/articles/am/v3/n1/ believe-in-ufos.

18. Chris Hedges, *American Fascists: The Christian Right and the War on America* (New York: Free, 2006), 146–149, 192–194. For examples of the vast literature linking Islam and the coming of the End Times, see Hal Lindsey, *The Everlasting Hatred: The Roots of Jihad* (Murietta, CA: Oracle House, 2002); Joel Richardson, *The Islamic Antichrist: The Shocking Truth about the Real Nature of the Beast* (Los Angeles: World Net Daily, 2009); Ellis H. Skolfield, *Islam in the End Times* (Fort Myers, FL: Fish House, 2007).

19. Barry A. Kosmin and Ariela Keysar, *American Religious Identification Survey 2008: Summary Report* (Hartford, CT: Trinity College Program of Public Values, 2009), http://www.americanreligionsurvey-aris.org; Jon Meacham, "The End of Christian America," *Newsweek*, April 13, 2009, http://www.thedaily beast.com/newsweek/2009/04/03/the-end-of-christian-america.html.

Meacham was certainly not the only commentator who prematurely suggested the demise of the Christian Right. See E. J. Dionne Jr., *Souled Out: Reclaiming Faith and Politics after the Religious Right* (Princeton, NJ: Princeton University Press, 2009).

20. David Barton, "A Foundation of Scripturecans [?]," *Answers* 1(July–September 2006), www.answersingenesis.org/articles/am/v1/n1/foundation-of -scripture; Mike Belknap, "Graffiti Alley and the Decay of the West," answersin genesis, July 1, 2013, http://www.answersingenesis.org/articles/2013/07/01/ graffiti-alley-decay-of-the-west; Ken Ham, "The State of the Nation: Addressing America's Real Foundational Problem," *answersingenesis*, March 8, 2010, http://www.answersingenesis.org/articles/2010/03/08/state-of-the-nation.

21. Ken Ham, "Separation of Christianity and State," *answersingenesis*, May

3,2010,http://www.answersingenesis.org/articles/au/separation-of-christianity
-and-state; Ken Ham, "As in the Days of Noah," *Answers* 8 (October–December
2013), http://answersingenesis.org/culture/as-in-the-days-of-noah; Ken Ham,
"'Wise Men' Remove Nativities," *Ken Ham* (blog), *answersingenesis*, December 22, 2014, https://answersingenesis.org/blogs/ken-ham/2014/12/22/wise
-men-remove-nativities; Ken Ham, "Getting Rid of *Christ*?," *answersingenesis*, December 11, 2007, http://www.answersingenesis.org/articles/2007/12/11/christ
-out-of-christmas; Mike Riddle, "The Stealing of America," *answersingenesis*,
September 23, 2008, http://www.answersingenesis.org/articles/2008/09/23/
stealing-of-america.

22. Barack Obama, *The Audacity of Hope: Thoughts on Reclaiming the
American Dream* (New York: Crown, 2006), 218; Ken Ham, "The End of Christian America," *Ken Ham* (blog), *answersingenesis*, April 13, 2009, http://blogs
.answersingenesis.org/blogs/ken-ham/2009/04/13/the-end-of-christian-amer
ica; Ken Ham, "Revealing the Unknown God," as quoted in "Foundations," Resource Preview, *Answers* 6 (July–September 2011), http://www.answersingene
sis.org/articles/am/v6/n3/foundations.

23. Ken Ham, "One Nation Under . . . ?," *Answers* 5 (October–December
2010), http://www.answersingenesis.org/articles/am/v5/n4/nation; William Vance
Trollinger, Jr., "Evangelicalism and Religious Pluralism in Contemporary America: Diversity Without, Diversity Within, and Maintaining the Borders," in *Gods
in America: Religious Pluralism in the United States*, ed. Charles Cohen and
Ronald L. Numbers (New York: Oxford University Press, 2013), 105–124; Paul
Maltby, *Christian Fundamentalism and the Culture of Disenchantment* (Charlottesville: University of Virginia Press, 2013), esp. 94–99.

24. Ken Ham, "The Chasm Is Widening: Are You on God's Side?," Letter
from Ken, *answersingenesis*, April 29, 2013, http://www.answersingenesis
.org/articles/2013/04/29/chasm-is-widening-are-you-on-gods-side; Don Landis, "Response to Persecution," *answersingenesis*, August 2, 2013, http://www
.answersingenesis.org/christianity/christian-life/response-to-persecution; Ken
Ham, *The Lie: Evolution/Millions of Years*, rev. ed. (Green Forest, AR: Master
Books, 2012), 33.

25. Ham, "Chasm Is Widening"; Ham, *The Lie*, 29–30, 34; Ken Ham, "The
Atheists' Real Motive," *Ken Ham* (blog), *answersingenesis*, June 6, 2013, http://
blogs.answersingenesis.org/blogs/ken-ham/2013/06/06/the-atheists-real
-motive.

26. Ham, *The Lie*, 30; Ham, "State of the Nation."

27. Bernadette C. Barton, *Pray the Gay Away: The Extraordinary Lives
of Bible Belt Gays* (New York: New York University Press, 2012), 157–169. The
one African American in the group found the museum less disorienting: as he
"listened closely to his classmates process their feelings about visiting the museum during our class discussion following the field trip . . . [he] chimed in with

amazement, 'What you are all describing is how I feel all the time here. It's even worse in the Morehead Wal-Mart'" (164–165).

28. Stephanie Pappas, "Creation Museum Creates Discomfort for Some Visitors," http://www.livescience.com/8501-creation-museum-creates-discom fort-visitors.html; Answers in Genesis-U.S. editors, "Feedback: Please Stop the Hate," *answersingenesis*, August 20, 2010, www.answersingenesis.org/articles /2010/08/20/feedback-please-stop-the-hate.

29. "Please Stop the Hate."

30. The biblical passage on the left placard comes from Matthew 19:4; the biblical passage on the right placard comes from Mark 10:7–9.

31. "Please Stop the Hate"; Randall Balmer, *Thy Kingdom Come: How the Religious Right Distorts the Faith and Threatens America* (New York: Basic Books, 2006), 5; Ken Ham and Stacia McKeever, "Seven C's of History," answersingen esis, May 20, 2004, www.answersingenesis.org/articles/2004/05/20/seven-cs -of-history. For more on AiG's obsession with opposing gay marriage, see Jeffrey Goldberg, "The Genesis Code: How Promoting Creationism Has Become a Way to Oppose Gay Marriage," *The Atlantic* (October 2014): 11–13.

32. Jeffrey J. Ventrella, "Genesis and Justice," *Answers* 1 (July–September 2006), http://www.answersingenesis.org/articles/am/v1/n1/genesis_and_justice; Travis Nuckolls, "When Did You Choose to Be Straight?, http:/www.youtube .com/watch?v=QJtjqLUHYoY; Roger Patterson, "Feedback: When Did You Choose to Be Straight?," *answersingenesis*, July 5, 2013, www.answersingen esis.org/articles/2013/07/05/feedback-choose-to-be-straight; Tim Chaffey, "Feed-back: God's Not Clear on Homosexuality?," *answersingenesis*, January 14, 2011, http://www.answersingenesis.org/articles/2011/01/14/feedback-gods-not-clear-on -homosexuality. While Chaffey provides no evidence for his assertion, the Ameri-can Psychological Association has repeatedly found that there is no evidence "to show that therapy aimed at changing sexual orientation . . . is safe or effective" and that the continued emphasis on "change therapies" appears to be particu-larly damaging to "lesbian, gay, and bisexual individuals who grow up in more conservative religious settings." American Psychological Association, "Answers to your Questions: For a Better Understanding of Sexual Orientation and Ho-mosexuality," *American Psychological Association* (2008), http://www.apa.org/ helpcenter/sexual-orientation.aspx. As regards two of the most prominent min-istries devoted to helping gays become straight, Love Won Out (a Focus on the Family organization) folded in 2009. More famously, in June 2013 Exodus In-ternational, the largest and oldest evangelical organization devoted to changing sexual orientation, shut down, in the process issuing an apology to the LGBTQ community.

33. Elizabeth Mitchell, "Homosexuality Ultimately a Result of Gene Regula-tion, Researchers Find," News to Note, Answers in Genesis, December 22, 2012, http://www.answersingenesis.org/articles/2012/12/22/news-to-note-12222012.

34. Seth Cline, "Chick-fil-A's Controversial Gay Marriage Beef," *U.S. News and World Report*, July 27, 2012, http://www.usnews.com/news/articles/2012/07/27/chick-fil-as-controversial-gay-marriage-beef; Ken Ham, "Our 'Chikin'* Day," *Ken Ham* (blog), *answersingenesis*, August 2, 2012, http://blogs.answersingenesis.org/blogs/ken-ham/2012/08/02/our-chikin-day; Ken Ham, "The Evolving Sexual Agenda," *Ken Ham* (blog), *answersingenesis*, March 27, 2013, http://blogs.answersingenesis.org/blogs/ken-ham/2013/03/27/the-evolving-sexual-agenda; Ken Ham, "eHarmony Compromises on Same Sex 'Marriage,'" *Ken Ham* (blog), *answersingenesis*, April 13, 2013, http://blogs.answersingenesis.org/blogs/ken-ham/2013/04/13/eharmony-compromises-on-same-sex-marriage; Ken Ham, "Southern Baptist Convention Opposes Boy Scouts of America (BSA) Decision," *Ken Ham* (blog), *answersingenesis*, June 16, 2013, http://blogs.answersingenesis.org/blogs/ken-ham/2013/06/16/southern-baptist-convention-opposes-boy-scouts-of-america-bsa-decision.

35. Adam Liptak, "Supreme Court Bolsters Gay Marriage with Two Major Rulings," *New York Times*, June 26, 2013, http://www.nytimes.com/2013/06/27/us/politics/supreme-court-gay-marriage.html; Elizabeth Mitchell, "Defining Marriage—*Supreme Court* v. *Sovereign God*," *answersingenesis*, June 27, 2013, https://answersingenesis.org/family/marriage/defining-marriagesupreme-court-v-sovereign-god.

36. Ken Ham, "The Great Delusion—The Spiritual State of the Nation," You-Tube video, 1:00:48, from the Answers Mega Conference, Sevierville, TN, July 24, 2013, https:// answersingenesis.org/media/video/worldview/great-delusion.

37. Ham's assertion that God is "withdrawing" and "giving [America] over to judgment" because of its tolerance of homosexuality bears striking resemblance to Jerry Falwell's infamous remarks two days after the September 11, 2001, attacks: "God continues to lift the curtain and allow the enemies of America to give us probably what we deserve . . . The pagans, and the abortionists, and the feminists, and the gays and the lesbians who are trying to make that an alternative lifestyle, the ACLU, People for the American Way—all of them who have tried to secularize America—I point the finger in their face and say, 'You helped this happen.'" Having quoted Falwell, Cynthia Burack goes on to document how such language is standard among Christian Right leaders. Burack, *Sin, Sex, and Democracy: Antigay Rhetoric and the Christian Right* (Albany: State University of New York Press, 2008), 109.

38. In a blog response to Obama's speech at a June 2014 LGBTQ fundraiser, entitled "Do Not Fret Because of Evildoers," Ham quoted Psalm 37:1–9, which includes these verses: "Do not fret because of evildoers, Nor be envious of the workers of iniquity. For they shall soon be cut down like the grass . . . Evildoers shall be cut off; But those who wait on the LORD, They shall inherit the earth." *Ken Ham* (blog), *answersingenesis*, June 20, 2014, http://blogs.answersingenesis.org/blogs/ken-ham/2014/06/20/do-not-fret-because-of-evildoers.

39. While the placard says this passage comes from Genesis 3:4, it actually comes from Genesis 3:1.

40. Georgia Purdom, "Genesis, Wifely Submission, and Modern Wives," *answersingenesis*, November 24, 2009, http://www.answersingenesis.org/arti cles/2009/11/24/genesis-wifely-submission-modern-wives; SteveGolden, "Feedback: Is Male Headship a 'Curse?,'" *answersingenesis*, August 31, 2012, http://www.answersingenesis.org/articles/2012/08/31/feedback-male-headship -curse.

41. Margaret Lamberts Bendroth, *Fundamentalism and Gender, 1875 to the Present* (New Haven, CT: Yale University Press, 1993), 44–45; C. I. Scofield, ed., *The Scofield Reference Bible* (King James Version) (New York: Oxford University Press, 1917), 9.

42. Bendroth, *Fundamentalism and Gender*, 124; W. Gary Phillips, "Adam's Headship before the Fall," Readers Respond, *Answers* 4 (October–December 2009), http://www.answersingenesis.org/articles/am/v4/n4/readers-respond; Golden, "Male Headship."

43. Purdom, "Wifely Submission"; Phillips, "Adam's Headship."

44. Wayne A. Grudem, *Evangelical Feminism and Biblical Truth: An Analysis of More Than One Hundred Disputed Questions* (Sisters, OR: Multnomah, 2004), 93, 95, 46; Bruce A. Ware, *Father, Son, and Holy Spirit: Relationships, Roles, and Relevance* (Wheaton, IL: Crossway Books, 2005): 21, 140–141, 145.

45. For a summary of this heated debate within conservative evangelicalism, see Millard J. Erickson, *Who's Tampering with the Trinity? An Assessment of the Subordination Debate* (Grand Rapids, MI: Kregel, 2009). Erickson concludes his analysis by asserting that while those who argue for subordination within the Trinity have "explicitly reject[ed] the idea of ontological subordination, [their] view actually implies it and thus contains an implicit ontological subordination" (257).

46. Ken Ham, Carl Wieland, and Don Batten, *One Blood: The Biblical Answer to Racism* (Green Forest, AR: Master Books, 1999), 29.

47. Ken Ham, ed., *Demolishing Supposed Bible Contradictions: Exploring Forty Alleged Contradictions*, vol. 1 (Green Forest, AR: Master Books, 2010); Ken Ham, Bodie Hodge, and Tim Chaffey, eds., *Demolishing Supposed Bible Contradictions: Exploring Forty Alleged Contradictions*, vol. 2 (Green Forest, AR: Master Books, 2012). The quote comes from the description of the second volume in the AiG online store.

48. *The World's Most Famous Court Trial: Tennessee Evolution Case*, 2nd rpt. ed. (Dayton, TN: Bryan College, 1990), 302.

49. Ken Ham, "Was Ham Cursed?," *Ken Ham* (blog), *answersingenesis*, January 26, 2013, http://blogs.answersingenesis.org/blogs/ken-ham/2013/01/26/ was-ham-cursed; Mark Looy, "Race around the Nation," *answersingenesis*, January 19, 2009, https://answersingenesis.org/racism/race-around-the-nation.

50. Stephen J. Gould, *Ontogeny and Phylogeny* (Cambridge, MA: Belknap

Press of Harvard University Press, 1977), 127. As Ronald L. Numbers mentioned to the authors, Gould may have simply been wrong here, given that polygenism generated much more controversy regarding race than did Darwinism. See, for example, Monte Harrell Hampton, *Storm of Words: Science, Religion, and Evolution in the Civil War Era* (Tuscaloosa: University of Alabama Press, 2014); G. Blair Nelson, "Infidel Science! Polygenism in the Mid-Nineteenth-Century American Weekly Religious Press" (PhD diss., University of Wisconsin–Madison, 2014).

51. Much has been written on the topic of colonial/settler "genocide," including two very helpful collections, ed. A. Dirk Moses, each of which contains essays on Tasmania and Australia: *Genocide and Settler Society: Frontier Violence and Stolen Indigenous Children in Australian History* (New York: Berghahn Books, 2004); *Empire, Colony, Genocide: Conquest, Occupation, and Subaltern Resistance in World History* (New York: Berghahn Books, 2008).

52. Ken Ham and A. Charles Ware, with Todd A. Hillard, *Darwin's Plantation: Evolution's Racist Roots* (Green Forest, AR: Master Books, 2007). It is not clear what Hillard contributed to the book; when the book was reissued under a new title Hillard's name was not included: Ken Ham and A. Charles Ware, *One Race One Blood: A Biblical Answer to Racism* (Green Forest, AR: Master Books, 2010). In the latter book, there is no explanation for the title change, and the only difference between the two books—besides the different titles—would seem to be the name of the first chapter (written by Ham), which was changed from "Darwin's Plantation" to "Darwin's Garden."

53. Ham, Ware, and Hillard, *Darwin's Plantation*, 15–34, 92. For more on the Ota Benga story, see Phillips Verner Bradford and Harvey Blume, *Ota: The Pygmy in the Zoo* (New York: St. Martin's, 1992); Nancy J. Parezo and Don D. Fowler, *Anthropology Goes to the Fair: The 1904 Louisiana Purchase Exposition* (Lincoln: University of Nebraska Press, 2007), esp. 200–210.

54. Ham, Ware, and Hillard, *Darwin's Plantation*, 23; Paul Weindling, "Genetics, Eugenics, and the Holocaust," in *Biology and Ideology: From Descartes to Dawkins*, ed. Denis R. Alexander and Ronald L. Numbers (Chicago: University of Chicago Press, 2010), 197; Henry M. Morris, *The Long War against God: The History and Impact of the Creation/Evolution Conflict* (Grand Rapids, MI: Baker Book House, 1989), 75. Morris's critique of the social effects of evolutionism was not limited to Nazism. In appendix 2 of *The Henry Morris Study Bible*, he lists thirty-two "Evolutionist Fruits," including Amoralism, Bestiality, Cannibalism, Communism, Homosexuality, Monopolism, Occultism, Pantheism, Pollution, and Racism. Henry M. Morris, ed., *The Henry Morris Study Bible* (King James Version) (1995; repr., Green Forest, AR: Master Books, 2012), 2065. In his politics as in his creationism, Morris was in keeping with George McCready Price. See Carl R. Weinberg, "'Ye Shall Know Them by Their Fruits': Evolution, Eschatology, and the Anticommunist Politics of George McCready Price," *Church History* 83 (September 2014): 684–722.

55. Jerry R. Bergman, "Biology," in *In Six Days: Why 50 Scientists Choose to Believe in Creation*, ed. John Ashton (Green Forest, AR: Master Books, 2001), http://creation.com/jerry-r-bergman-biology-in-six-days; Ronald L. Numbers, *The Creationists: From Scientific Creationism to Intelligent Design*, expanded ed. (Cambridge, MA: Harvard University Press, 2006), 317; Jerry Bergman, *The Dark Side of Charles Darwin: A Critical Analysis of an Icon of Science* (Green Forest, AR: Master Books, 2011); Jerry Bergman, *Hitler and the Nazi Darwinian Worldview: How the Nazi Eugenic Crusade for a Superior Race Caused the Greatest Holocaust in World History* (Kitchener, ONT: Joshua, 2012); Richard Weikart, *From Darwin to Hitler: Evolutionary Ethics, Eugenics, and Racism in Germany* (New York: Palgrave Macmillan, 2004); and Richard Weikart, *Hitler's Ethic: The Nazi Pursuit of Evolutionary Progress* (New York: Palgrave Macmillan, 2009). The other major Darwin-to-Hitler author is Daniel Gasman, *The Scientific Origins of National Socialism* (1971; reprint, New Brunswick, NJ: Transaction, 2004).

56. Weindling, "Genetics, Eugenics, and the Holocaust," 195, 197, 198; Robert J. Richards, "The Narrative Structure of Moral Judgments in History: Evolution and Nazi Biology" (Nora and Edward Ryerson Lecture, University of Chicago, Chicago, IL, April 12, 2005), http://home.uchicago.edu/~rjr6/articles/Ryerson%20Lecture—%20Moral%20Judgment%20in%20History.pdf. For a detailed dismantling of the Darwin-to-Hitler trope, see Robert J. Richards, *Was Hitler a Darwinian? Disputed Questions in the History of Evolutionary Theory* (Chicago: University of Chicago Press, 2013), 192–242.

57. Karl Giberson, *Saving Darwin: How to Be a Christian and Believe in Evolution* (New York: HarperOne, 2008), 77, 79; Anti-Defamation League, "Intelligent Design: It's Not Science," 2012, http://www.adl.org/assets/pdf/civil-rights/religiousfreedom/religfreeres/ID-NotSci-docx.pdf; Martin Luther, "On the Jews and Their Lies" (1543), as quoted in Robert Michael, *Holy Hatred: Christianity, Antisemitism, and the Holocaust* (New York: Palgrave Macmillan, 2006), 114–115.

58. For more on the Curse of Ham and modern slavery, see Stephen R. Haynes, *Noah's Curse: The Biblical Justification of American Slavery* (Oxford: Oxford University Press, 2002); and David M. Whitford, *The Curse of Ham in the Early Modern Era: The Bible and the Justifications for Slavery* (Farnham, England: Ashgate, 2009).

59. Paul S. Taylor, "William Wilberforce: A Leader for Biblical Equality," *Answers* 2 (January–March 2007), https://answersingenesis.org/ministry-news/ministry/william-wilberforce; Bodie Hodge and Paul S. Taylor, "The Bible and Slavery," *answersingenesis*, February 2, 2007, https://answersingenesis.org/bible-history/the-bible-and-slavery.

60. Molly Oshatz, *Slavery and Sin: The Fight against Slavery and the Rise of Liberal Protestantism* (Oxford: Oxford University Press, 2012), 5–9; Robert P. Forbes, "Slavery and the Evangelical Enlightenment," in *Religion and the Ante-*

bellum Debate over Slavery, ed. John R. McKivigan and Mitchell Snay (Athens: University of Georgia Press, 1998), 70.

61. Oshatz, *Slavery and Sin*, 59; Elizabeth Fox-Genovese and Eugene D. Genovese, *The Mind of the Master Class: History and Faith in the Southern Slaveholders' Worldview* (Cambridge: Cambridge University Press, 2005), 490; J. Albert Harrill, "The Use of the New Testament in the American Slave Controversy: A Case History in the Hermeneutical Tension between Biblical Criticism and Christian Moral Debate," *Religion and American Culture* 10 (Summer 2000):163–173; Mark A. Noll, *The Civil War as a Theological Crisis* (Chapel Hill: University of North Carolina Press, 2006), 50.

62. Mark Newman, *Getting Right with God: Southern Baptists and Desegregation, 1945–1995* (Tuscaloosa: University of Alabama Press, 2001), 54; "Segregation Resolution," Baptist Bible Fellowship, November 28, 1957, as quoted in Jeffrey D. Lavoie, *Segregation and the Baptist Bible Fellowship: Integration, Anti-Communism, and Religious Fundamentalism, 1950s–1970s* (Bethesda, MD: Academica, 2013), 9; Paul Harvey, *Freedom's Coming: Religious Culture and the Shaping of the South from the Civil War through the Civil Rights Era* (Chapel Hill: University of North Carolina Press, 2005), 229–245.

63. Carolyn Renee Dupont, *Mississippi Praying: Southern White Evangelicals and the Civil Rights Movement, 1945–1975* (New York: New York University Press, 2013), 231, 96; Lavoie, *Segregation and the Baptist Bible Fellowship*, 83; Todd S. Purdum, *An Idea Whose Time Has Come: Two Presidents, Two Parties, and the Battle for the Civil Rights Act of 1964* (New York: Henry Holt, 2014), 101–103, 231–232; Randall Balmer, *Redeemer: The Life of Jimmy Carter* (New York: Basic Books, 2014), 102–108; Michael W. Fuquay, "Civil Rights and the Private School Movement in Mississippi, 1964–1971," *History of Education Quarterly* 42 (Summer 2002): 159–180; Joseph Crespino, "Civil Rights and the Religious Right," in *Rightward Bound: Making America Conservative in the 1970s*, ed. Bruce J. Schulman and Julian E. Zelizer (Cambridge, MA: Harvard University Press, 2008), 91–105.

64. Goldberg, *Kingdom Coming*, 70; Ham, Ware, and Hillard, *Darwin's Plantation*, 126; A. Charles Ware, *Prejudice and the People of God: How Revelation and Redemption Lead to Reconciliation* (Grand Rapids, MI: Kregel, 2001); "Our Mission and History," *Crossroads Bible College*, http://www.crossroads.edu/aboutcbc/whoweare/missionhistory.php.

65. Susan Friend Harding, *The Book of Jerry Falwell: Fundamentalist Language and Politics* (Princeton, NJ: Princeton University Press, 2000), 180.

66. Daniel T. Rodgers, *Age of Fracture* (Cambridge, MA: Belknap Press of Harvard University Press, 2011), 143.

67. For all of these examples in one article, see Ken Ham, "Are There Really Different Races?," *answersingenesis*, November 29, 2007, https://answersingenesis.org/racism/are-there-really-different-races.

68. Ham and Ware, *Darwin's Plantation*, 47–49, 165–183. Carolyn Renee Dupont has observed that, for evangelicals and fundamentalists, if there is racism, it is to be found in individuals. The result is that, "no matter how vigorously evangelicals decry racism, the continued location of it in the individual—where they locate all social problems—hinders, rather than enhances, any understanding of the racial past and, as a consequence, the country's racial present." *Mississippi Praying*, 238.

69. Adam Liptak, "Supreme Court Invalidates Key Part of Voting Rights Act," *New York Times*, June 25, 2013, http://www.nytimes.com/2013/06/26/us/supreme-court-ruling.html?pagewanted=all&_r=1. As one indication of the AiG response to the Supreme Court ruling regarding gay marriage, Ken Ham published ten blog entries on this topic between June 26 and July 13, 2015, with titles such as "Gay 'Marriage' Decision Will Fundamentally Change the Culture," "Supreme Court, Gay 'Marriage,' and Pandora's Box," and "What Has the Supreme Court Unleashed on the United States?," *Ken Ham* (blog), *answersingen esis*, https://answersingenesis.org/blogs/ken-ham.

70. Mark Looy, "$1.5 Million Dinosaur Exhibit Dedicated Today at the Creation Museum," *answersingenesis*, May 23, 2014, https://answersingenesis.org/ministry-news/creation-museum/15-million-dinosaur-exhibit-dedicated-today; Michael Peroutka, "Impeachment Is Appropriate: An Open Letter to Congress," *Freedom Outpost*, February 5, 2014, http://freedomoutpost.com/2014/02/impeachment-appropriate-open-letter-congress; Michael Peroutka, "Has Your State Legislature Forfeited Its Validity?," *American Clarion*, June 25, 2014, http://www.americanclarion.com/state-legislature-forfeited-validity-31757; Michael Peroutka, "Government Schools: Children Are the Path to Successful Tyranny," *Freedom Outpost*, January 15, 2014, http://freedomoutpost.com/2014/01/government-schools-children-path-successful-tyranny.

71. Michael Peroutka, "Policies Based on Skin Color Are Sinful and Shameful," *Freedom Outpost*, January 20, 2014, http://freedomoutpost.com/2014/01/policies-based-skin-color-sinful-shameful-2; "Michael Peroutka Elected to the League of the South Board of Directors," *Independent Political Report*, June 22, 2013, http://www.independentpoliticalreport.com/2013/06/michael-peroutka-appointed-to-the-league-of-the-south-board-of-directors; Warren Throckmorton, "League of the South President Michael Hill Defines Southern People as White," *Patheos*, September 24, 2013, http://www.patheos.com/blogs/warren throckmorton/2013/09/24/league-of-the-south-president-michael-hill-defines-southern-people-as-white-2. For more on the League of the South—which is not subtle in its "Southern Nationalist" commitments—see http://www.dixienet.org. For a fascinating article on Hill and other white nationalists and their support for Donald Trump's presidential campaign, see Evan Osnos, "The Fearful and the Frustrated," *New Yorker*, August 31, 2015, 50–59.

72. Rachel Maddow, "Dinosaur Find Cited as Proof of Bible Story," MSNBC,

May 27, 2014, http://www.msnbc.com/rachel-maddow-show/watch/dinosaur
-find-cited-as-proof-of-bible-story-268002371823; Ken Ham, "Rachel's Rant on
MSNBC against Answers in Genesis," *answersingenesis*, May 28, 2014, https://
answersingenesis.org/ministry-news/creation-museum/media-coverage
/rachels-rant-msnbc.

Chapter 5. Judgment

1. While the Last Adam Theater was a Creation Museum mainstay from its
opening in 2007, sometime in 2015 the museum's website announced that the
theater would be devoted to a new exhibit, with *The Last Adam* now to be shown
in the Special Effects Theater.

2. *The Last Adam*, directed by John Grooters (Hebron, KY: Answers in Genesis in association with Grooter Productions, 2007), DVD. In our comments
below we are using the DVD version, which obviously does not give the three-
screen effect one gets at the museum, but it does include bonus material, including Michael Matthews's "Content Manager's Commentary."

3. Ken Ham, "Saved at the Creation Museum," *Ken Ham* (blog), answers
ingenesis, April 26, 2008, http://blogs.answersingenesis.org/blogs/ken-ham/
2008/04/26/saved-at-the-creation-museum; Ken Ham, "Saved at the Museum," *Ken Ham* (blog), *answersingenesis*, October 29, 2009, http://blogs.an
swersingenesis.org/blogs/ken-ham/2009/10/29/saved-at-the-museum.

4. Matthews, "Content Manager's Commentary."

5. Julia Parks has observed that the story of Mary being raised in the temple,
with frequent exposure to the sacrifices, has deep roots in the Eastern Church.
"Savior Slain as Sacrifice for Sinners: A Look at the Creation Museum's *The Last
Adam*" (paper for REL 620, University of Dayton, April 28, 2014).

6. Matthews, "Content Manager's Commentary."

7. Regarding the Behemoth and the Leviathan, the "Not the Stuff of Legends!" signboard provides the following biblical text: Job 39: 1, 5, 19, 26–27; Job
40: 15–24; Job 41: 1–2, 9, 18–28.

8. Bill Cooper, *After the Flood: The Early Post-Flood History of Europe
Traced Back to Noah* (Chichester: New Wine, 1993); Eve Siebert, "Monsters and
Dragons and Dinosaurs, Oh My: Creationist Interpretations of *Beowulf*," *Skeptical Inquirer* 37 (January–February, 2013): 43–48 (quote: 44). According to the
blurb accompanying *After the Flood* in the AiG online store, "Bill Cooper shares
fascinating revelations that the earliest Europeans recorded their descent from
Noah through Japheth and knew all about the Flood, and had encounters with
creatures we would call dinosaurs." *answersingenesis*, https://answersingenesis
.org/store/product/after-flood/?sku=10-2-055.

9. All quotes come from product descriptions at the online AiG store: *an
swersingenesis*: https://answersingenesis.org/store/product/america-beginning/
?sku=40-1-403; https://answersingenesis.org/store/product/history-revealed

-full-curriculum/?sku=90-7-618; https://answersingenesis.org/store/product/
biology-101-dvd-based-curriculum/?sku=30-9-177;https://answersingenesis.org
/store/product/gods-design-life-complete-set/?sku=40-1-324; https://answers
ingenesis.org/store/product/ken-ham-foundations-curriculum-kit/?sku=40
-1-381.

10. Sebastian Adams, *Adams' SynChronological Chart or Map of History*
(1871: reprint, Green Forest, AR: Master Books, 2007); *answersingenesis*, https://
answersingenesis.org/store/product/adams-chart-history/?sku=10-2-301. According to the *Oregon Encyclopedia*, at the time he was working on the *Synchronological Chart* Adams was pastor of the First Christian Church in Salem,
Oregon; upon finishing the *Chart*, Adams "spent the next six years traveling, successfully selling the publication at a low cost." Given how his chart is used by AiG,
it is interesting to note that Adams's "personal philosophy and studies led him
to the Unitarian faith." Virginia Green, "Sebastian C. Adams (1825–1898)," *Oregon Encyclopedia*, http://www.oregonencyclopedia.org/articles/adams_sebas
tian_c_1825_1898_/#.U9kgAEjPa6A.

11. *answersingenesis*, https://answersingenesis.org/store/curriculum/courses;
answersingenesis, https://answersingenesis.org/education/online-courses.

12. Ken Ham, "A Spiritual War Is Raging," Letter from Ken, *answersingen
esis*, August 26, 2013, https://answersingenesis.org/apologetics/a-spiritual-war
-is-raging; Ken Ham, "Are You Aware of the Battles against AiG and the Truth of
God's Word?," Letter from Ken, *answersingenesis*, November 14, 2011, https://
answersingenesis.org/ministry-news/core-ministry/are-you-aware-of-the-bat
tles; Ken Ham, "A Spiritual War Is Raging," Letter from Ken, *answersingene
sis*, August 26, 2013, https://answersingenesis.org/apologetics/a-spiritual-war
-is-raging; Ken Ham, "The Creation/Evolution Debate Continues," *Around the
World with Ken Ham* (blog), June 7, 2014, http://blogs.answersingenesis.org/
blogs/ken-ham/2014/06/07/the-creationevolution-debate-continues.

13. Randall J. Stephens and Karl W. Giberson, *The Anointed: Evangelical
Truth in a Secular Age* (Cambridge, MA: Belknap Press of Harvard University
Press, 2011), 43.

14. The following is based on the extensive notes that each of us took during
the presentations.

15. "Bob Gillespie," *answersingenesis*, https://answersingenesis.org/outreach
/speakers/bob-gillespie.

16. For discussion of how dinosaurs became part of the young Earth creationist narrative, see Ronald L. Numbers and T. Joe Willey, "Baptizing Dinosaurs: How Once-Suspect Evidence of Evolution Came to Support the Biblical
Narrative," *Spectrum* 43 (Winter 2015): 57–68.

17. "Aliens and the Creation of Man," *Ancient Aliens*, season 3, episode 16,
History Channel, aired November 16, 2011 (Los Angeles: Prometheus Entertainment, 2011).

18. *answersingenesis*, https://answersingenesis.org/store/product/check-out /?sku=30-9-381&; *answersingenesis*, "Evolution? Impossible!," https://answers ingenesis.org/evidence-against-evolution/evolution-impossible. "Evolution Refuted" is included in *Check This Out!*, which—according to the AiG online store, is a "creative, cutting-edge new movie . . . loaded with six mini-videos, each covering a distinct 'hot topic' in the creation/evolution debate and each answering the controversy at warp speed!"

19. Brian Switek, "The Idiocy, Fabrications, and Lies of *Ancient Aliens*," *Smithsonian.Com*, May 11, 2012, http://www.smithsonianmag.com/science-na ture/the-idiocy-fabrications-and-lies-of-ancient-aliens-86294030/?no-ist. Switek notes that *Ancient Aliens* often uses young Earth creationists as their scientific experts and uses the young Earth creationist tactic—made infamous by Duane Gish and referred to as the "Gish Gallop"—of burying "opponent[s] under an avalanche of fictions and distortions."

20. *NW Creation Network*, http://www.nwcreation.net/presentations; "Heinz Lycklama," *CreationWiki*, http://www.creationwiki.org/Heinz_Lycklama; "Russell Humphreys," *CreationWiki*, http://creationwiki.org/Russell_Humphreys; D. Russell Humphreys, "Evidence for a Young World," *Impact* (June 2005), http://www.icr.org/i/pdf/imp/imp-384.pdf; D. Russell Humphreys, "Our Galaxy Is the Centre of the Universe, 'Quantized' Red Shifts Show," *TJ* 16(2002), *answersingenesis*, https://cdn-assets.answersingenesis.org/doc/articles/2002/ TJv16n2_CENTRE.pdf; Roger Patterson, *Evolution Exposed: Earth Science* (Hebron, KY: Answers in Genesis, 2008), 58.

21. Ham's notion that "why we wear clothes" is a "major [Christian] doctrine" is peculiar, to say the least. But this is a topic that clearly occupies his thinking. In his chapter on "important Christian doctrines" in *The Lie* he devotes three pages to "Why Clothes?," arguing that clothing is "a reminder of our sin problem"; more than this, because "men are very easily sexually aroused," it is incumbent upon women to dress modestly to ensure that men do not commit "adultery in their hearts—adultery for which they and the women will have to answer." Ken Ham, *The Lie: Evolution/Millions of Years*, rev. ed. (Green Forest, AR: Master Books, 2012), 104.

22. Ken Ham, "More Sad Compromise in the Church," *Ken Ham* (blog), *an swersingenesis*, March 30, 2011, http://blogs.answersingenesis.org/blogs/ken -ham/2011/03/30/more-sad-compromise-in-the-church; Ken Ham, *Six Days: The Age of the Earth and the Decline of the Earth* (Green Forest, AR: Master Books, 2013), 9, 12, 29–30. For more on Ham's assertion—which he claims is backed by polling research—that allowing for the possibility of an old Earth is a major factor in youth leaving the church, see Ken Ham and Britt Beemer, with Todd Hillard, *Already Gone: Why Your Kids Will Quit Church and What You Can Do to Stop It* (Green Forest, AR: Master Books, 2009).

23. Ham, *Six Days*, 42, 218.

24. Bodie Hodge, "Institution Questionnaire," in Ken Ham and Greg Hall with Britt Beemer, *Already Compromised: Christian Colleges Took a Test on the State of Their Faith and the Final Exam Is In* (Green Forest, AR: Master Books, 2011), appendix D, 228–236.

25. "Creation Colleges," *answersingenesis*, https://answersingenesis.org/col leges; "College-Expo [2014]," *answersingenesis*, https://answersingenesis.org/outreach/event/College-Expo. The list of topics is from the promotional information for the November 2014 conference.

26. "About Creation Colleges," *answersingenesis*, https://answersingenesis .org/colleges/about; "Colleges in Agreement with AiG's Statement of Faith," *answersingenesis*, Creation Colleges, https://answersingenesis.org/colleges/all.

27. "Answers in Genesis: Creation Museum Grand Opening," *Cedarville University*, https://www.cedarville.edu/cf/calendar/viewsingleevent/id/ bd81e473-fd2d -f790-5c42-6381d4b2137e; "Planning a Successful College Visit," *Cedarville University*, October 21, 2011, http://www.cedarville.edu/eNews/ParentPrep/2012/~/ link.aspx?_id=7291B0F3826E4A95A6DC502EA9A29E26&_z=zKenHam,"Register Today for Creation College 4," *Ken Ham* (blog), *answersingenesis*, June 22, 2014, http://blogs.answersingenesis.org/blogs/ken-ham/2014/06/22/register-to day-for-creation-college-4.

28. Georgia Purdom, "Answers for Women 4: Women Living after Eve," February 13, 2014, *Georgia Purdom* (blog), *answersingenesis*, http://blogs.an swersingenesis.org/blogs/georgia-purdom/2014/02/13/answers-for-women -2014-women-living-after-eve; "Alumni Awards," Cedarville University, http:// www.cedarville.edu/Alumni/Awards.aspx; Creation Biology Society, "CBS Conferences," http://www.creationbiology.org/content.aspx?page_id=22&club_id= 201240&module_id=36812; Roger Patterson, "Creation Geologists Meet at Cedarville," *answersingenesis*, July 31, 2007, https://answersingenesis.org/ministry -news/ministry/creation-geologists-meet-at-cedarville.

29. "First-of-Its-Kind Program for Creation Geologists," *Answers* 4 (July–September 2009), https://answersingenesis.org/ministry-news/ministry/first -of-its-kind-program-for-creation-geologists; "Faculty," Cedarville University, http://www.cedarville.edu/Academics/Science-and-Mathematics/Geology/ Faculty.aspx; John Whitmore, "Aren't Millions of Years Required for Geological Processes?," in *The New Answers Book 2*, ed. Ken Ham (Green Forest, AR: Master Books, 2008), 229–244; Ken Ham, "Interact Live Tonight with Dr. Andrew Snelling and Dr. John Whitmore," *Ken Ham* (blog), *answersingenesis*, April 15, 2014, http://blogs.answersingenesis.org/blogs/ken-ham/2014/04/15/ interact-live-tonight; "2015 Grand Canyon Raft Trips," Answers in Genesis, https://answersingenesis.org/events/grand-canyon-raft-trips; Mark Weinstein, "Geology Students Assist with New Allosaurus Exhibit at Creation Museum," Cedarville University, May 23, 2014, http://www.cedarville.edu/Offices/Public

-Relations/CampusNews/2014/Geology-Students-Assist-with-New-Allo
saurus-Exhibit-at-Creation-Museum.aspx.

30. Ken Ham, "Special Visitors at AiG / Creation Museum," *Ken Ham* (blog), *answersingenesis*, May 14, 2008, http://blogs.answersingenesis.org/blogs/ken
-ham/2008/05/14/special-visitors-at-aigcreation-museum; "CDR Radio Network Sold," Cedarville University, http://www.cedarville.edu/Advancement/
WCDR.aspx; Andrea Speros, "Debate on Creation Live-Streamed at the Dixon Ministry Center," Cedarville University, http://www.cedarville.edu/Offices/Pub
lic-Relations/CampusNews/2014/Debate-on-Creation-Live-Streamed-At
-Cedarville-University.aspx; John Whitmore, "Principles of Earth Science, GSCI 1010," Cedarville University, http://www.cedarville.edu/Academics/Dual-Enroll
ment/~/media/Files/PDF/Dual-Enrollment/Courses/Fall/gsci_1010.ashx;
John Whitmore, "Principles of Earth Science, GSCI 1010," Cedarville University, http://www.cedarville.edu/Academics/Dual-Enrollment/~/media/Files/PDF/
Dual-Enrollment/Courses/Spring/gsci_1010.ashx.

31. Public Relations Office, "Worldview Weekend Coming to CU," Cedarville University, http://www.cedarville.edu/Offices/Public-Relations/Campus-
News/2006/Worldview-Weekend-Coming-to-CU.aspx; "Pastor's Appreciation Day Schedule," Cedarville University, http://www.cedarville.edu/Event/Pastors
-Appreciation-Day.aspx; Kaitlyn Coughlin, "Cedarville to Host Answers in Genesis Conference," Cedarville University, http://www.cedarville.edu/Offices/
Public-Relations/CampusNews/2012/Cedarville-to-Host-Answers-in-Genesis
-Conference.aspx; Ken Ham, "Speaking at a Creationist College," *Ken Ham* (blog), *answersingenesis*, September 1, 2010, http://blogs.answersingenesis.org/
blogs/ken-ham/2010/09/01/speaking-at-a-creationist-college; "Creation Museum Founder Speaks at Cedarville," Cedarville University, http://www.cedarville
.edu/Offices/Public-Relations/CampusNews/2011/Creation-Museum-Founder
-Speaks-at-Cedarville.aspx.

32. E-mail to authors from anonymous informant, August 28, 2014.

33. "Doctrinal Statement" and "Doctrinal Statement White Papers," Cedarville University, http://www.cedarville.edu/About/Doctrinal-Statement.aspx;
Melissa Steffan, "Crisis of Faith Statements," *Christianity Today*, October 29, 2012, http://www.christianitytoday.com/ct/2012/november/crisis-of-faith-state
ments.html?paging=off.

34. Mark Oppenheimer, "An Ohio Christian College Struggles to Further Define Itself," *New York Times*, February 16, 2013; Melissa Steffan, "Administrators' Resignations Fuel Fears at Cedarville University," *Christianity Today*, January 23, 2013, http://www.christianitytoday.com/ct/2013/january-web-only/
resignations-of-top-administrators-fuel-fears-at-cedarville.html?paging=off;
Sarah Pulliam Bailey, "Leadership changes at Cedarville University point to conservative direction," Religion News Service, December 13, 2013, http://www

.religionnews.com/2013/12/13/reports-conservative-shakeup-ohio-christian
-university-hits-women.

35. E-mail to authors from anonymous informant, August 31, 2014; "What's Happened at Cedarville?," *Let There Be Light* (blog), http://fiatlux125.word press.com/2014/05/05/whats-happened-at-cedarville-villefeedbackforum; "Faculty and Staff," *Cedarville University*, http://www.cedarville.edu/Academics /Biblical-and-Theological-Studies/Faculty-Staff.aspx.

36. Bailey, "Leadership changes"; Ruth Moon, "Christian College Solidifies Complementarian Stance," *Christianity Today* ("Gleanings"), March 21, 2014, http://www.christianitytoday.com/gleanings/2014/march/christian-college -solidifies-complementarian-cedarville.html?paging=off; Sarah Jones, "I Will Not Suffer a Woman," *AnthonyBSusan* (blog), March 18 2014, http://anthony bsusan.wordpress.com/2014/03/18/i-will-not-suffer-a-woman. Jones is a frequent and—in some quarters—controversial critic of Cedarville's turn to doctrinaire fundamentalism. For her scathing attack on Cedarville's decision to build a $6 million firing range on campus, see "University or Militia? Cedarville Jumps the Shark," *AnthonyBSusan* (blog), July15, 2013, http://anthonybsusan.word press.com/2013/07/15/university-or-militia-cedarville-jumps-the-shark.

37. Ken Ham, "Partnering with a Creationist University," *Ken Ham* (blog), *answersingenesis*, August 18, 2015, https://answersingenesis.org/blogs/ken-ham /2015/08/18/partnering-with-a-creationist-university.

38. "Doctrinal Statement White Papers."

39. Ham and Hall, *Already Compromised*, 19, 22, 55, 82. "Members and Affiliates," Council for Christian Colleges and Universities, http://www.cccu.org/ members_and_affiliates; "Creation Colleges," *answersingenesis*, https://legacy -cdn-assets.answersingenesis.org/assets/pdf/colleges/colleges-list.pdf.

40. Ham and Hall, *Already Compromised*, 35, 90–91.

41. Ken Ham, "Poll Shows Nazarene Scholars Embracing Evolution," *Ken Ham* (blog), *answersingenesis*, June 24, 2014, http://blogs.answersingenesis.org/ blogs/ken-ham/2014/06/24/poll-shows-nazarene-scholars-embracing-evolu tion; Ken Ham, "Nazarene Churches—What Is Really Happening?," *Ken Ham* (blog), *answersingenesis*, August 23, 2011, http://blogs.answersingenesis.org/ blogs/ken-ham/2011/08/23/nazarene-churceswhat-is-really-happening; Ken Ham, "Evolutionist Christian College Professor Admits What He's Doing to Students," *Ken Ham* (blog), *answersingenesis*, September 4, 2012, http://blogs .answersingenesis.org/blogs/ken-ham/2012/09/04/evolutionist-christian-college -professor-admits; Ken Ham, "Nazarenes Defending Evolution," *Ken Ham* (blog), *answersingenesis*, June 22, 2013, http://blogs.answersingenesis.org/ blogs/ken-ham/2013/06/22/nazarenes-defending-evolution; Ken Ham, "Many Nazarenes Helping Evolution's Destructive Effects," *Ken Ham* (blog), *answers ingenesis*, October 16, 2013, http://blogs.answersingenesis.org/blogs/ken-ham/ 2013/10/16/nazarenes-helping-evolutions-destructive-effects.

42. Bob Smietana, "Young Evangelical Writer: 'Move On' from Evolution-Creationism Debate," *USA Today*, July 23, 2010, http://www.usatoday.com/news/religion/2010-07-23-scopes23_ST_N.htm.

43. Ken Ham, "'Move On from Evolution-Creationism Debate?'—No, It Is Heating Up!," *Ken Ham* (blog), *answersingenesis*, July 29, 2010, http://blogs.answersingenesis.org/blogs/ken-ham/2010/07/29/move-on-from-evolution-creationism-debate-no-it-is-heating-up.

44. Kevin Hardy, "Bryan College takes stand on creation that has professors worried for their jobs," *Times Free Press*, March 2, 2014, http://www.timesfreepress.com/news/2014/mar/02/bryan-college-draws-takes-stand-creation-has-profe; Kevin Hardy, "Discord roils Bryan College as creation flap brings host of issues," *Times Free Press*, March 9, 2014, http://timesfreepress.com/news/2014/mar/09/discord-roils-bryan-college.

45. Ken Ham, "What Is Happening at Bryan College?," *Ken Ham* (blog), *answersingenesis*, http://blogs.answersingenesis.org/blogs/ken-ham/2014/03/04/what-is-happening-at-bryan-college; Kevin Hardy, "Bryan College faculty prepare to leave over evolution controversy," *Times Free Press*, April 4, 2014, http://www.timesfreepress.com/news/2014/apr/04/bryan-faculty-exodus-looms; "2 Fired Professors Sue Bryan College," *Diverse: Issues in Higher Education*, May 15, 2014, http://diverseeducation.com/article/64293; Scot McKnight, "Board Resignations, Board Support at Bryan College," *Patheos*, Jesus Creed, July 21, 2014, http://www.patheos.com/blogs/jesuscreed/2014/07/21/board-resignations-board-support-at-bryan-college; Kevin Hardy, "Bryan College to cut 20 positions," *Times Free Press*, May 31, 2014, http://www.timesfreepress.com/news/2014/may/31/bryan-college-to-cut-20-positions; "Colleges in Agreement with AiG's Statement of Faith"; "College-Expo [2014]"; Ken Ham, "College Expo 2013—Still Time to Register," *Ken Ham* (blog), *answersingenesis*, October 25, 2013, http://blogs.answersingenesis.org/blogs/ken-ham/2013/10/25/college-expo-2013-still-time-to-register. We do not know when exactly the Bryan College administration signed on to AiG's Statement of Faith, but we first spotted the listing in August 2014.

46. This section includes discussion of a late 1980s creation–evolution controversy at Calvin. For a summary of this controversy, see Harry Boonstra, *Our School: Calvin College and the Christian Reformed Church* (Grand Rapids, MI: Eerdmans, 2001), 117–133. For a less measured appraisal, see Ken Ham, "Warning! Rampant Compromise—but Isn't It Really Heresy?," *Ken Ham* (blog), *answersingenesis*, May 16, 2013, http://blogs.answersingenesis.org/blogs/ken-ham/2013/05/16/warning-rampant-compromise-but-isnt-it-really-heresy.

47. Ken Ham, "Send Students to Calvin College to Be Indoctrinated in Evolution," *Ken Ham* (blog), *answersingenesis*, October 5, 2009, http://blogs.answersingenesis.org/blogs/ken-ham/2009/10/05/send-students-to-calvin-college-to-be-indoctrinated-in-evolution; Ken Ham, "Writer for Calvin Col-

lege Newspaper Lashes out at Answers in Genesis," *Ken Ham* (blog), *answers ingenesis*, March 6, 2010, http://blogs.answersingenesis.org/blogs/ken-ham/ 2010/03/06/writer-for-calvin-college-newspaper-lashes-out-at-answers-in -genesis; Ken Ham, "You Won't Believe What Came Out of Calvin College Last Week," *Ken Ham* (blog), *answersingenesis*, March 15, 2010, http://blogs.answers ingenesis.org/blogs/ken-ham/2010/03/15/you-wont-believe-what-came-out -of-calvin-college-last-week; Ken Ham, "Calvin College and USA Today," *Ken Ham* (blog), *answersingenesis*, March 20, 2010, http://blogs.answersingenesis .org/blogs/ken-ham/2010/03/20/calvin-college-and-usa-today.

48. Scott Jaschik, "Fall from Grace," *Inside Higher Education*, August 15, 2011, https://www.insidehighered.com/news/2011/08/15/a_professor_s_depar ture_raises_questions_about_freedom_of_scholarship_at_calvin_college; Daniel C. Harlow, "After Adam: Reading Genesis in an Age of Evolutionary Science," *Perspectives on Science and Christian Faith* 62 (September 2010), 179–195; John R. Schneider, "Recent Genetic Science and Christian Theology on Human Origins: An 'Aesthetic Supralapsarianism,'" *Perspectives on Science and Christian Faith* 62 (September 2010): 196–212; Ken Ham, "False Teaching Rife at Calvin College," *Ken Ham* (blog), *answersingenesis*, February 17, 2011, http://blogs.answersingenesis.org/blogs/ken-ham/2011/02/17/false-teaching -rife-at-calvin-college; Ken Ham, "I Agree with the Atheists!," *Ken Ham* (blog), *an swersingenesis*, June 1, 2011, http://blogs.answersingenesis.org/blogs/ken-ham /2011/06/01/i-agree-with-the-atheists.

49. Jaschik, "Fall from Grace"; Napp Nazworth, "Calvin College Professor Claims Administration Not Truthful over Colleague's Resignation," *The Christian Post*, August 17, 2011, http://www.christianpost.com/news/calvin-college -professor-claims-administration-not-truthful-over-colleagues-resignation -54046; Ken Ham, "The Secularization of a Christian College," *Ken Ham* (blog), *answersingenesis*, July 6, 2014, http://blogs.answersingenesis.org/blogs/ken -ham/2014/07/06/the-secularization-of-a-christian-college.

50. "The Language of God: BioLogos Website and Workshop," *John Templeton Foundation*, http://www.templeton.org/what-we-fund/grants/the-lan guage-of-god-biologos-website-and-workshop; "About the BioLogos Foundation," *BioLogos*, biologos.org/about.

51. "About the BioLogos Foundation."

52. Ken Ham, "More 'Uncertain' Sounds from the Church," *Ken Ham* (blog), *answersingenesis*, June 25, 2012, http://blogs.answersingenesis.org/blogs/ken -ham/2012/06/25/more-uncertain-sounds-from-the-church; Ken Ham, "Did Eve Come from Adam or an 'Ape-Woman?,'" *Ken Ham* (blog), *answersingenesis*, April 21, 2014, http://blogs.answersingenesis.org/blogs/ken-ham/2014/04/21/ did-eve-come-from-adam-or-an-ape-woman.

53. Ken Ham, "Who Teaches This? You May Be Surprised!," *Ken Ham* (blog), *answersingenesis*, May 11, 2009, http://blogs.answersingenesis.org/blogs/ken

-ham/2009/05/11/who-teaches-this-you-may-be-surprised; Ken Ham, "Calvin and Westmont Christian College Professors Leading Charge against Biblical Authority," *Ken Ham* (blog), *answersingenesis*, January 31, 2013, http://blogs.answersingenesis.org/blogs/ken-ham/2013/01/31/calvin-and-westmont-christian-college-professors-leading-charge-against-biblical-authority; Ken Ham, "Peter Enns—Mutilating God's Word," *Ken Ham* (blog), *answersingenesis*, December 14, 2012, http://blogs.answersingenesis.org/blogs/ken-ham/2012/12/14/peter-enns-mutilating-gods-word.

54. Ken Ham, "Should I Have Dinner with BioLogos?," *Ken Ham* (blog), *answersingenesis*, October 14, 2014, http://blogs.answersingenesis.org/blogs/ken-ham/2014/10/14/should-i-have-dinner-with-biologos; Ken Ham, "BioLogos Targets Children and Teens with Theistic Evolution," *Ken Ham* (blog), *answersingenesis*, July 2, 2014, http://blogs.answersingenesis.org/blogs/ken-ham/2014/07/02/biologos-targets-children-and-teens-with-theistic-evolution.

55. Ham and Hall, *Already Compromised*, 128–129.

56. Ken Ham, "Wheaton College and False Teaching in Tennessee," *Ken Ham* (blog), *answersingenesis*, February 18, 2011, http://blogs.answersingenesis.org/blogs/ken-ham/2011/02/18/wheaton-college-and-false-teaching-in-tennessee; Ken Ham, "The Refreshing Faith of a Child," *Ken Ham* (blog), *answersingenesis*, February 13, 2013, http://blogs.answersingenesis.org/blogs/ken-ham/2013/02/13/the-refreshing-faith-of-a-child; Ken Ham, "Peter Enns Responds with 'Ken Ham Clubs Baby Seals,'" *Ken Ham* (blog), *answersingenesis*, September 24, 2012, http://blogs.answersingenesis.org/blogs/ken-ham/2012/09/24/peter-enns-responds-with-ken-ham-clubs-baby-seals; Ham, "I Agree with the Atheists."

57. Ken Ham, "Never-Ending List of Christians Who Compromise," *Ken Ham* (blog), *answersingenesis*, September 20, 2013, http://blogs.answersingenesis.org/blogs/ken-ham/2013/09/20/never-ending-list-of-christians-who-compromise; Ken Ham, "Wolves in Sheep's Clothing?," *Ken Ham* (blog), *answersingenesis*, August 28, 2012, http://blogs.answersingenesis.org/blogs/ken-ham/2012/08/28/wolves-in-sheeps-clothing; Ham, "Evolutionist Christian College Professor"; Ken Ham, "It Is a Fearful Thing to Fall into the Hands of the Living God," *Ken Ham* (blog), *answersingenesis*, June 1, 2014, http://blogs.answersingenesis.org/blogs/ken-ham/2014/06/01/a-fearful-thing-to-fall-into-the-hands-of-the-living-god.

58. Ken Ham, "Many Nazarenes Helping Evolution's Destructive Effects," *Ken Ham* (blog), *answersingenesis*, October 16, 2013, http://blogs.answersingenesis.org/blogs/ken-ham/2013/10/16/nazarenes-helping-evolutions-destructive-effects; Ken Ham, "Pastor Calls for More Active Teaching of Evolution and Rejection of Genesis," *Ken Ham* (blog), *answersingenesis*, June 14, 2014, http://blogs.answersingenesis.org/blogs/ken-ham/2014/06/14/pastor-calls-for-more-active-teaching-of-evolution-and-rejection-of-genesis; Ken Ham, "Why

Not Edit the Bible?," *Ken Ham* (blog), *answersingenesis*, September 19, 2012, http://blogs.answersingenesis.org/blogs/ken-ham/2012/09/19/why-not-edit-the-bible.

59. Henry M. Morris, ed., *The Henry Morris Study Bible* (King James Version) (1995; repr., Green Forest, AR: Master Books, 2012), 1738–1739, 2038–2039; Ken Ham, "The Leaders of the Presbyterian Church USA Need to Fear God," *Ken Ham* (blog), *answersingenesis*, June 26, 2014, http://blogs.answersingenesis.org/blogs/ken-ham/2014/06/26/the-leaders-of-the-presbyterian-church-usa-need-to-fear-god.

60. Doug Frank, *A Gentler God: Breaking Free of the Almighty in the Company of the Human Jesus* (Menangle, Australia: Albatross Books, 2010), 105–132; Tim Challies, "What Kind of God Would Condemn People to Eternal Torment?," *Answers* 7 (July–September 2012), https://answersingenesis.org/who-is-god/god-is-good/what-kind-of-god-would-condemn-people-to-eternal-torment. In his *Study Bible* Henry Morris added another two reasons for hell as the place where sinners suffer eternal, conscious torment: people in hell will be "less miserable" than if "they were forced to be in His presence in heaven forever," and anyway, "they must exist forever somewhere, since they had been created in God's image, which by definition is eternal" (2039).

61. Morris, ed., *Henry Morris Study Bible*, 1445. Here it would seem Morris is proposing a "third" judgment day, in addition to the Great White Throne judgment and the Judgment Seat of Christ.

62. Michael J. Himes, *Doing the Truth in Love: Conversations about God, Relationships, and Service* (New York: Paulist, 1995), 51; Richard T. Hughes, *Christian America and the Kingdom of God* (Urbana: University of Illinois Press, 2009), 68–71; Meghan Henning, *Educating Early Christians through the Rhetoric of Hell: "Weeping and Gnashing of Teeth" as Paideia in Matthew and the Early Church* (Tübingen: Mohr Siebeck, 2014), 229–230.

Epilogue

1. "Bond Offering Succeeds for Full-Size Ark," press release, *answersingenesis*, February 27, 2014, https://answersingenesis.org/ministry-news/ark-encounter/bond-offering-succeeds-for-full-size-ark/. For more on the Ark Encounter project, see James S. Bielo, "Literally Creative: Intertextual Gaps and Artistic Agency," in *Scripturalizing the Human: The Written as the Political*, ed. Vincent L. Wimbush (New York: Routledge, 2015), 20–34.

2. "Ark Encounter Construction Begins," video embedded in "Bond Offering Succeeds for Full-Size Ark"; Ken Ham, "An (Intriguing) Ark Construction Update: Then and Now," *Ken Ham* (blog), *answersingenesis*, January 7, 2015, http://blogs.answersingenesis.org/blogs/ken-ham/2015/01/07/an-intriguing-ark-construction-update-then-and-now/. "The Ark Is Happening!" video embedded in "The Ark Is Happening!," *Ken Ham* (blog), Gary Cole speaking, Ark

Encounter project manager, December 18, 2014, http://blogs.answersingenesis
.org/blogs/ken-ham/2014/12/18/the-ark-is-happening/; "Ark Encounter Museum Staff Update," video; Ken Ham speaking, April 7, 2015, https://arkencoun
ter.com/media/.

3. The description here is based on information provided in "Bond Offering Succeeds for Full-Size Ark," press release and embedded video, Patrick Marsh speaking. See also three diagrams of the interior of the ark at the Ark Encounter website: https//arkencounter.com/about/.

4. "About," Ark Encounter, no date (accessed February 4, 2015), https://ark
encounter.com/about/park-map; "Ark Encounter Construction Begins."

5. Ken Ham, "The Ark Needed Now More Than Ever!," fundraising e-mail received by William Trollinger, June 30, 2015; "Donate," Ark Encounter, https://
arkencounter.com/donate/; "Boarding Passes," Ark Encounter, https://ark
encounter.com/support-the-ark/. We note that the Ark Encounter website indicates that they are trying to raise $29.5 million for the project. We are unable to determine how that figure fits with the figures from Ken Ham's fundraising e-mail.

6. "The Building Begins," Ark Encounter, December 1, 2014, https://ark
encounter.com/blog/2014/12/01/the-building-begins/.

7. Ark Encounter, "Lehman Brothers Visit the Ark Site," Ark Encounter, June 11, 2014, https://arkencounter.com/blog/2014/06/11/lehman-brothers
-visit-the-ark-site/; "About," Ark Encounter, no date (accessed May 9, 2015).

8. "Ark Encounter Construction Begins," Ken Ham speaking.

9. "Ark Encounter Construction Begins," Patrick Marsh speaking.

10. Ark Encounter, "The Exit," Ark Encounter (blog), November 11, 2011, https://arkencounter.com/blog/2011/11/11/the-exit/.

11. Ken Ham, "Was There Really a Noah's Ark and Flood?," *answersingenesis*, October 23, 2012, video, https://answersingenesis.org/media/video/bible/really
-noahs-ark-flood/.

12. Amanda Petrusich, "A Boat of Biblical Proportions," *The Atlantic*, December 2012, http://www.theatlantic.com/magazine/archive/2012/12/a-boat-of-biblical
-proportions/309173/.

13. "Bond Offering Succeeds for Full-Size Ark." See also "United States District Court Eastern District of Kentucky Central Division at Frankfort" on behalf of Ark Encounter, LLC, Crosswater Canyon, Inc., and Answers in Genesis, Inc., against Bob Stewart (secretary of the Kentucky Tourism, Arts and Heritage Cabinet) and Steven Beshear (governor, KY) that is posted in pdf form at https://
answersingenesis.org/religious-freedom/religious-discrimination-lawsuit
-filed/, 16–17.

14. Laurie Goodstein, "In Kentucky, Noah's Ark Theme Park Is Planned," *New York Times*, December 5, 2010, http://www.nytimes.com/2010/12/06/
us/06ark.html?_r=0.

15. According to the *New York Times*, "Erwin Chemerinsky, a constitutional scholar and founding dean of the School of Law at the University of California, Irvine, said: 'If this is about bringing the Bible to life, and it's the Bible's account of history that they're presenting, then the government is paying for the advancement of religion. And the Supreme Court has said that the government can't advance religion.'" Laurie Goodstein, "In Kentucky, Noah's Ark Theme Park Is Planned," *New York Times*, December 5, 2010, http://www.nytimes.com/2010/12/06/us/06ark.html?_r=0. The *New York Times* editorial page appeared to agree with Chemerinsky when it published the following statement: "granting tax incentives to the explicitly Christian enterprise clearly clashes with the First Amendment's prohibition on government establishment of religion. Public money is not supposed to pay to advance religion. Kentucky's citizens should certainly ask themselves if this is really the best use of taxpayer dollars." "Crossing the Church-State Divide by Ark," *New York Times* (editorial), Lexis-Nexis Academic, May 31, 2011.

16. Tom Loftus, "Ark Park Tax Incentives in Limbo over Hiring," *Courier Journal*, October 7, 2014, http://www.courier-journal.com/story/news/politics/2014/10/07/ark-park-hiring-issue-jeopardizes-tax-incentives/16854657/. "About," Ark Encounter, no date (accessed May 9, 2015); "United States District Court," 1.

WORKS CITED

Adams, Sebastian C. *Adams' SynChronological Chart or Map of History*. 1871. Reprint, Green Forest, AR: Master Books, 2007.

Akenson, Donald Harman. *Surpassing Wonder: The Invention of the Bible and the Talmuds*. New York: Harcourt Brace, 1998.

Ammerman, Nancy Tatom. *Bible Believers: Fundamentalists in the Modern World*. New Brunswick, NJ: Rutgers University Press, 1987.

Andermann, Jens, and Silke Arnold-de Simine. "Introduction: Memory, Community, and the New Museum." *Theory, Culture & Society* 29, no. 1 (2012): 3–13. doi:10.1177/0263276411423041.

Answers in Genesis. *Journey through the Creation Museum: Prepare to Believe*. Green Forest, AR: Master Books, 2008.

Asma, Stephen T. "Solomon's House: The Deeper Agenda of the New Creation Museum in Kentucky." *Skeptic*, May 23, 2007. http://www.skeptic.com /eskeptic/07-05-23.

Bailey, Lloyd R. *Genesis, Creation, and Creationism*. New York: Paulist, 1993.

Balmer, Randall. *Redeemer: The Life of Jimmy Carter*. New York: Basic Books, 2014.

———. *Thy Kingdom Come: How the Religious Right Distorts the Faith and Threatens America*. New York: Basic Books, 2006.

Balthrop, V. William, Carole Blair, and Neil Michel. "The Presence of the Present: Hijacking 'The Good War'?" *Western Journal of Communication* 74, no. 2 (2010): 170–210. doi:10.1080/10570311003614500.

Barr, James. *Escaping from Fundamentalism*. London: SCM, 1984.

———. *Fundamentalism*. Philadelphia: Westminster, 1978.

Barry, Andrew. "On Interactivity: Consumers, Citizens, and Culture." In *The Politics of Display: Museums, Science, Culture*, edited by Sharon MacDonald, 98–117. New York: Routledge, 1998.

Barry, Judith. "Dissenting Spaces." In *Thinking about Exhibitions*, edited by Reesa Greenberg, Bruce W. Ferguson, and Sandy Nairne, 307–312. New York: Routledge, 1996.

Barton, Bernadette C. *Pray the Gay Away: The Extraordinary Lives of Bible Belt Gays*. New York: New York University Press, 2012.

Bendroth, Margaret Lamberts. *Fundamentalism and Gender, 1875 to the Present*. New Haven, CT: Yale University Press, 1993.

Bennett, Tony. *The Birth of the Museum: History, Theory, Politics*. New York: Routledge, 1995.

———. "Speaking to the Eyes: Museums, Legibility, and the Social Order." In *The Politics of Display: Museums, Science, Culture*, edited by Sharon MacDonald, 25–35. New York: Routledge, 1998.

Bergman, Jerry. "Biology." In *In Six Days: Why 50 Scientists Choose to Believe in Creation*, edited by John Ashton. Green Forest, AR: Master Books, 2001.

———. *The Dark Side of Charles Darwin: A Critical Analysis of an Icon of Science*. Green Forest, AR: Master Books, 2011.

———. *Hitler and the Nazi Darwinian Worldview: How the Nazi Eugenic Crusade for a Superior Race Caused the Greatest Holocaust in World History*. Kitchener, ONT: Joshua, 2012.

Bielo, James S. "Literally Creative: Intertextual Gaps and Artistic Agency." In *Scripturalizing the Human: The Written as the Political*, edited by Vincent L. Wimbush, 20–34. New York: Routledge, 2015.

Biesecker, Barbara A. "Remembering World War II: The Rhetoric and Politics of National Commemoration at the Turn of the 21st Century." *Quarterly Journal of Speech* 88, no. 4 (2002): 393–409.

Blair, Carole. "Contemporary US Memorial Sites as Exemplars of Rhetoric's Materiality." In *Rhetorical Bodies*, edited by Jack Selzer and Sharon Crowley, 16–57. Madison: University of Wisconsin Press, 1999.

Blair, Carole, Marsha S. Jeppeson, and Enrico Pucci, Jr. "Public Memorializing in Postmodernity: The Vietnam Veterans Memorial as Prototype." *Quarterly Journal of Speech* 77, no. 3 (1991): 263–288.

Blair, Carole, and Neil Michel. "Reproducing Civil Rights Tactics: The Rhetorical Performances of the Civil Rights Memorial." *Rhetoric Society Quarterly* 30, no. 2 (2000): 31–55. http://www.jstor.org/stable/3886159.

Boone, Kathleen C. *The Bible Tells Them So: The Discourse of Protestant Fundamentalism*. Albany: State University of New York Press, 1989.

Boonstra, Harry. *Our School: Calvin College and the Christian Reformed Church*. Grand Rapids, MI: Eerdmans, 2001.

Bouw, Gerardus D. *Geocentricity*. Cleveland, OH: Association for Biblical Astronomy, 1992.

Boyer, Paul. *When Time Shall Be No More: Prophecy Belief in Modern American Culture*. Cambridge, MA: Belknap Press of Harvard University Press, 1992.

Bozeman, Theodore Dwight. *Protestants in an Age of Science: The Baconian Ideal and Antebellum American Religious Thought*. Chapel Hill: University of North Carolina Press, 1977.

Bradford, Phillips Verner, and Harvey Blume. *Ota: The Pygmy in the Zoo*. New York: St. Martin's, 1992.

Brown, William P. *The Seven Pillars of Creation: The Bible, Science, and the Ecology of Wonder*. New York: Oxford University Press, 2010.

Brueggemann, Walter. *Genesis*. Atlanta, GA: John Knox, 1982.

Buchanan, Margaret. *Salem Revisited*. Brisbane, Australia: Margaret Buchanan, 1990.

Burack, Cynthia. *Sin, Sex, and Democracy: Antigay Rhetoric and the Christian Right*. Albany: State University of New York Albany Press, 2008.

Butler, Ella. "God Is in the Data: Epistemologies of Knowledge at the Creation Museum." *Ethnos* 75, no. 3 (2010): 229–251. http://dx.doi.org/10.1080.001 41844.2010.507907.

Caldwell, Bob. "'If I Profess': A Spurious, If Consistent, Luther Quote?" *Concordia Journal* 35 (Fall 2009): 356–359.

Campbell, Gordon. *Bible: The Story of the King James Version, 1611–2011*. Oxford: Oxford University Press, 2010.

Carpenter, Joel A. *Revive Us Again: The Reawakening of American Fundamentalism*. New York: Oxford University Press, 1997.

Carr, Nicholas. *The Shallows: What the Internet Is Doing to Our Brains*. New York: W. W. Norton, 2010.

Caudill, Edward. *Intelligently Designed: How Creationists Built the Campaign against Evolution*. Urbana: University of Illinois Press, 2013.

Chaffey, Tim, and Jason Lisle. *Old-Earth Creationism on Trial: The Verdict Is In*. Green Forest, AR: Master Books, 2008.

Charles, Elizabeth Rundle. *Chronicles of the Schonberg-Cotta Family by Two of Themselves*. New York: M. W. Dodd, 1864.

Clark, Joseph. "Specters of a Young Earth." *Triplecanopy*, December 1, 2008. http://canopycanopycanopy.com/contents/specters_of_a_young_earth.

Conkin, Paul K. *When All the Gods Trembled: Darwinism, Scopes, and American Intellectuals*. Lanham, MD: Rowman and Littlefield, 1998.

Cooper, Bill. *After the Flood: The Early Post-Flood History of Europe Traced Back to Noah*. Chichester: New Wine, 1993.

Crespino, Joseph. "Civil Rights and the Religious Right." In *Rightward Bound: Making America Conservative in the 1970s*, edited by Bruce J. Schulman and Julian E. Zelizer, 91–105. Cambridge, MA: Harvard University Press, 2008.

Cullen, Dave. *Columbine*. New York: Twelve, 2009.

Dickinson, Greg. "Joe's Rhetoric: Finding Authenticity at Starbucks." *Rhetoric Society Quarterly* 32, no. 4 (2002): 5–27. http://www.jstor.org/stable/3886018.

———. "Memories for Sale: Nostalgia and the Construction of Identity in Old Pasadena. *Quarterly Journal of Speech* 83, no. 1 (1997): 1–27.

Dickinson, Greg, Brian Ott, and Eric Aoki. "Spaces of Remembering and Forgetting: The Reverent Eye/I at the Plains Indian Museum." *Communication and Critical/ Cultural Studies* 3, no. 1 (2006): 27–47. doi:10.1080/14791420500505619.

Dionne, E. J., Jr. *Souled Out: Reclaiming Faith and Politics after the Religious Right.* Princeton, NJ: Princeton University Press, 2008.

Dochuk, Darren. *From Bible Belt to Sunbelt: Plain-Folk Religion, Grassroots Politics, and the Rise of Evangelical Conservatism.* New York: W. W. Norton, 2011.

Dupont, Carolyn Renee. *Mississippi Praying: Southern White Evangelicals and the Civil Rights Movement, 1945–1975.* New York: New York University Press, 2013.

Efron, Noah J. *A Chosen Calling: Jews in Science in the Twentieth Century.* Baltimore: Johns Hopkins University Press, 2014.

Enns, Peter. *The Evolution of Adam: What the Bible Does and Doesn't Say about Human Origins.* Grand Rapids, MI: Brazos, 2012.

Erickson, Millard J. *Who's Tampering with the Trinity? An Assessment of the Subordination Debate.* Grand Rapids, MI: Kregel, 2009.

Faulkner, Danny. "Interpreting Craters in Terms of the Day Four Cratering Hypothesis." *Answers Research Journal* 7 (2014): 11–25.

Fedo, Michael. *The Lynchings in Duluth.* St. Paul: Minnesota Historical Society Press, 2000.

Ferguson, Bruce W. "Exhibition Rhetorics: Material Speech and Utter Sense." In *Thinking about Exhibitions,* edited by Reesa Greenberg, Bruce W. Ferguson, and Sandy Nairne, 175–190. New York: Routledge, 1996.

Finocchiaro, Maurice A. "That Galileo Was Imprisoned and Tortured for Advocating Copernicanism." In *Galileo Goes to Jail: And Other Myths about Science and Religion,* edited by Ronald L. Numbers, 68–78. Cambridge, MA: Harvard University Press, 2009.

———, ed. and trans. *The Galileo Affair: A Documentary History.* Berkeley: University of California Press, 1989.

Forbes, Robert P. "Slavery and the Evangelical Enlightenment." In *Religion and the Antebellum Debate over Slavery,* edited by John R. McKivigan and Mitchell Snay, 68–103. Athens: University of Georgia Press, 1998.

Fox-Genovese, Elizabeth, and Eugene D. Genovese. *The Mind of the Master Class: History and Faith in the Southern Slaveholders' Worldview.* Cambridge, MA: Cambridge University Press, 2005.

Frank, Doug. *A Gentler God: Breaking Free of the Almighty in the Company of the Human Jesus.* Menangle, Australia: Albatross Books, 2010.

———. *Less than Conquerors: How Evangelicals Entered the Twentieth Century.* Grand Rapids, MI: Eerdmans, 1986.

Fretheim, Terence E. "The Book of Genesis: Introduction, Commentary, and Reflections." In vol. 1, *The New Interpreter's Bible: A Commentary in Twelve*

Volumes, edited by Leander E. Keck et al., 319–674. Nashville: Abingdon, 1994.

Friedman, Richard Elliott. *The Bible with Sources Revealed: A New View into the Five Books of Moses*. San Francisco: HarperSanFrancisco, 2003.

——. *The Hidden Book in the Bible*. San Francisco: HarperSanFrancisco, 1998.

——. *Who Wrote the Bible?* New York: Summit Books, 1987.

Fuquay, Michael W. "Civil Rights and the Private School Movement in Mississippi, 1964–1971." *History of Education Quarterly* 42 (Summer 2002): 159–180.

Galilei, Galileo. "Letter to Madame Christina of Lorraine, Grand Duchess of Tuscany, Concerning the Use of Biblical Quotations in Matters of Science [1615]." In *Discoveries and Opinions of Galileo*, edited and translated by Stillman Drake, 175–216. Garden City, NY: Doubleday Anchor Books, 1957.

Gallagher, Victoria J. "Memory and Reconciliation in the Birmingham Civil Rights Institute." *Rhetoric and Public Affairs* 2, no. 2 (1999): 303–320.

——. "Remembering Together: Rhetorical Integration and the Case of the Martin Luther King, Jr. Memorial." *Southern Communication Journal* 60, no. 2 (1995): 109–119. doi:10.1080/10417949509372968.

Gasman, Daniel. *The Scientific Origins of National Socialism*. 1971. Reprint, New Brunswick, NJ: Transaction, 2004.

Gerlach, Justin, et al. "Giant Tortoise Distribution and Abundance in the Seychelles Islands: Past, Present, and Future." *Chelonian Conservation and Biology* 12, no. 1 (2013): 70–83.

Giberson, Karl. *Saving Darwin: How to Be a Christian and Believe in Evolution*. New York: HarperOne, 2008.

Gieryn, Thomas F. "Boundary Work and the Demarcation of Science from Non-Science: Strains and Interests in Professional Ideologies of Scientists." *American Sociological Review* 48, no. 6 (1983): 781–795.

Gloege, Timothy E. W. *Guaranteed Pure: The Moody Bible Institute, Business, and the Making of Modern Evangelicalism*. Chapel Hill: University of North Carolina Press, 2015.

Goldberg, Michelle. *Kingdom Coming: The Rise of Christian Nationalism*. New York: W. W. Norton, 2006.

Gould, Stephen Jay. *Ontogeny and Phylogeny*. Cambridge, MA: Belknap Press of Harvard University Press, 1977.

Grudem, Wayne A. *Evangelical Feminism and Biblical Truth: An Analysis of More than One Hundred Disputed Questions*. Sisters, OR: Multnomah, 2004.

Habermehl, Anne. "Where in the World Is the Tower of Babel?" *Answers Research Journal* 4 (2011): 25–53.

Ham, Ken., ed. *Demolishing Supposed Bible Contradictions: Exploring Forty Alleged Contradictions*. Vol. 1. Green Forest, AR: Master Books, 2010.

———. *The Lie: Evolution/Millions of Years*. Revised edition. Green Forest, AR: Master Books, 2012.

———, ed. *The New Answers Book 2*. Green Forest, AR: Master Books, 2008.

———. *Six Days: The Age of the Earth and the Decline of the Church*. Green Forest, AR: Master Books, 2013.

Ham, Ken, and Britt Beemer, with Todd Hillard. *Already Gone: Why Your Kids Will Quit Church and What You Can Do to Stop It*. Green Forest, AR: Master Books, 2009.

Ham, Ken, and Greg Hall, with Britt Beemer. *Already Compromised: Christian Colleges Took a Test on the State of Their Faith and the Final Exam Is In*. Green Forest, AR: Master Books, 2011.

Ham, Ken, Bodie Hodge, and Tim Chaffey, eds. *Demolishing Supposed Bible Contradictions: Exploring Forty Alleged Contradictions*. Vol. 2. Green Forest, AR: Master Books, 2012.

Ham, Ken, and Stacia McKeever. *The Seven C's of History*. Hebron, KY: Answers in Genesis, 2004.

Ham, Ken, and A. Charles Ware. *One Race One Blood: A Biblical Answer to Racism*. Green Forest, AR: Master Books, 2010.

Ham, Ken, and A. Charles Ware, with Todd A. Hillard. *Darwin's Plantation: Evolution's Racist Roots*. Green Forest, AR: Master Books, 2007.

Ham, Ken, Carl Wieland, and Don Batten. *One Blood: The Biblical Answer to Racism*. Green Forest, AR: Master Books, 1999.

Hamilton, Victor P. *The Book of Genesis: Chapters 1–17*. Grand Rapids, MI: Eerdmans, 1990.

Hampton, Monte Harrell. *Storm of Words: Science, Religion, and Evolution in the Civil War Era*. Tuscaloosa: University of Alabama Press, 2014.

Hankins, Barry. *God's Rascal: J. Frank Norris and the Beginnings of Southern Fundamentalism*. Lexington: University Press of Kentucky, 1996.

Haraway, Donna. "Teddy Bear Patriarchy: Taxidermy in the Garden of Eden, 1908–1936." *Social Text* 11 (Winter 1984–85): 20–64. http://www.jstor.org/stable/466593.

Harding, Susan Friend. *The Book of Jerry Falwell: Fundamentalist Language and Politics*. Princeton, NJ: Princeton University Press, 2000.

Harlow, Daniel C. "After Adam: Reading Genesis in an Age of Evolutionary Science." *Perspectives on Science and Christian Faith* 62 (September 2010): 179–195.

Harrill, J. Albert. "The Use of the New Testament in the American Slave Controversy: A Case History in the Hermeneutical Tension between Biblical Criticism and Christian Moral Debate." *Religion and American Culture* 10 (Summer 2000): 149–186.

Harris, Harriet A. *Fundamentalism and Evangelicals*. Oxford: Oxford University Press, 1998.

Hart, D. G. *Defending the Faith: J. Gresham Machen and the Crisis of Conservative Protestantism in Modern America.* Baltimore: Johns Hopkins University Press, 1994.

Harvey, Paul. *Freedom's Coming: Religious Culture and the Shaping of the South from the Civil War through the Civil Rights Era.* Chapel Hill: University of North Carolina Press, 2005.

Hasian, Marouf, Jr. "Remembering and Forgetting the 'Final Solution': A Rhetorical Pilgrimage through the US Holocaust Memorial Museum." *Critical Studies in Media Communication* 21, no. 1 (2004): 64–92. doi:10.1080/07 39318042000184352.

Haynes, Stephen R. *Noah's Curse: The Biblical Justification of American Slavery.* Oxford: Oxford University Press, 2002.

Hedges, Chris. *American Fascists: The Christian Right and the War on America.* New York: Free, 2006.

Hennigan, Tom. "An Initial Estimate toward Identifying and Numbering the Ark Turtle and Crocodile Kinds." *Answers Research Journal* 7 (2014): 1–10.

Henning, Meghan. *Educating Early Christians through the Rhetoric of Hell: "Weeping and Gnashing of Teeth" as Paideia in Matthew and the Early Church.* Tübingen: Mohr Siebeck, 2014.

Hentschel, Jason. "Evangelicals, Inerrancy, and the Quest for Certainty: Making Sense of Our Battles for the Bible." PhD diss., University of Dayton, 2015.

Hetherington, Kevin. "The Unsightly: Touching the Parthenon Frieze." *Theory, Culture & Society* 19, no. 5/6 (2002): 187–205.

Himes, Michael J. *Doing the Truth in Love: Conversations about God, Relationships, and Service.* New York: Paulist, 1995.

Homchick, Julie. "Displaying Controversy: Evolution, Creation, and Museums." PhD diss., University of Washington, 2009. UMI Microform (3377287).

Hughes, Richard T. *Christian America and the Kingdom of God.* Urbana: University of Illinois Press, 2009.

Jackson, Brian. "Jonathan Edwards Goes to Hell (House): Fear Appeals in American Evangelism." *Rhetoric Review* 26, no. 1 (2007): 42–59. doi:10.1207/ s15327981rr2601_3.

Johnson, Davi. "Psychiatric Power: The Post-Museum as a Site of Rhetorical Alignment." *Communication and Critical/Cultural Studies* 5, no. 4 (2008): 344–362. doi:10.1080/14791420802412423.

Joubert, Callie. "The Unbeliever at War with God: Michael Ruse and the Creation-Evolution Controversy." *Answers Research Journal* 5 (2012): 125–39.

Kelly, Casey Ryan, and Kristen E. Hoerl. "Genesis in Hyperreality: Legitimizing Disingenuous Controversy at the Creation Museum." *Argumentation and Advocacy* 48, no. 3 (2012): 123–141.

Kirshenblatt-Gimblett, Barbara. "Objects of Ethnography." In *Exhibiting Cultures: The Poetics and Politics of Museum Display,* edited by Ivan Karp and Steven D. Lavine, 386–443. Washington, DC: Smithsonian Institution Press, 1991.

Kosmin, Barry A., and Ariela Keysar. *American Religious Identification Survey 2008: Summary Report*. Hartford, CT: Trinity College Program of Public Values, 2009. http://www.americanreligionsurvey-aris.org.

Krauss, Lawrence. "Museum of Misinformation." *New Scientist*, May 26, 2007, 23.

Kraybill, Donald B., Steven M. Nolt, and David Weaver-Zercher. *Amish Grace: How Forgiveness Transcended Tragedy*. San Francisco: Jossey-Bass, 2007.

Kruse, Kevin M. *One Nation under God: How Corporate America Invented Christian America*. New York: Basic Books, 2015.

Laats, Adam. *Fundamentalism and Education in the Scopes Era: God, Darwin, and the Roots of America's Culture Wars*. New York: Palgrave Macmillan, 2010.

Laclau, Ernesto, and Chantal Mouffe. *Hegemony and Socialist Strategy: Towards a Radical Democratic Politics*. 2nd ed. New York: Verso, 2001.

Lamoureux, Denis O. *I Love Jesus & I Accept Evolution*. Eugene, OR: Wipf and Stock, 2009.

Larkin, Ralph W. *Comprehending Columbine*. Philadelphia: Temple University Press, 2007.

Larson, Edward J. *Summer for the Gods: The Scopes Trial and America's Continuing Debate over Science and Religion*. Cambridge, MA: Harvard University Press, 1997.

———. *Trial and Error: The American Controversy over Creation and Evolution*. 3rd ed. Oxford: Oxford University Press, 2003.

Lavoie, Jeffrey D. *Segregation and the Baptist Bible Fellowship: Integration, Anti-Communism, and Religious Fundamentalism, 1950s–1970s*. Bethesda, MD: Academica, 2013.

Lessl, Thomas M. *Rhetorical Darwinism: Religion, Evolution, and the Scientific Identity*. Waco, TX: Baylor University Press, 2012.

Lindberg, David C. "Galileo, the Church, and the Cosmos." In *When Science and Christianity Meet*, edited by David C. Lindberg and Ronald L. Numbers, 33–60. Chicago: University of Chicago Press, 2003.

Lindsey, Hal. *The Everlasting Hatred: The Roots of Jihad*. Murietta, CA: Oracle House, 2002.

Loucks, Ira S. [pseud.]. "Fungi from the Biblical Perspective: Design and Purpose in the Original Creation." *Answers Research Journal* 2 (2009): 123–131.

Luther, Martin. *Luther's Works*. Vol. 1, *Lectures on Genesis: Chapters 1–5*, edited by Jaroslav Pelikan, translated by George V. Schick. St. Louis, MO: Concordia, 1958.

———. *Luther's Works*. Vol. 54, *Table Talk*, edited by Helmut T. Lehmann, edited and translated by Theodore G. Tappert. Philadelphia: Fortress, 1967.

Lynch, John. "'Prepare to Believe': The Creation Museum as Embodied Conversion Narrative." *Rhetoric and Public Affairs* 16, no. 1 (2013): 1–28.

MacDonald, Sharon. "Afterword: From War to Debate." In *The Politics of Display: Museums, Science, Culture*, edited by Sharon MacDonald, 229–35. New York: Routledge, 1998.

———. "Exhibitions of Power and Powers of Exhibition: An Introduction to the Politics of Display." In *The Politics of Display: Museums, Science, Culture*, edited by Sharon MacDonald, 1–24. New York: Routledge, 1998.

———. "Supermarket Science? Consumers and 'the Public Understanding of Science.'" In *The Politics of Display: Museums, Science, Culture*, edited by Sharon MacDonald, 118–137, New York: Routledge, 1998.

Maltby, Paul. *Christian Fundamentalism and the Culture of Disenchantment.* Charlottesville: University of Virginia Press, 2013.

Marsden, George M. *Fundamentalism and American Culture: The Shaping of Twentieth-Century Evangelicalism: 1870–1925.* New York: Oxford University Press, 1980.

———. *Understanding Fundamentalism and Evangelicalism.* Grand Rapids, MI: Eerdmans, 1991.

Martin, Rod. "A Proposed Bible-Science Perspective on Global Warming." *Answers Research Journal* 3(2010): 91–106.

Martin, William. *With God on Our Side: The Rise of the Religious Right in America.* New York: Broadway Books, 1996.

Matzke, Nicholas J. "The Evolution of Creationist Movements." *Evolution: Education and Outreach* 3 (June 2010): 145–162. http://link.springer.com/article/10.1007%2Fs12052-010-0233-1.

May, Gerhard. *Creatio ex Nihilo: The Doctrine of "Creation out of Nothing" in Early Christian Thought*, translated by A. S. Worrall. Edinburgh: T & T Clark, 1994.

McGrath, Alister E. *Christianity's Dangerous Idea: The Protestant Revolution—a History from the Sixteenth Century to the Twenty-First.* New York: HarperOne, 2007.

McKeever, Stacia, Gary Vaterlaus, and Diane King. *The Creation Museum: Behind the Scenes!* Hebron, KY: Answers in Genesis, 2008.

Michael, Robert. *Holy Hatred: Christianity, Antisemitism, and the Holocaust.* New York: Palgrave Macmillan, 2006.

Moran, Jeffrey P. *American Genesis: The Antievolution Controversies from Scopes to Creation Science.* Oxford: Oxford University Press, 2012.

Moreton, Bethany. *To Serve God and Wal-Mart: The Making of Christian Free Enterprise.* Cambridge, MA: Harvard University Press, 2009.

Morris, Henry M., ed. *The Henry Morris Study Bible* (King James Version). 1995. Reprint, Green Forest, AR: Master Books, 2012.

———. *The Long War against God: The History and Impact of the Creation/ Evolution Conflict*. Grand Rapids, MI: Baker Book House, 1989.

Moses, A. Dirk. *Empire, Colony, Genocide: Conquest, Occupation, and Subaltern Resistance in World History*. New York: Berghahn Books, 2008.

———. *Genocide and Settler Society: Frontier Violence and Stolen Indigenous Children in Australian History*. New York: Berghahn Books, 2004.

Nelson, G. Blair. "Infidel Science! Polygenism in the Mid-Nineteenth-Century American Weekly Religious Press." PhD diss., University of Wisconsin–Madison, 2014.

Newman, Mark. *Getting Right with God: Southern Baptists and Desegregation, 1945–1995*. Tuscaloosa: University of Alabama Press, 2001.

Newsom, Carol A. "The Book of Job: Introduction, Commentary, and Reflections." In vol. 4, *The New Interpreter's Bible: A Commentary in Twelve Volumes*, edited by Leander E. Keck et al., 317–637. Nashville, TN: Abingdon, 1996.

Noll, Mark A. *The Civil War as a Theological Crisis*. Chapel Hill: University of North Carolina Press, 2006.

———. *Jesus Christ and the Life of the Mind*. Grand Rapids, MI: Eerdmans, 2011.

———. *The Scandal of the Evangelical Mind*. Grand Rapids, MI: Eerdmans, 1994.

Numbers, Ronald L. *The Creationists: From Scientific Creationism to Intelligent Design*. Expanded edition. Cambridge, MA: Harvard University Press, 2006.

———. *Darwinism Comes to America*. Cambridge, MA: Harvard University Press, 1998.

———. "Ironic Heresy: How Young-Earth Creationists Came to Embrace Rapid Microevolution by Means of Natural Selection." In *Darwinian Heresies*, edited by Abigail Lustig, Robert J. Richards, and Michael Ruse, 84–100. Cambridge: Cambridge University Press, 2004.

Numbers, Ronald L., and Rennie Schoepflin. "Science and Medicine." In *Ellen Harmon White: American Prophet*, edited by Terrie Dopp Aamodt, Gary Land, and Ronald L. Numbers, 196–223. Oxford: Oxford University Press, 2014.

Numbers, Ronald L., and T. Joe Willey. "Baptizing Dinosaurs: How Once-Suspect Evidence of Evolution Came to Support the Biblical Narrative." *Spectrum* 43 (Winter 2015): 57–68.

Obama, Barack. *The Audacity of Hope: Thoughts on Reclaiming the American Dream*. New York: Crown, 2006.

Oberlin, Kathleen Curry. "Mobilizing Epistemic Conflict: The Creation Museum and the Creationist Social Movement." PhD diss., Indiana University, 2014.

Oppenheimer, Frank. "The Exploratorium: A Playful Museum Combines Perception and Art in Science Education." *American Journal of Physics* 40,

no. 7 (1972). http://www.exploratorium.edu/files/about/our_story/history/
frank/pdfs/playful_museum.pdf.

——. "Rationale for a Science Museum." *Curator: The Museum Journal* 11, no.
3 (1968): 206–9. http://www.exploratorium.edu/files/about/our_story/his
tory/frank/pdfs/rationale.pdf.

Oshatz, Molly. *Slavery and Sin: The Fight against Slavery and the Rise of Liberal Protestantism.* Oxford: Oxford University Press, 2012.

Ott, Brian L., Eric Aoki, and Greg Dickinson. "Ways of (Not) Seeing Guns: Presence and Absence at the Cody Firearms Museum." *Communication and
Critical/Cultural Studies* 8, no. 3 (2011): 215–239. doi:10.1080/14791420
.2011.594068.

Pahl, Michael W. *The Beginning and the End: Rereading Genesis's Stories and
Revelation's Visions.* Eugene, OR: Cascade Books, 2011.

Parezo, Nancy J., and Don D. Fowler. *Anthropology Goes to the Fair: The 1904
Louisiana Purchase Exposition.* Lincoln: University of Nebraska Press,
2007.

Patterson, Roger. *Evolution Exposed: Earth Science.* Hebron, KY: Answers in
Genesis, 2008.

Peters, John Durham. *Speaking into the Air: A History of the Idea of Communication.* Chicago: University of Chicago Press, 1999.

Pierce, Charles P. *Idiot America: How Stupidity Became a Virtue in the Land of
the Free.* New York: Anchor Books, 2009.

Plato. *Phaedrus.* In *Plato on Rhetoric and Language*, edited by Jean Nienkamp,
translated by Alexander Nehamas and Paul Woodruff, 165–214. New York:
Routledge, 1999.

Purdum, Todd S. *An Idea Whose Time Has Come: Two Presidents, Two Parties,
and the Battle for the Civil Rights Act of 1964.* New York: Henry Holt, 2014.

Radosh, Daniel. *Rapture Ready! Adventures in the Parallel Universe of Christian Pop Culture.* New York: Soft Skull, 2010.

Richards, Robert J. "The Narrative Structure of Moral Judgments in History:
Evolution and Nazi Biology." The 2005 Nora and Edward Ryerson Lecture.
University of Chicago, April 12, 2005. http://home.uchicago.edu/~rjr6/ar
ticles/Ryerson%20Lecture—%20Moral%20Judgment%20in%20History
.pdf.

——. *Was Hitler a Darwinian? Disputed Questions in the History of Evolutionary Theory.* Chicago: University of Chicago Press, 2013.

Richardson, Joel. *The Islamic Antichrist: The Shocking Truth about the Real
Nature of the Beast.* Los Angeles: World Net Daily, 2009.

Rios, Christopher M. *After the Monkey Trial: Evangelical Scientists and a New
Creationism.* New York: Fordham University Press, 2014.

Roberts, Michael B. "Genesis Chapter 1 and Geological Time from Hugo Grotius
and Marin Mersenne to William Conybeare and Thomas Chalmers (1620–

1825)." In *Myth and Geology*, edited by Luigi Piccardi and W. Bruce Masse, 39–50. London: Geological Society of London, 2007.

Rodgers, Daniel T. *Age of Fracture*. Cambridge, MA: Belknap Press of Harvard University Press, 2011.

Rosenhouse, Jason. *Among the Creationists: Dispatches from the Anti-Evolutionist Front Line*. New York: Oxford University Press, 2012.

Rothstein, Jandos. "Graphic Displays of Faith." *Print* 62, no. 1 (2008): 97–101.

Sandeen, Ernest R. *The Roots of Fundamentalism: British & American Millenarianism, 1800–1930*. Chicago: University of Chicago Press, 1970.

Sarna, Nahum M. *Genesis = Be-reshit: The Traditional Hebrew Text with New JPS Translation*, with commentary by Nahum M. Sarna. Philadelphia: Jewish Publication Society, 1989.

Schneider, John R. "Recent Genetic Science and Christian Theology on Human Origins: An 'Aesthetic Supralapsarianism.'" *Perspectives on Science and Christian Faith* 62 (September 2010): 196–212.

Scofield, C. I., ed. *The Scofield Reference Bible* (King James Version). New York: Oxford University Press, American Branch, 1909.

———. *The Scofield Reference Bible* (King James Version). 2nd ed. New York: Oxford University Press, 1917.

Shapiro, Adam R. *Trying Biology: The Scopes Trial, Textbooks, and the Antievolution Movement in American Schools*. Chicago: University of Chicago Press, 2013.

Siebert, Eve. "Monsters and Dragons and Dinosaurs, Oh My: Creationist Interpretations of *Beowulf*." *Skeptical Inquirer* 37 (January–February 2013), 43–48.

Skolfield, Ellis H. *Islam in the End Times*. Fort Myers, FL: Fish House, 2007.

Smith, Christian. *The Bible Made Impossible: Why Biblicism Is Not a Truly Evangelical Reading of Scripture*. Grand Rapids, MI: Brazos, 2011.

Smith, Robert O. *More Desired than Our Owne Salvation: The Roots of Christian Zionism*. New York: Oxford University Press, 2013.

Stephens, Randall J., and Karl W. Giberson. *The Anointed: Evangelical Truth in a Secular Age*. Cambridge, MA: Belknap Press of Harvard University Press, 2011.

Storr, Will. *The Unpersuadables: Adventures with the Enemies of Science*. New York: Overlook, 2014.

Sutton, Matthew Avery. *American Apocalypse: A History of Modern Evangelicalism*. Cambridge, MA: Belknap Press of Harvard University Press, 2014.

Teslow, Tracy Lang. "Reifying Race: Science and Art in *Races of Mankind* at the Field Museum of Natural History." In *Politics of Display: Museums, Science, Culture*, edited by Sharon MacDonald, 53–76. New York: Routledge, 1998.

Thuesen, Peter Johannes. *In Discordance with the Scriptures: American Protestant Battles over Translating the Bible*. New York: Oxford University Press, 1999.

Thurs, Daniel P. "Scientific Methods." In *Wrestling with Nature: From Omens to Science*, edited by Peter Harrison, Ronald L. Numbers, and Michael H. Shank, 307–335. Chicago: University of Chicago Press, 2011.

Trollinger, Susan. "From Reading to Revering the Good Book: How the Bible Became Fossil at the Creation Museum." *The Maryville Symposium: Conversations of Faith and the Liberal Arts: The Proceedings*, 29–49. Maryville, TN: Maryville College, 2011.

Trollinger, William Vance, Jr. "Evangelicalism and Religious Pluralism in Contemporary America: Diversity Without, Diversity Within, and Maintaining the Borders." In *Gods in America: Religious Pluralism in the United States*, edited by Charles Cohen and Ronald L. Numbers, 105–124. New York: Oxford University Press, 2013.

———. *God's Empire: William Bell Riley and Midwestern Fundamentalism*. Madison: University of Wisconsin Press, 1990.

———. "An Outpouring of 'Faithful' Words: Protestant Publishing in the United States." In *Print in Motion: The Expansion of Publishing and Reading in the United States, 1880–1940*, edited by Carl Kaestle and Janice Radway, vol. 4, *History of the Book in America*, 359–75. Chapel Hill: University of North Carolina Press, 2009.

"Trouble in Paradise: Answers in Genesis Splinters." *Reports of the National Center for Science Education* 26 (2006), http://ncse.com/rncse/26/6/trouble-paradise.

Waltke, Bruce K., with Cathi J. Fredricks. *Genesis: A Commentary*. Grand Rapids, MI: Zondervan, 2001.

Walton, John H. *Ancient Near Eastern Thought and the Old Testament: Introducing the Conceptual World of the Hebrew Bible*. Grand Rapids, MI: Baker Academic, 2006.

———. *The Lost World of Genesis One: Ancient Cosmology and the Origins Debate*. Downers Grove, IL: InterVarsity, 2009.

Ware, A. Charles. *Prejudice and the People of God: How Revelation and Redemption Lead to Reconciliation*. Grand Rapids, MI: Kregel, 2001.

Ware, Bruce A. *Father, Son, and Holy Spirit: Relationships, Roles, and Relevance*. Wheaton, IL: Crossway, 2005.

Watkins, Steven Mark. "An Analysis of the Creation Museum: Hermeneutics, Language, and Information Theory." PhD diss., University of Louisville, 2014.

Weber, Timothy P. "The Two-Edged Sword: The Fundamentalist Use of the Bible." In *The Bible in America: Essays in Cultural History*, edited by Nathan O. Hatch and Mark A. Noll, 101–120. New York: Oxford University Press, 1982.

Weikart, Richard. *From Darwin to Hitler: Evolutionary Ethics, Eugenics, and Racism in Germany*. New York: Palgrave Macmillan, 2004.

———. *Hitler's Ethic: The Nazi Pursuit of Evolutionary Progress*. New York: Palgrave Macmillan, 2009.

Weinberg, Carl R. "'Ye Shall Know Them by Their Fruits': Evolution, Eschatology, and the Anticommunist Politics of George McCready Price." *Church History* 83 (September 2014): 684–722.

Weindling, Paul. "Genetics, Eugenics, and the Holocaust." In *Biology and Ideology: From Descartes to Dawkins*, edited by Denis R. Alexander and Ronald L. Numbers, 192–214. Chicago: University of Chicago Press, 2010.

Wenham, Gordon J. *Genesis 1–15*. Waco, TX: Word, 1987.

Whitcomb, John C., and Henry M. Morris. *The Genesis Flood: The Biblical Record and Its Scientific Implications*. Phillipsburg, NJ: Presbyterian and Reformed, 1961.

Whitford, David M. *The Curse of Ham in the Early Modern Era: The Bible and the Justifications for Slavery*. Farnham, England: Ashgate, 2009.

Williams, Daniel K. *God's Own Party: The Making of the Christian Right*. Oxford: Oxford University Press, 2010.

Wolf, Maryanne. "Our 'Deep Reading' Brain: Its Digital Evolution Poses Questions." *Nieman Reports* 64 (Summer 2010): 7–8.

The World's Most Famous Court Trial: Tennessee Evolution Case. 1925. 2nd reprint edition. Dayton, TN: Bryan College, 1990.

Worthen, Molly. *Apostles of Reason: The Crisis of Authority in American Evangelicalism*. Oxford: Oxford University Press, 2014.

Zagacki, Kenneth S., and Victoria J. Gallagher. "Rhetoric and Materiality in the Museum Park at the North Carolina Museum of Art." *Quarterly Journal of Speech* 95, no. 2 (2009): 171–191. doi:10.1080/00335630902842087.

Zelizer, Barbie. "The Voice of the Visual in Memory." In *Framing Public Memory*, edited by Kendall R. Phillips, 157–186. Tuscaloosa: University of Alabama Press, 2004.

SUGGESTIONS FOR FURTHER READING

Contemporary scholarship on fundamentalism has its origins in two ground-breaking works: Ernest R. Sandeen, *The Roots of Fundamentalism: British and American Millenarianism, 1800–1930* (Chicago: University of Chicago Press, 1970); and, George M. Marsden, *Fundamentalism and American Culture: The Shaping of Twentieth-Century Evangelicalism: 1870–1925* (New York: Oxford University Press, 1980). In their wake came a number of historical studies, including Margaret Lamberts Bendroth, *Fundamentalism and Gender, 1875 to the Present* (New Haven, CT: Yale University Press, 1993); Joel A. Carpenter, *Revive Us Again: The Reawakening of American Fundamentalism* (New York: Oxford University Press, 1997); Barry Hankins, *God's Rascal: J. Frank Norris and the Beginnings of Southern Fundamentalism* (Lexington: University Press of Kentucky, 1996); D. G. Hart, *Defending the Faith: J. Gresham Machen and the Crisis of Conservative Protestantism in Modern America* (Baltimore: Johns Hopkins University Press, 1994); William Vance Trollinger, Jr., *God's Empire: William Bell Riley and Midwestern Fundamentalism* (Madison: University of Wisconsin Press, 1990).

In these same years a number of scholars addressed the question of how fundamentalists make use of the Bible. The best of these include: Nancy Tatom Ammerman, *Bible Believers: Fundamentalists in the Modern World* (New Brunswick, NJ: Rutgers University Press, 1987); James Barr, *Escaping from Fundamentalism* (London: SCM, 1984) and *Fundamentalism* (Philadelphia: Westminster, 1978); Kathleen C. Boone, *The Bible Tells Them So: The Discourse of Protestant Fundamentalism* (Albany: State University of New York Press, 1989); Susan Friend Harding, *The Book of Jerry Falwell: Fundamentalist Language and Politics* (Princeton, NJ: Princeton University Press, 2000); Harriet A. Harris, *Fundamentalism and Evangelicals* (Oxford: Oxford University Press, 1998); Mark A. Noll, *The Scandal of the Evangelical Mind* (Grand Rapids, MI: William B. Eerdmans, 1994), which is actually an incisive examination of evangelical intellectual life in general, and which has as its "sequel" Noll's *Jesus Christ and the Life of the Mind*

313

(Grand Rapids, MI: William B. Eerdmans, 2011); Peter Johannes Thuesen, *In Discordance with the Scriptures: American Protestant Battles over Translating the Bible* (New York: Oxford University Press, 1999); and Timothy P. Weber, "The Two-Edged Sword: The Fundamentalist Use of the Bible," in *The Bible in America: Essays in Cultural History*, ed. Nathan O. Hatch and Mark A. Noll (New York: Oxford University Press, 1982), 101–120. The best study of dispensational premillennialism in American life remains Paul Boyer's *When Time Shall Be No More: Prophecy Belief in Modern American Culture* (Cambridge, MA: Belknap Press of Harvard University Press, 1992).

In the twenty-first century, the study of American fundamentalism has really come into its own, with a surfeit of outstanding works, many of which pay close attention to economics and politics. At the top of the list are Darren Dochuk, *From Bible Belt to Sunbelt: Plain-Folk Religion, Grassroots Politics, and the Rise of Evangelical Conservatism* (New York: W. W. Norton, 2011); Timothy E. W. Gloege, *Guaranteed Pure: The Moody Bible Institute, Business, and the Making of Modern Evangelicalism* (Chapel Hill: University of North Carolina Press, 2015); Kevin M. Kruse, *One Nation under God: How Corporate America Invented Christian America* (New York: Basic Books, 2015); Paul Maltby, *Christian Fundamentalism and the Culture of Disenchantment* (Charlottesville: University of Virginia Press, 2013); Bethany Moreton, *To Serve God and Wal-Mart: The Making of Christian Free Enterprise* (Cambridge, MA: Harvard University Press, 2009); Matthew Avery Sutton, *American Apocalypse: A History of Modern Evangelicalism* (Cambridge, MA: Belknap Press of Harvard University Press, 2014); and Molly Worthen, *Apostles of Reason: The Crisis of Authority in American Evangelicalism* (Oxford: Oxford University Press, 2014).

Scholars and journalists have written much about the Christian Right, with mixed results; works worth reading include Randall Balmer, *Thy Kingdom Come: How the Religious Right Distorts the Faith and Threatens America* (New York: Basic Books, 2006); Cynthia Burack, *Sin, Sex, and Democracy: Antigay Rhetoric and the Christian Right* (Albany: State University of New York Press, 2008); Joseph Crespino, "Civil Rights and the Religious Right," in *Rightward Bound: Making America Conservative in the 1970s*, ed. Bruce J. Schulman and Julian E. Zelizer (Cambridge, MA: Harvard University Press, 2008), 91–105; Michelle Goldberg, *Kingdom Coming: The Rise of Christian Nationalism* (New York: W. W. Norton, 2006); Chris Hedges, *American Fascists: The Christian Right and the War on America* (New York: Free, 2006); William C. Martin, *With God on Our Side: The Rise of the Religious Right in America* (New York: Broadway Books, 1996), which was the best general study until the publication of Daniel K. Williams' *God's Own Party: The Making of the Christian Right* (Oxford: Oxford University Press, 2010).

Regarding the history of creationism, Ronald L. Numbers dominates the field; his output is prodigious, and the best place to start is *The Creationists:*

From Scientific Creationism to Intelligent Design, expanded edition (Cambridge, MA: Harvard University Press, 2006). Nicholas Matzke nicely delineates the various forms of creationism in "The Evolution of Creationist Movements," *Evolution: Education and Outreach* 3(June 2010): 145–162. Carl R. Weinberg makes the connection between creationism and right-wing politics: "'Ye Shall Know Them by Their Fruits': Evolution, Eschatology, and the Anticommunist Politics of George McCready Price," *Church History* 83 (September 2014): 684–722.

There is a host of excellent books on the Scopes trial and its aftermath, including Paul K. Conkin, *When All the Gods Trembled: Darwinism, Scopes, and American Intellectuals* (Lanham, MD: Rowman and Littlefield, 1998); Adam Laats, *Fundamentalism and Education in the Scopes Era: God, Darwin, and the Roots of America's Culture Wars* (New York: Palgrave Macmillan, 2010); Edward J. Larson, *Summer for the Gods: The Scopes Trial and America's Continuing Debate over Science and Religion* (Cambridge, MA: Harvard University Press, 1997), and *Trial and Error: The American Controversy over Creation and Evolution*, 3rd ed. (Oxford: Oxford University Press, 2003); Jeffrey P. Moran, *American Genesis: The Antievolution Controversies from Scopes to Creation Science* (Oxford: Oxford University Press, 2012); Christopher M. Rios, *After the Monkey Trial: Evangelical Scientists and a New Creationism* (New York: Fordham University Press, 2014); Adam R. Shapiro, *Trying Biology: The Scopes Trial, Textbooks, and the Antievolution Movement in American Schools* (Chicago: University of Chicago Press, 2013).

While this book is the first book on the Creation Museum, a number of scholars and commentators have visited and written about the museum, in articles and as part of larger works. For example, see Stephen T. Asma, "Solomon's House: The Deeper Agenda of the New Creation Museum in Kentucky," *Skeptic*, May 23, 2007, http://www.skeptic.com/eskeptic/07-05-23; Bernadette C. Barton, *Pray the Gay Away: The Extraordinary Lives of Bible Belt Gays* (New York: New York University Press, 2012); Ella Butler, "God Is in the Data: Epistemologies of Knowledge at the Creation Museum," *Ethnos* 75(2010): 229–251; Casey Ryan Kelly and Kristen E. Hoerl, "Genesis in Hyperreality: Legitimizing Disingenuous Controversy at the Creation Museum," *Argumentation and Advocacy* 48 (2012): 123–141; John Lynch, "'Prepare to Believe': The Creation Museum as Embodied Conversion Narrative," *Rhetoric and Public Affairs* 16 (2013): 1–28; Daniel Phelps, "The Anti-Museum: An Overview and Review of the Answers in Genesis Creation 'Museum,'" *National Center for Science Education*, December 17, 2008, http://ncse.com/creationism/general/anti-museum-overview-review-genesis-creation-museum; Daniel Radosh, *Rapture Ready! Adventures in the Parallel Universe of Christian Pop Culture* (New York: Soft Skull, 2010); Jason Rosenhouse, *Among the Creationists: Dispatches from the Anti-Evolutionist Front Line* (New York: Oxford University Press, 2012); Jandos Rothstein, "Graphic Displays of Faith," *Print* 62 (2008): 97–101; Randall J. Stephens and

Karl W. Giberson, *The Anointed: Evangelical Truth in a Secular Age* (Cambridge, MA: Belknap Press of Harvard University Press, 2011), which is the best analysis of Ken Ham as evangelical guru.

Not surprisingly, PhD dissertations are starting to appear. For two good examples, see: Kathleen Curry Oberlin, "Mobilizing Epistemic Conflict: The Creation Museum and the Creationist Social Movement" (PhD diss., Indiana University, 2014), and Steven Mark Watkins, "An Analysis of the Creation Museum: Hermeneutics, Language, and Information Theory" (PhD diss., University of Louisville, 2014). For the battles over inerrancy which provide the context for young Earth creationism, see Jason Hentschel, "Evangelicals, Inerrancy, and the Quest for Certainty: Making Sense of Our Battles for the Bible" (PhD diss., University of Dayton, 2015).

Finally, for a sympathetic but critical treatment of Ken Ham's old creationist ally, John Mackay, see Will Storr, *The Unpersuadables: Adventures with the Enemies of Science* (New York: Overlook, 2014), 1–20; for an ethnographical examination of the creation of Ark Encounter, see James S. Bielo, "Literally Creative: Intertextual Gaps and Artistic Agency," in *Scripturalizing the Human: The Written as the Political*, ed. Vincent L. Wimbush (New York: Routledge, 2015), 20–34.

INDEX

Page numbers in *italics* indicate photographs and tables.

Christian Right: "color-blind" society of, 188–89; creationism and, 6–7; institutions of, 234–35; Muslims and, 160; origins of, 187; Schiavo case and, 159–60; works on, 314

"Christ-the-Door Theater," 231

church age, 3, 47

church conferences, AiG presentations at, 202–7

Church of the Nazarene, 216

Clark, Joseph, 36

colleges: screening, 209–10; shootings at, 158–59. *See also* "Creation Colleges"

Collins, Francis, 219, 220, 221

"color-blind" society, 188–89, 190

Columbine school shooting, 154–55, 158

commonsensical reading of Bible, 112–13, 133–34, 186, 188. *See also* biblical inerrancy, doctrine of; literal reading of Bible; "plain sense" reading of Bible

"Communities" video, 78–79

Concerned Women of America, 6

contemporary natural history museums, 25–27, 36–38

content analysis, 14

conversion and *The Last Adam* film, 194

Cooper, Bill, 198

Copernicus, 103, 144

Cornerstone Community Church, Millersburg, Ohio, 202–3, 207

cosmic binary, 149, 163–64

cosmology of Bible: geocentrism, 144–47; museum and, 103–8, 144

Council for Christian Colleges and Universities, 215

Coyne, Jerry A., 64

Created Cosmos film, 71, 73–75, 76, 106

creatio ex nihilo, doctrine of, 132–33

creation: accounts of, use of biblical text in, 128–33; day-age theory of, 7, 140, 141–43; "gap theory" of, 138–39, 140; science as confirming biblical account of, 71, 94–96. *See also* creationism; Genesis; young Earth creationism

"Creation Colleges," 209–11, 218

creationism: Darwin-to-Hitler narrative in, 182–84; definition of, 241n2; works on history of, 314–15. *See also* young Earth creationism

Creation Museum: appearance as museum, 16–17; argument presented by, 17; Christian Right and, 234–35; exterior of, *27*, 28; institutional status as museum, 16; interior of, 28–30; material speech of, 19; object of, 59–63; outdoor dining area of, *28*; purpose of, 1, 17; revenue generated by, 13; as rhetorical, 18–19; tour of, 27–36; writings on, 315–16. *See also* Answers in Genesis; Bible; Ham, Ken; judgment; politics; science; *specific rooms*

Creation Research Society, 8–9

Creation Science Foundation, 9, 10

"Creator Clearly Seen" video, 77–78

cultural binary, 207

cultural decline: from Christian America to pluralistic society, 161–64; Graffiti Alley and, 157–61; homosexuality, same-sex marriage, and, 168–70, 189, 224

Culture in Crisis room, 32, 49, 52–54, 118

culture war: AiG, Creation Museum, and, 15, 190; in America, 2; Dragon Hall Bookstore as arsenal for, 198–201; Ham on, 10, 207

curiosity cabinets, 20–21

"Curse of Ham," 184, 185

"Curse" placard, 130, 172–73

damnation, emphasis on, 196, 224–27

Darby, John Nelson, 44, 45, 46, 47

Darrow, Clarence, 141–42, 177–78

Darwin, Charles, 2, 148, 181, 184

Darwinism, 2, 4, 5, 22–23, 181. *See also* evolution

Darwin's Plantation (Ham, Ware, and Hillard), 181–82, 184, 187–88, 189

Darwin-to-Hitler narrative, 182–84

dating methods, 84

day-age theory of creation, 7, 140, 141–43

"day," meaning of, 131–32

deep reading, 120

DeLay, Tom, 159, 160

Demolishing Supposed Bible Contradictions books, 177

"Designed for Flight" video, 78

detail, attention to, 40, 61–62

Dig Site room, 69–71, *70*, 115